A Series of Books in Geology

EDITORS:
James Gilluly
A. O. Woodford

PRINCI

Paleon

PRINCIPLES OF
Paleontology

David M. Raup
UNIVERSITY OF ROCHESTER

Steven M. Stanley
THE JOHNS HOPKINS UNIVERSITY

W. H. FREEMAN AND COMPANY
San Francisco

Printed in the United States of America

Library of Congress Catalog Card Number: 79-120302

International Standard Book Number: 0-7167-0247-9

3   4   5   6   7   8   9

Contents

Preface

Our goal has been to write a book that presents the principles of paleontology at a level suitable for an undergraduate course. This book is not designed to provide the entire content of such a course; rather, it is meant to provide a conceptual background for the course and essential information and ideas that may be only partially presented — or presented from a different point of view — in lectures and laboratories. Many teachers of undergraduate paleontology courses follow a phylum-by-phylum taxonomic format in their lectures, which we believe makes assignment of extensive reading on principles and approaches all the more important.

Paleontology is an exciting field of study. It is currently moving rapidly forward, and many of the ideas and assumptions presented here will be refined or supplanted during the next several years. We have tried to include as many new and promising ideas and approaches as possible to spark the imaginations of potential paleontologists and other scientists — to plant seeds for future harvest.

There are twelve chapters in our volume; an average assignment of one chapter per week allows for coverage of the entire book in a typical semester. Early chapters are more abstract and contain fewer biologic and paleontologic examples than later chapters, so the student should have been introduced to most taxonomic groups through some combination of laboratory work, lectures, and reading of reference books by the time the groups are cited.

In Part I of the text, Description and Classification of Fossils, we give some general information about the fossil record in an introductory chapter. In subsequent chapters we discuss the ways in which fossils are studied as specimens within species, as species, and as hierarchical groups of species. Part I might have been called "Taxonomy" or "Systematics," but is slightly different in scope from these broad disciplines.

We explain in Part II, The Uses of Fossil Data, ways in which information derived from fossil study is applied to various geologic and biologic problems. In addition to chapters on the traditionally recognized study areas of paleoecology, evolution, and biostratigraphy, we have included chapters entitled "Adaptation and Functional Morphology" and "The Uses of Paleontologic Data in Geophysics and Geochemistry." Functional morphology, though sometimes considered a subdivision of paleoecology, is not concerned exclusively with ecologic inferences; we believe it deserves status as a distinct branch of paleontology. Inasmuch as paleontology has traditionally resided within the administrative framework of geology departments, contributions of paleontology to geophysics and geochemistry are of particular interest to many students.

Our omission of topics that might together be called "biogeochemistry" may draw criticism, but we believe that the many isotopic and trace-element approaches undertaken during the past two decades have thus far contributed little to general paleontologic knowledge, largely because of the thorny problems imposed by diagenetic alteration. Perhaps study of non-carbonate fossil organic compounds will be more profitable, but relatively little work has been done with them to date.

Finally, we considered as titles for our book "Paleobiology" and "Geobiology," but decided that textbook titles should follow, rather than precede, name changes of major disciplines. Adoption by the scientific community of "paleobiology" or "geobiology" (the latter complementing geophysics and geochemistry) is worth considering, however, to mark the progressive change in paleontologic research since the midpoint of the twentieth century.

We gratefully acknowledge the able photographic work of our steadfast assistant, Robert Eaton, and the thorough and helpful evaluation of the manuscript by our colleague Zeddie P. Bowen. We would greatly appreciate hearing from readers who detect errors or have constructive suggestions for improving the text.

David M. Raup
Steven M. Stanley

January 1970

PART 1

DESCRIPTION AND CLASSIFICATION OF FOSSILS

CHAPTER 1

Preservation and the Fossil Record

The fossil record — far from being complete — represents only a small sample of past life. Furthermore, this is not a random sample but is highly distorted and biased by a variety of biologic and geologic factors. Any study of fossils or use of paleontologic data must be based on a clear understanding of the strengths and weaknesses of the record. We must learn what can be done through the use of fossils and what cannot.

Not all plants and animals have an equal chance of being preserved as fossils, and not all geologic environments are equally favorable for preservation. Figure 1-1 shows several examples of unusually good preservation — all are from the Solnhofen Limestone (Jurassic) of southern Bavaria.

The skeleton of the crustacean (*Penaeus*) shown in Figure 1-1, A is preserved essentially intact with the many skeletal elements articulated in natural position. This in itself would not be so striking were it not for the preservation of the additional features indicated by black arrows. The dead animal was apparently upside down when it hit the surface of the soft sediment, and its impact made marks, which are seen as depressions in the rock surface beside it. The impact marks were formed by the dorsal median ridge of the carapace (a), the rostrum (b), and the eyes (c). The body subsequently fell over into the position in which we find the fossil.

Also shown in Figure 1-1 is a fossil horseshoe crab (*Mesolimulus*). The preservation of such a relatively hard skeleton is not unusual, but this specimen also shows the animal's tracks — apparently his last! Quarrymen in the Solnhofen region learned years ago that the best way to find arthropod fossils is simply to follow their tracks along bedding surfaces in the limestone.

The other fossil illustrated in Figure 1-1 is that of a small bird (*Archaeopteryx*). The delicate feathers are preserved as impressions. It is said that the fossil squids from the Solnhofen were first illustrated by drawings made with the ink taken from preserved ink sacs of the fossils. A surprising feature of the Solnhofen limestone is that excellent preservation is the rule rather than the exception. Only occasionally are broken or fragmental fossils found.

Although preservation as good as that in the Solnhofen is not typical of the fossil record, it is certainly not unique to the Solnhofen: the Hunsrück Shale (Devonian of Germany), the Burgess Shale (Cambrian of British

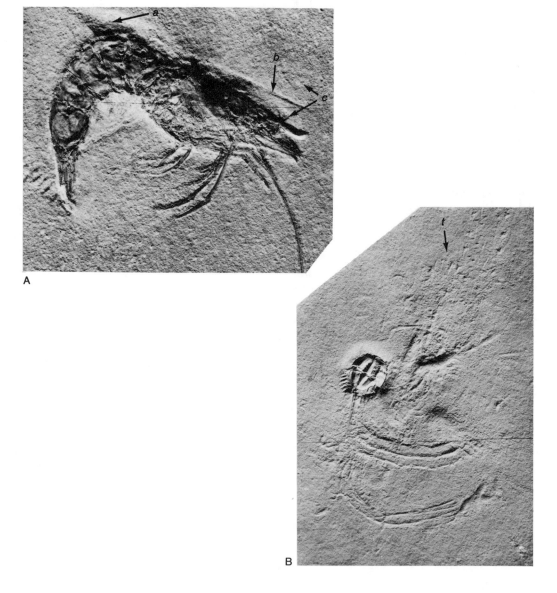

A

B

C

FIGURE 1-1
Fossils from the Solnhofen Limestone (Jurassic, Bavaria.) A: *Penaeus speciosus* (×½). The arrow marked *a* shows the impact mark made by the carapace; *b* that made by the rostrum; *c* those made by the eyes. B: *Mesolimulus walchi* (×⅓). Its final walking tracks are preserved (arrow *t*); the curved grooves below were made by the animal sliding upside-down over the sediment before reaching its final resting place. C: *Archaeopteryx lithographica* (×½). Marks made by its feathers are preserved. (A and B provided by Helmut Leich, B from Leich, 1965; C courtesy of the U.S. National Museum.)

Columbia), the La Brea tar deposits (Pleistocene of southern California), the Mazon Creek Shale (Pennsylvanian of Illinois), and the Baltic amber (Oligocene of Germany) are a few of the many parts of the stratigraphic record that rank with the Solnhofen in quality of fossil preservation. Each of these examples differs from the others, however, in many biologic and geologic factors.

Examples of poor preservation are more difficult to deal with. When a rock is totally barren of fossils, it may mean that plants and animals did not live at or near the locality, that they lived there but were not preserved, or that they were preserved but subsequently destroyed. The Permian reptile tracks shown in Figure 1-2 are spectacular examples of good preservation. The puzzling feature is that the reptiles themselves have never been found in these rocks. The conditions were evidently such that the tracks could be preserved but not the skeletons.

Later in this chapter, after looking at more of the broad characteristics of the fossil record, we will consider the process of preservation in more detail.

Number of Species

About 1.5 million different kinds of plants and animals are known to be living today (Grant, 1963). Since this number is based on species that have actually been found, described, and classified, the real number is probably considerably higher. New species are being discovered each year, and thus the number is constantly growing. The rate of this growth varies from one group of plants and animals to another. Insects, for example, are being described at the rate of approximately ten thousand per year but it has been estimated that no more than one hundred species of birds remain to be described. The difference stems partly from the much greater diversity shown by the insects—there are simply more species (about 850,000 known at present compared with about 8,600 bird species)—and partly from the fact that birds are larger and more readily observed and have long attracted the interest of amateurs and professionals alike.

The list of recognized species has probably been increased somewhat by instances in which a single species has inadvertently been given more than one name. This factor is minor, however. Some authorities have estimated that when the job of description and classification of living species is complete, as many as 4.5 million plants and animals will be known (Grant, 1963).

By contrast, only about 130,000 fossil species have been described and named! This is about 8.7 percent of the number of *known* living species and less than 3 percent of the *probable total* of living organisms. These comparisons are particularly striking when we consider that the fossil record covers many hundreds of million years and the living fauna and flora represent only an instant of geologic time. If overall preservation of fossils were even reasonably good, we would expect the number of fossil species to outnumber by far the number of living species.

FIGURE 1-2
Tracks of a Permian reptile from the Grand Canyon. Although the tracks are well preserved, skeletal remains of the animals have not been found in these rocks. (Courtesy of Raymond Alf Museum.)

There are several possible explanations for the relative paucity of fossil species. One is that there may have been a marked increase in biologic diversity in the course of time. That is, the limited extent of the fossil record may simply reflect a lack of variety in the geologic past.

What does the fossil record itself tell us about changes in diversity throughout the earth's history? According to the most recent information, the fossil record reaches back at least 3.2 billion years (Barghoorn and Schopf, 1966; Engel et al., 1968). That is, the oldest known fossils, those from the Onverwacht Series in South Africa, are found in rocks that are at least this old. The record may be even somewhat longer — older fossils may remain to be discovered.

Fossils are not evenly distributed throughout the rocks that accumulated during 3.2 billion years. More than 99 percent of the 130,000 recognized species come from rocks deposited since the beginning of the Cambrian (that is, rocks younger than approximately 0.6 billion years). The increase in number and variety of fossils at the beginning of the Cambrian is abrupt and may indicate a marked evolutionary radiation, or increase in diversity. The fossil record since the beginning of the Cambrian is usually interpreted as showing a gradual increase in diversity. But is this interpretation correct? Let us look at the evidence.

Diversity is usually measured by the number of *taxa* (species, genera, families, etc.) known to have lived during a given time interval. Figure 1-3, A (taken from data given by Harland et al., 1967) shows changes in apparent diversity through time, from Cambrian through Cenozoic. The width of the shaded area is proportional to the number of taxa (families, for the most part) known for successive intervals of geologic time. According to this diagram (and others like it that have been constructed by various workers), diversity increased steadily through the Paleozoic, decreased in the Triassic, and then increased markedly during the Mesozoic and Tertiary. The decrease in the Triassic is generally interpreted as the result of many forms becoming extinct. Plots more detailed than that shown in Figure 1-3 show the decrease in diversity starting in mid-Permian time.

There is a basic difficulty with this picture of change in amount of diversity. Figure 1-3, A does not take into account the fact that some parts of the geologic column are better known than others. Our knowledge of the fossil record is substantially better in young rocks than in older rocks. The younger rocks are more widely exposed because they are closer to the "top of the stack." Figure 1-4 shows the relative areas of exposed rocks of various ages in the United States. Note that the area of exposed Cretaceous rocks is about four times that of Ordovician rocks even though the time durations of the two periods were about the same.

Figure 1-5, which shows *thickness* of rocks per million years plotted against the age of the rocks, might suggest, at first glance, that sedimentation rates have increased since the Cambrian. There is no evidence for this, however. Rather, it is generally agreed that the plot reflects the fact that younger rocks are better preserved and exposed and thus yield apparently greater thickness of sediment accumulation (Gilluly, 1949).

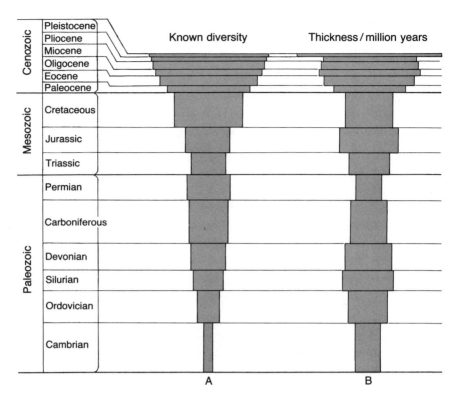

FIGURE 1-3
Relative numbers of taxa of fossil plants and animals. A: Known diversity, as estimated from literature surveys, expressed by width of shaded area. (Data from Harland et al., 1967.) B: Thickness exposed of sediments per million years. (Data from Holmes, 1960.)

As a result of the several factors just considered, we have a much more complete picture of the fossil record in the later periods of the earth's history — and have inevitably found and described more species for these periods. Figure 1-3, B shows the data for thickness of rocks per million years plotted on the same format as the raw diversity data in Figure 1-3, A. (Thickness is probably more important in determining relative fossil diversity than area.) For many of the time periods there is a positive correspondence between thickness and amount of diversity. This suggests that present-day diversity (the estimated 4.5 million species) may not be appreciably higher than the average for the time since the Cambrian.

Let us return to the problem of interpreting the 130,000 known fossil species as a sample of past life. It is evident from Figure 1-3 that the smallness of this number cannot be entirely explained by the idea that diversity increased with the progress of evolution.

Species become extinct and are replaced by others during the course of geologic time. Some estimates of the average rate of species turnover have been made. Teichert (1956), for example, concluded that the average time

FIGURE 1-4
Geographic extent (shown by "surface of step" area) of exposures of rocks of various ages in the United States. (After Gilluly, 1949.)

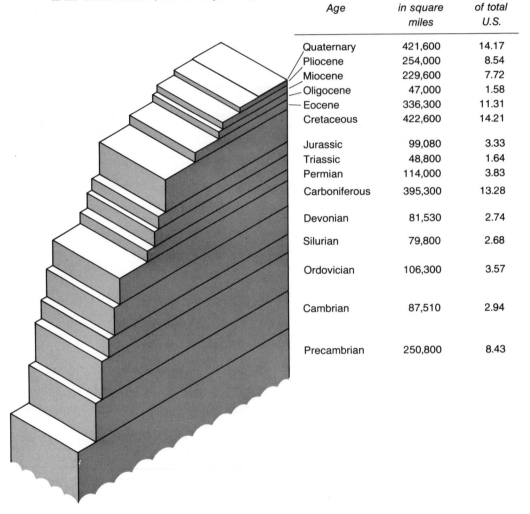

Age	Area in square miles	Percent of total U.S.
Quaternary	421,600	14.17
Pliocene	254,000	8.54
Miocene	229,600	7.72
Oligocene	47,000	1.58
Eocene	336,300	11.31
Cretaceous	422,600	14.21
Jurassic	99,080	3.33
Triassic	48,800	1.64
Permian	114,000	3.83
Carboniferous	395,300	13.28
Devonian	81,530	2.74
Silurian	79,800	2.68
Ordovician	106,300	3.57
Cambrian	87,510	2.94
Precambrian	250,800	8.43

FIGURE 1-5
Relation between the age of rocks and their exposed thickness per million years.
(Data from Holmes, 1960.)

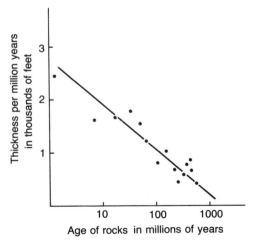

for complete replacement of all species is about 12 million years. Simpson (1952) expressed an estimate in terms of the average life-span of a single species, his best guess being that an average of 2.75 million years elapses between the origin and extinction of a species. He acknowledged, however, that the actual average may be as low as 0.5 million years or as high as 5.0 million years, which demonstrates the uncertainty of estimates of this sort.

If we assume that there are 4.5 million species of plants and animals living today, and that this number is reasonably close to the average diversity since the beginning of the Cambrian, and if we accept Simpson's 2.75 million years as the average duration of a species, we may calculate an estimated number of species that have lived in the 600 million years since the beginning of the Cambrian, as follows:

$$(4.5 \times 10^6) \frac{600 \times 10^6}{2.75 \times 10^6} = 982 \times 10^6 \text{ species.}$$

If, using this estimate of 982 million species, we compare the total with the 130,000 known fossil species, we see that only about .013 of one percent of the species that have lived during this 600 million year period have been recognized in the fossil record.

Many other estimates of the number of species have been made. Simpson's calculations were directed at estimating the total species since life on earth began, and he concluded that the number is somewhere between 50 million and 4 billion. Grant (1963) estimated that there have been at least 1.6 billion species since the beginning of the Cambrian. More recently, Durham (1967) calculated that since the beginning of the Cambrian there have been at least 4 million and probably as many as 10 million species of *preservable, marine* organisms alone.

Regardless of which estimate we accept, we may agree that the paleontologist has discovered only a tiny fraction of the species that have lived.

Kinds of Species

Several textbooks contain tabulations of the taxonomic distribution of species. Among the most recent are those by Easton (1960), which lists fossil and living animals, and Grant (1963), which includes all known living organisms. The numbers of species in various biologic groups are estimates based on surveys of the taxonomic literature and, as such, are subject to considerable uncertainty. Clearly, however, the numbers of species are not randomly distributed among the various phyla and classes. It is usually true that a given phylum or class contains many more living species than fossil species, but certain groups are more heavily represented in the fossil record than among living organisms. For example, there are about 10,000 fossil cephalopod species known but only 400 living species.

Basic questions concerning all tabulations of species abundance are: To what extent do the fossils actually represent life in the past? How has the composition of the biologic world changed? To what extent is the fossil record biased by preferential preservation?

About three-quarters of all known living species are animals, and of these about three-quarters (850,000) are insects. But only about 12,000 fossil insect species are known (the oldest being of Devonian age)—not because the actual number of insect species was formerly so small, but because insects generally occupy habitats that are not conducive to preservation. About 180 species of insects are preserved in the Solnhofen Limestone. (This is, incidentally, about one-quarter of the total number of Solnhofen species.) Many of the other known fossil insects are from a few source-beds of amber, which serves as an unusually favorable medium for preservation of insect fossils. Amber itself is, however, rare.

The important point is that where we do find a well-preserved insect fauna, it is diverse, containing insects that show a wide variety of structural adaptations to a wide variety of environments. In an authoritative review of the insect fossil record, Carpenter (1953) made the following general comments:

> From a survey of the Carboniferous fauna it is apparent that the insects had acquired surprising diversity. . . . I am convinced that we have not yet begun to appreciate the extent of the Upper Carboniferous insect fauna. . . . If the same number of living species were collected at a few isolated localities over the world, we could not expect to obtain from them a good idea of the complexity of the world fauna as it exists today. It is not beyond the realm of possibility, therefore, that the extinct orders of Carboniferous insects were in their time comparable in extent to the major orders now living.

We may conclude, therefore, that the sparse fossil record of insects is due primarily to the unlikeliness of their preservation. The virtual nonexistence of a Cretaceous insect record in all probability stems from a lack—by pure chance—of insect environments suitable for preservation during the Cretaceous.

As noted earlier, known fossil species actually outnumber living species in a few biologic groups. This is most striking when we consider cephalopods and crinoids but is also true for brachiopods, bivalves, and echinoids. Clearly, there has been an evolutionary *decrease* in diversity of cephalopods in the course of time. One of the most common fossils in some parts of the record (particularly in the Mesozoic of Western Europe) is the shelled cephalopod. Throughout much of the Mesozoic, ammonoid and nautiloid cephalopods constitute a large part of the fossil record in terms of number of species, morphologic types, and numbers of individuals. Yet comparable cephalopods are represented in modern seas by only one genus containing four species. Without question the Mesozoic was truly an "age of cephalopods."

Similarly, marine environments of the Paleozoic era supported thousands of species of crinoids and brachiopods. The present-day sparsity of these groups can be explained only by evolutionary decline (extinction).

The bivalves and the echinoids show a somewhat different pattern. The fossil record of both of these groups exhibits a progressive *increase* in number of fossil species through time and suggests that the number of species living today should exceed the number living *at any single point in time* in the geologic past. The fact that the *total* number of fossil species for these groups exceeds the number of living species simply reflects the fact that the fossil record is a composite of a long period of time.

Another important point concerning the groups in which the number of fossil species exceeds the number of living species is that these are groups that, by the nature of their skeletons, have a relatively high probability of being preserved. They are also groups that contain many species that lived in environments that were relatively favorable for fossil preservation.

Figure 1-6 shows the observed diversity through geologic time for selected animal groups. We have already discussed the decline in brachiopod diversity from the mid-Paleozoic onward. The echinoderm diversity data show a peak during the Paleozoic, a pronounced reduction in the Triassic, and a steady increase between Jurassic and Recent. The bimodal pattern reflects the composite nature of the sample: the Paleozoic peak stems from pelmatozoan echinoderms (crinoids, in particular); the post-Triassic peak documents the evolutionary radiation of eleutherozoan echinoderms (e.g., echinoids, starfish).

The reptile diversity pattern in Figure 1-6 shows a high in the Late Paleozoic and another in the Cretaceous. The abrupt drop at the end of the Cretaceous corresponds to the extinction of the dinosaurs and complements the dramatic rise in mammal diversity following the Cretaceous.

For a given biologic group, a fossil record that is meager may mean either that preservation of the organisms was poor or that they expressed little evolutionary diversity. A rich fossil record may result from better-than-average preservability, decrease in evolutionary diversity through time, or the simple fact that the fossil record is the composite of a long period of evolutionary turnover. One of the most important areas of research open to the paleontologist is the further evaluation of the fossil record as a sample of former life.

Numbers of Individuals

Our interpretation of the fossil record may be assisted by our considering living plants and animals as potential fossils. As we do this we should take into account numbers of individuals (abundance) as well as numbers and kinds of species (diversity). The number of individuals per species ranges from a few hundred to many millions or even billions. Each is a potential fossil.

From the biologist we get an occasional glimpse of the magnitude of the numbers involved. Figure 1-7, for example, shows the larger invertebrate organisms typically found in $\frac{1}{4}$ sq m of sea bottom off the Kii Peninsula, Japan, at depths from 10–60 fathoms. This assemblage represents the "standing crop" of individuals of three species: two bivalves and one

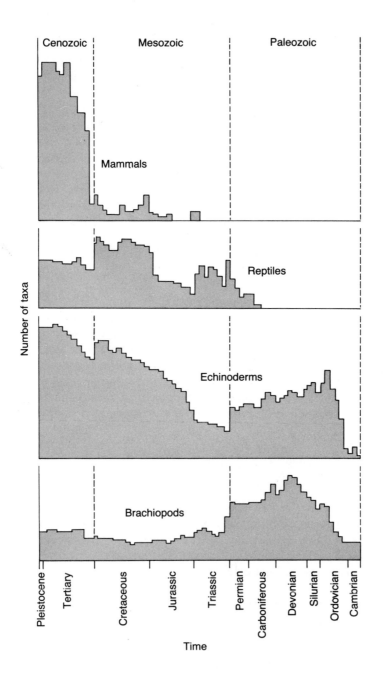

FIGURE 1-6
Diversity throughout time for four animal groups.
(Data from Harland et al., 1967.)

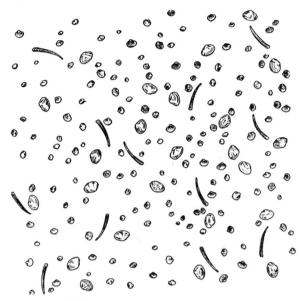

FIGURE 1-7
Animals obtained by dredging ¼ square meter of level
sea bottom off the Kii Peninsula, Japan. Species are the
bivalves *Macoma incongrua* (larger) and *Cardium
hungerfordi* (smaller) and the scaphopod *Dentalium
octangulatum*. (From Thorson, 1957.)

scaphopod. There are about 25 individuals of the larger bivalve (*Macoma
incongrua*), about 160 of the smaller bivalve (*Cardium hungerfordi*), and
12 of the scaphopod (*Dentalium octangulatum*). All this in one-quarter of
a square meter! The average age of the specimens is approximately two
years. This rate of production would yield 1,000 potential fossils in ten
years, or one hundred million in a million years. If these calculations were
extended to include a larger geographic area and longer periods, the number
of specimens — and the tonnage of potential fossils — would be truly staggering. In just the tiny area of sea bottom that we have been discussing would
probably be produced in a million years more individuals than the total
number of fossil specimens (of all species) that have ever been studied.

The sample calculation just carried out is, of course, subject to many errors and many tacit assumptions. The density of life varies greatly from
place to place and from time to time. Many sites have higher densities
than the sample area we used; many sites have lower. The calculation does
give us some impression, however, that the fossils we have to work with
represent an almost infinitesimal fraction of the total life of the geologic
past. If the fossil record were a random sample of the plants and animals
that have lived, there would be less difficulty. But it clearly is not random.
As one of the biases in the fossil record, we may note that those species
represented by many individuals are more likely to be preserved and found
(other things being equal) than those species having fewer individuals.

16

Preservability

To be a part of the fossil record, all or part of an organism or some trace of its activity must be preserved in the rock. The reptile tracks shown in Figure 1-2 are a part of the fossil record and may give us information not given by the preserved body or skeleton. If an organism or some trace of its activity is preserved, the wide variety of destructive processes that operate on dead organisms and their surroundings have been, at least partly, unsuccessful.

BIOLOGIC DESTRUCTION

Biologic agents of destruction are present in nearly all environments. Predators and scavengers are ubiquitous in the biologic world. Some are larger than the organisms they feed on; some are much smaller. Hardly a biologic structure fails to attract scavengers or other biologic agents of destruction. We tend to think, for example, that the shell of an oyster is almost a fossil as soon as it is formed by the animal. The shell is quite sturdy and is made up largely of calcium carbonate. The structure of the oyster shell is not continuous or solid, however, but rather is constructed of tiny needles

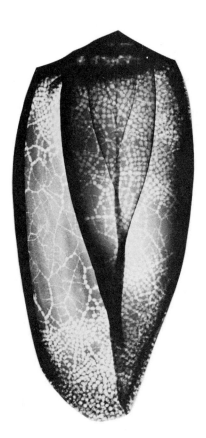

FIGURE 1-8
Radiograph of a Recent gastropod, *Conus geographicus,* showing the destructive effects of the boring sponge *Cliona.* (From Ginsburg, 1957.)

or lamellae of calcite held together by a network of organic tissue commonly referred to as the organic matrix. The strength of the shell is thus partly dependent on the integrity of the organic matrix.

As soon as an oyster or other mollusc dies, its shell is subject to deterioration resulting from the attack by a great variety of boring organisms, including worms, sponges, other molluscs, and algae (see Figure 1-8). Most sea bottoms on which living shelled organisms are abundant have surprisingly few empty shells.

If an organism is buried by sedimentation shortly after death, it is partially insulated from destructive biologic processes. The importance of this insulating effect is often exaggerated, however. The unconsolidated sediment immediately below the sediment-water interface in a normal aquatic environment is anything but biologically inert. In fact, much of the bacterial decay of biologic tissue is concentrated in the upper few inches of the sediment. A shell may survive long enough to be buried, only to be destroyed or fragmented beneath the sediment-water interface.

Although biologic destruction of potential fossils is generally recognized as an important factor in limiting the fossil record, our knowledge of the process is woefully small. Little research has been done on the process, particularly on bacterial activity and effects of bacteria on the chemistry of aqueous solutions in sediments. Yet understanding biologic destruction is vitally important if we are to understand fully the preservation process and its effect on the fossil record.

MECHANICAL DESTRUCTION

Several important studies have been made on mechanical breakage and abrasion of skeletal material of potential fossils. As has been known for a long time, organisms whose early post-mortem history takes place in a high-energy environment may be abraded beyond recognition or completely destroyed by the action of wind, waves, and currents. Also, some kinds of skeletons are known to be more susceptible to mechanical destruction than others, which of course contributes to bias in the fossil record.

One of the simplest and most meaningful studies of mechanical destruction was carried out by Chave (1964). Particle-against-particle abrasion was tested by placing shells and other skeletal parts of various marine invertebrates with chert pebbles in tumbling barrels. The time required for various degrees of destruction by abrasion was carefully noted. Some of the results are shown in Figure 1-9. In this illustration, time is shown on the horizontal axis and the percentage of the sample larger than 4 millimeters in diameter is indicated on the vertical axis. This particular experiment was performed on skeletons of gastropods, corals, echinoids, and bryozoans. The differences in durability are striking. After more than a hundred hours of tumbling, more than 60 percent of the material of one gastropod species still remained as particles larger than 4 millimeters and, therefore, as potentially recognizable fragments. By contrast, all of the bryozoans and calcareous algae were gone after one hour of tumbling. In view of this, it is not surprising

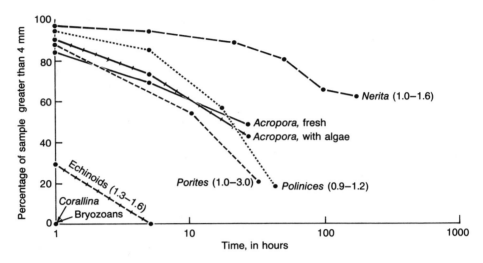

FIGURE 1-9
Experimental abrasion of skeletal material by tumbling with chert pebbles. The gastropod *Nerita* is most durable; the corals *Acropora* and *Porites* and the gastropod *Polinices* are intermediate; the calcareous alga *Corallina*, the bryozoans, and the echinoids are least durable under these conditions. The numbers in parentheses after the names refer to the original size, in inches, of the specimens. (From Chave, 1964.)

that bryozoans make up a relatively small fraction of the fossil record and gastropods a relatively large fraction.

Figure 1-10 shows the results of another tumbling barrel experiment carried out by Chave. Six different kinds of bivalve shells were used, and the variation in durability among them was almost as great as that among the groups of organisms in the first experiment. The relative durability of most of these shells can be interpreted in terms of such differences as their size, thickness, and internal fabric.

Studies such as Chave's can be used to evaluate preservability bias in the fossil record. They also are of value in providing material for comparison with actual fossils.

A further example of the way in which mechanical destruction may bias the fossil record is shown in Figure 1-11, which is also drawn from Chave's experimental work. A representative assemblage of shells from Corona del Mar, California, was placed with sand in a tumbling barrel. The illustration shows the effect on the composition of the assemblage; the bryozoans and calcareous algae were completely removed from the assemblage, in the sense that particle size was reduced to less than 2 mm, and the relative abundance of the remaining forms was considerably altered.

Figure 1-12 shows a photograph of a coarse sandstone slab containing an assemblage of molluscan fossils that Chave suggests may have survived mechanical attack similar to that to which his tumbled Corona del Mar sample was subjected. This emphasizes the fact that when we look at a

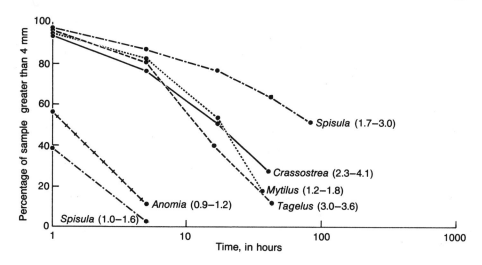

FIGURE 1-10
Experimental abrasion of skeletal material by tumbling with chert pebbles.
Species are all marine bivalves. (From Chave, 1964.)

FIGURE 1-11
Experimental abrasion of skeletal material. Starting sample (upper left)
contained fresh specimens of the bivalve *Mytilus*, the gastropods *Aletes*,
Haliotis, and *Tegula*, various species of limpets and echinoids, the starfish
Pisaster, and the calcareous alga *Corallina*. The series of diagrams shows
selective destruction of the assemblage by tumbling. (From Chave, 1964.)

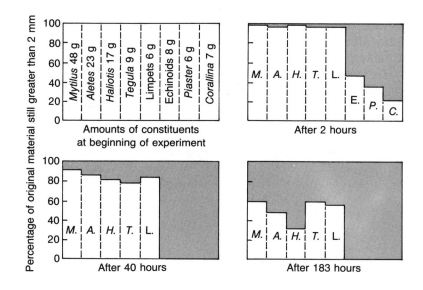

fossil assemblage in a rock we must attempt to look beyond the assemblage, even though the general level of preservation may seem to be good, in order to deduce what organisms may have lived at the site but have not been preserved.

CHEMICAL DESTRUCTION

The skeleton of an animal may withstand biologic and mechanical destructive processes yet not become part of the permanent fossil record. Simple chemical solution is one of the most important reasons why we do not have more identifiable fossils than we do. Chemical solution can take place at any time after the death of the animal — even after its skeleton has been a fossil for a long time. Solution may occur not only on the sea bottom but also in soft sediment. Solution of an organism's remains by ground water millions of years after its burial is common.

The ability of a fossil to survive solution depends on its chemical composition and on the composition and physical characteristics of the waters to which it is exposed. The chemistry of local waters may, in turn, be influenced strongly by the biologic environment, particularly by its bacterial activity.

If a shell is disintegrated by chemical solution after being imbedded in rock, the chances are good that the cavity may remain as a fossil. For many paleontological purposes such a fossil, which gives evidence about morphology, may be adequate.

Biologic Structures Most Likely To Be Preserved

Skeletons containing a high percentage of mineral matter are most readily preserved; soft tissue not intimately connected with skeletal parts is least likely to be preserved. This means that the fossil record contains a biased selection not only of types of organisms but also of parts of organisms. Much of our paleontological knowledge about mammals, for example, is based on teeth alone, the teeth being much more durable than other parts of the skeleton.

A striking effect of the difference in preservability of so-called hard parts and soft parts is that only in rare instances do we have any knowledge of the color of extinct organisms. Pigments, except in rare situations, are simply not preserved with fossils.

Evolutionary change may greatly alter the fossil record of a group. If, for example, an evolving animal group developed resistant skeletal elements, its preservability would suddenly have increased, and it would have begun

FIGURE 1-12
Slab of fossiliferous sandstone from the Pliocene of California. Condition and composition of the assemblage are comparable to the results of the tumbling experiments illustrated in Figure 1-11. The assemblage is dominated by worn specimens of gastropods and bivalves. (From Woodring et al., 1940.)

to make a greater impact on the fossil record. Arthropods probably took such an evolutionary step. Nearly all arthropods have a skeleton. The chemical composition of their skeletons varies widely, however, as does their strength and resistance to decay, solution, and abrasion. The skeleton of a crab, for example, is less fully calcified than that of a trilobite. The trilobite skeleton when it was part of a living organism was made up of a denser, firmer structure containing more pure calcium carbonate and proportionately less organic material than that of the crab. The trilobite was thus more likely to be fossilized, and, indeed, the fossil record of trilobites is much more complete than that of crabs. Many paleontologists have suggested that the rather abrupt beginning of the fossil record of trilobites and those of other organisms in the Cambrian stems not from the sudden evolution of many forms but rather from the sudden evolution of calcification.

Environment and Preservability

The example of excellent preservation of fossils in the Solnhofen Limestone emphasizes the importance of the physical environment in determining the preservability of organisms. Biologically identical organisms may be excellently preserved in one environment but destroyed in another. Obviously an environment in which burial is rapid is most conducive to fossilization, but other factors are also involved. For example, the Solnhofen Limestone is an extremely fine grained, well-bedded, pure limestone showing none of the features that the geologist normally associates with rapid sedimentation. The source of the limestone is not known for certain, but in all probability it was formed by very slow accumulation of inorganically precipitated calcium carbonate.

We can get a bit closer to evaluating an environment's effect on the preservability of fossils by distinguishing between areas receiving sediment, which tend to be conducive to preservation, and areas being eroded. Generally speaking, parts of the earth below sea level are more apt to be accumulating sediment than those above sea level. Thus, preservation of fossils below sea level is more common than their preservation above sea level, and the marine fossil record is infinitely more complete than the terrestrial record.

In areas that are receiving a large, steady supply of sediment, such as deltas of major rivers, sedimentation certainly assists fossil preservation. Often, however, *normal* rates of sedimentation are not high enough to afford the prompt burial of potential fossils that protects them from many processes of destruction.

Some of the most geologically unusual parts of the earth's crust provide much of the diversity in the fossil record. The spectacular preservation of vertebrates in the tar pits of southern California is a notable example. Not only are the tar pits relatively rare and unusual in the earth's history, but they also represent the occurrence of catastrophe, which may be more significant in fossil preservation than is generally acknowledged. Further-more, a tar pit is anything but the normal habitat of the organisms preserved there. The preservation of insects in amber is similar in many respects to that of the vertebrates in the tar pits.

A common denominator in examples of excellent fossil preservation is an environment that was biologically inert as was that of the California tar pits and also that of the Baltic amber. In such an environment, the remains of animals and plants are protected from the normal biologic agents of destruction as well as from most physical and chemical agents. We do not know what the environment was in which the Solnhofen Limestone formed, but preservation shows that it too must have been biologically inert. Preser-vation of soft tissues where there is no indication of an extremely high sedi-mentation rate, as, for example, in the Middle Cambrian Burgess Shale (Figure 1-13), argues for an almost complete absence of scavengers or bac-terial action. Individuals of some plant and animal species seem to be pre-served as fossils only when their remains are transported out of the organisms' normal environment. We cannot say at this time how common or how im-portant transport to burial sites is in the formation of the fossil record as a whole.

Habitat heavily influences preservability of an organism. A mountain sheep is less likely to be preserved than a hippopotamus because the sheep normally lives farther from an area of sedimentation. Likewise, within the marine environment, large differences exist. One of the most important from a paleontologic viewpoint is the distinction between those animals that live *in* the sediment on the sea bottom and those that live *on* the surface of the sediment or swim or float in the overlying water. For the organism living in the sediment, sedimentation rates may be less important because the or-ganism is wholly buried in the sediment even when it is alive and is thus somewhat protected against scavengers and to a large extent against mechanical abrasion and breakage.

Post-mortem Transport

The remains of different organisms are transported in many different ways; furthermore, similar organisms are transported in different ways in different environments. One of the simplest forms of transport is illustrated by free-swimming organisms that upon dying sink to the sea bottom — an entirely different environment, which may in fact be biologically inert. This pos-sibility has led to one of the most widely accepted interpretations of the excellent preservation of fossils in the Solnhofen.

Nearly all organisms preserved in the Solnhofen Limestone are either pelagic marine organisms or terrestrial organisms that might have drifted or fallen into the Solnhofen Sea. The general lack of bottom-dwelling marine organisms is evidenced not only by the absence of their actual remains in the fossil record, but also by the absence of any indication of their activities. The

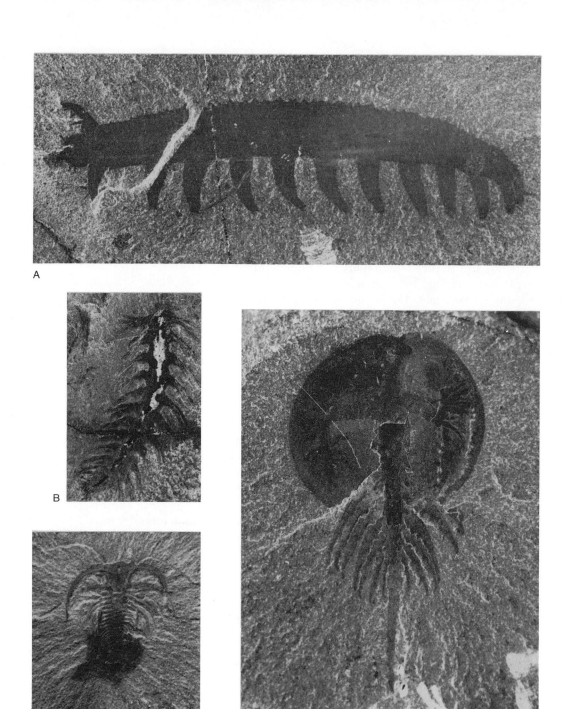

A

B

C

D

FIGURE 1-13
Fossils from the Burgess Shale of British Columbia. A: *Aysheaia pedunculata,* which belongs to the Onychophora, a phylum of animals that are morphologically intermediate between annelids and arthropods. B: *Canadia setigera,* an annelid worm. C and D: *Marella splendens* and *Burgessia bella,* trilobitomorph arthropods. E and F: Right-lateral

E

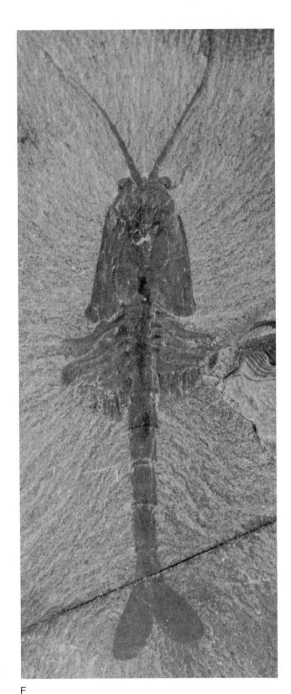

F

and dorsal views of *Waptia fieldensis*, another trilobitomorph. Most known trilobitomorph species have been collected only from the Burgess Shale. Preserved remains of internal organs are seen as dark areas in C and E. All photographs are two to three times actual size, except E, which is ×6. (Courtesy of the U.S. National Museum.)

tracks and trails that *are* preserved would not long persist in an environment of high biologic activity. The tracks recorded in the Solnhofen are the last tracks made by the organisms, and most, if not all, were made by organisms that were pelagic. This has led to the contention that the bottom waters of the Solnhofen Sea were inhospitable to most organisms and that the fossil record exhibits the organisms that either drifted in after death or inadvertently swam in and were killed by toxicity.

Some living pelagic organisms, for example most crustaceans, remain suspended in the water by their own swimming activity; others are suspended in the water or float at the surface because of their buoyancy. Usually when organisms of the first sort die, their remains settle to the bottom. With the second sort of organism, there may be considerable transport of the buoyant cadaver by currents or wind. The coiled cephalopod furnishes a prime example. In life, the gas-filled chambers of the shell provide buoyancy. Upon death this buoyancy may actually increase because decay of the soft body of the organism makes the shell lighter. The shell may float in the surface water for days, weeks, or even months before becoming so damaged that it sinks to the bottom. Post-mortem transport is probably the rule rather than the exception for coiled cephalopods and similar organisms.

Most post-mortem transport of bottom-dwelling marine organisms depends on currents. In relatively shallow water in which currents are strong, the skeleton of an organism may be carried a considerable distance to its final site of burial. The possibility also exists that bottom-dwelling organisms may float after death because the decay of the soft tissue produces gas that may become trapped inside the skeleton. Unfortunately it is not known how common this phenomenon is. Certainly fish and other soft-bodied animals do become bloated during decay and rise to the surface. Occurrences of the same phenomenon have been recorded for rather heavy-shelled invertebrates but not enough research has been done to assess this possibility as a major factor in fossil preservation.

Pollen and spores of terrestrial plants, which make up an important part of the fossil record, can be carried long distances by wind. Also, rivers and streams carry a variety of plant and animal fragments. Very commonly, terrestrial organisms, particularly plants, are carried to the ocean and float for hundreds of miles before sinking to the bottom. An interesting result of this is that several sea urchin species living in very deep water off the coast of New Guinea depend, for food, upon terrestrial plant material brought into the ocean by the rivers of the New Guinea coast.

It is exceedingly difficult to say just how important a role is played by post-mortem transport. Physical deterioration of fossils or their discovery where the organisms are known not to have lived gives clear evidence of post-mortem transport. On the other hand some fossils show no evidence of having been disturbed after death.

The significance of post-mortem transport as a biasing factor in the fossil record depends entirely on the use to which the fossil record is put. If the concern is with tracing the evolutionary development of a biologic group on a broad scale, which is done by studying comparative morphology and distribution in time, movement of a few miles or even a few hundreds of miles is negligible. Similarly, if fossil evidence is to be used to reconstruct regional climate, movement of a few miles or tens of miles may be insignificant. On the other hand, movement of a short distance may be vitally important when the problem is to reconstruct local environmental conditions.

Conclusion

We have seen that preservation of an organism's remains or of evidence of its activity is a rare event. The more we investigate the difficulties of fossil preservation, the more surprised we become that the fossil record is as good as it is. But the number of potentially fossilizable plants and animals is so enormous that even such an unlikely event as preservation becomes a relatively common phenomenon. Although we have a generally accurate understanding of some of the basic problems and processes in fossil preservation there still are enormous areas of ignorance. For example, it has been suggested in this chapter that geologically unusual or even catastrophic conditions contribute to the preservation of fossils. But to what degree? We do not have enough information yet to answer this question.

The best way for us to proceed is never to accept a fossil assemblage at face value; it probably represents a strongly biased picture of past life. A corollary is that lack of fossils in given rocks should not be taken to indicate that animals or plants were definitely absent.

Although the fossil record is limited in many ways, it still contains an extraordinary amount of information. There are perhaps two reasons why the rather poor overall sample of past life in the fossil record is often adequate for the types of study that characterize paleontology:

1. In *large-scale* studies of rates, trends, and patterns of evolution and of evolutionary relationships within major plant and animal groups, paleontologists tend to restrict themselves to groups that have relatively good fossil records — fossil records that are adequate statistical samples.

2. In interpreting fossil faunas and floras of *local* rock units, the paleontologist attempts to deal with rocks in which a large proportion of potentially fossilizable species have been preserved.

Thus, to undertake a particular research project, an attempt is made to choose certain fossil groups or rock units that lend themselves to a particular approach. Our main purpose in the remainder of this book will be to explore the methods by which fossil information, carefully evaluated in light of preservational biases, can be interpreted and applied to geologic and biologic problems.

Supplementary Reading

Durham, J. W. (1967) The incompleteness of our knowledge of the fossil record. *Jour. Paleont.,* **41:**559–565. (A presidential address to the Paleontological Society on the nature of the fossil record, particularly with regard to marine invertebrates.)

Easton, W. H. (1960) *Invertebrate Paleontology.* New York, Harper, 701 p. (Chapter 1 contains data and discussion on representation of animal taxa in the fossil record.)

Gilluly, J. (1949) Distribution of mountain building in geologic time. *Geol. Soc. Amer. Bull.,* **60:**561–590. (Includes a classic discussion of the imperfections of the rock record.)

Grant, V. (1963) *The Origin of Adaptations.* New York, Columbia University Press, 606 p. (Chapter 4 contains an authoritative treatment of present and past diversity.)

Harland, W. B., et al. (1967) *The Fossil Record.* London, Geol. Soc. London, 827 p. (The most complete compilation of diversity data thus far published for the fossil record.)

Newell, N. D. (1959) Adequacy of the fossil record. *Jour. Paleont.,* **33:**488– 499. (A general discussion of the size of the fossil record in relation to its study by paleontologists.)

Simpson, G. G. (1952) How many species? *Evolution,* **6:**342. (Calculated estimates of the number of species that have lived on the earth.)

Teichert, C. (1956) How many fossil species? *Jour. Paleont.,* **30:**967–969.

CHAPTER 2

Describing a
Single Specimen

Description of fossils is fundamental to nearly all paleontologic research. The question for a given study is what to describe and how to describe it. If the objective is to define and classify a new fossil species, the description should be as complete as possible, but at the same time be such that comparison with other fossils is facilitated. Comparison is usually based on selected attributes. For any fossil, there are literally thousands that *could* be described. How then are attributes chosen? Furthermore, once they have been chosen, a practical and meaningful scheme of expressing differences in them must be developed.

Problems of description are not limited to the definition and classification of new species. Let us suppose that a paleontologist wishes to test the evolutionary hypothesis that *size* in a group of organisms increases with time. This seems simple enough, but several problems soon arise. What is the best measure of size? Should he use total weight, total volume, maximum length or width, surface area, or some combination of these? Should the whole organism be considered, or can parts of the organism be assumed to reflect overall size? Choosing attributes is in part a biologic problem, but in part a uniquely paleontologic one. Many attributes that the biologist would measure are not available to the paleontologist, who must operate within the restrictions of fossilized material. For example, characteristics such as weight and volume are greatly affected by fossilization. The paleontologist must arrive at a selection of attributes that is both biologically sound and paleontologically reasonable.

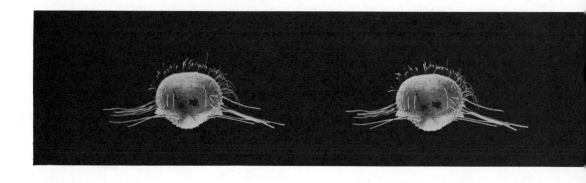

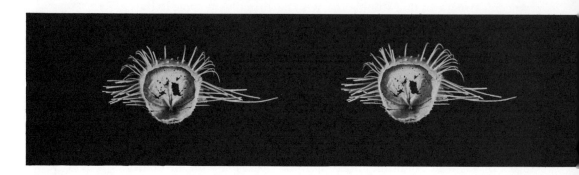

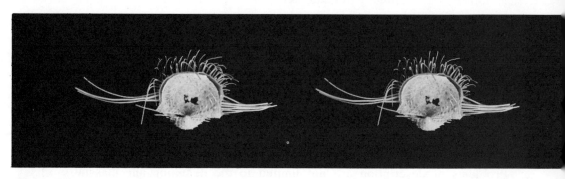

FIGURE 2-1
Stereophotographs of the brachiopod *Echinauris opuntia* from the Permian of Pakistan.
Three-dimensional images can be seen when the pairs of photographs are viewed with a
pocket stereoscope. (From Grant, 1968.)

There is no single format that can be followed to produce a perfect or
all-purpose description. In this chapter we will explore some of the general
problems of selection of attributes and methods of description. We will
assume that we have at hand fossils that are collected, prepared, and ready
to be described, bypassing the important problems of fossil collection and
preparation. (There are references at the end of the chapter to several
comprehensive works on collection and preparation.)

The Photographic Approach

In some ways, a good photograph of a fossil is the best description. It is both objective and comprehensive. For this reason, the photograph has become an indispensable part of the formal definition of a fossil species. The photograph cannot, of course, record certain attributes like chemical composition and some minute or internal structures.

A high point in the paleontologic use of photography is illustrated by the pairs of stereophotographs that are reproduced as Figure 2-1. Appropriate preparation for photography, such as coating, and appropriate lighting may bring out structures that would otherwise be virtually invisible to the camera, as they are to the naked eye.

Photographic methods other than the conventional ones that use visible light may be used to bring out characteristics that are not ordinarily visible. Figure 2-2 shows several examples. X-radiography shows many internal structures not otherwise apparent. The stereographic pair of X-ray photographs of a brachiopod in Figure 2-2 shows the internal lophophore support structure characteristic of this group of brachiopods. The spiral lophophore is completely embedded in solid sediment that fills the shell. Only with elaborate sectioning techniques and reconstruction of the structure from a series of sections could the lophophore be observed with the clarity of the X-ray photograph. This illustrates another advantage of photographic description: it is essentially nondestructive.

Figure 2-2 also shows both an example of the effectiveness of ultraviolet photography, which depends upon the fluorescence of small amounts of organic materials to outline skeletal structures not evident in visible light, and an example of photography with infrared film. Differential absorption of heat by infrared film brings out structures not otherwise visible. Electron microscopy has become important in recent years for the examination and photography of ultramicrofossils and of the fine details of structure in larger fossils. Examples are shown in Figure 2-3.

Several attempts have been made recently to develop further the use of photography in fossil description by putting photographic information into "machine-recognizable form." This development has followed logically from the advances made in high-speed computers. There is the tantalizing possibility that photographic information can be read directly into a computer and that the assessment and comparison of attributes can be made precise and automatic. This field is still in its infancy, and most of the work so far has been outside of paleontology, but several of the methods developed are obviously applicable to fossils.

The simplest method of putting photographic information into machine-recognizable form is *digitizing*. The usual procedure is to superimpose on a photograph a conventional orthogonal coordinate system (that is, a grid in X and Y). The X and Y coordinate values of points on the photograph are noted and recorded on punched cards, punched paper tape, or magnetic tape that can be read into the memory of the computer. The photographic quality of the image recorded depends only on the number of points digitized.

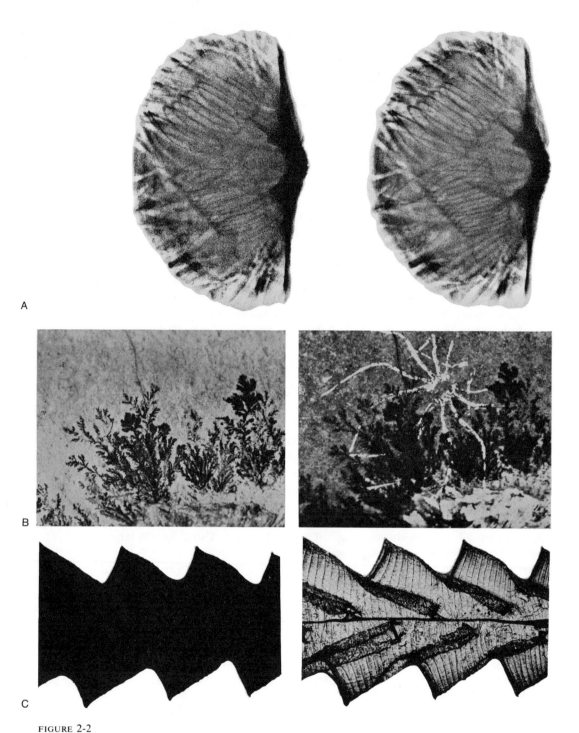

FIGURE 2-2
A: X-ray photographs (stereo pair) of the brachiopod *Spinocyrtia euruteines* from
the Devonian of Ohio, showing the internal lophophore support structure. (From Zangerl,
1965.) B: Bedding surface of Solnhofen Limestone in visible light (*left*) showing dendrites
but no fossils and in ultraviolet light (*right*) showing larva of a decapod crustacean. (From
Leon, 1933.) C: Graptolite *Diplograptus gracilis* in visible light (*left*) and in infrared
light (*right*). (From Kraft, 1932.)

This digitizing method assumes that we are dealing with a two-dimensional structure or a structure that can be expressed as a plane projection. For the computer study of most fossils we need instead a method of three-dimensional digitizing. The most straightforward method is to digitize a stereographic pair of photographs. Each point on the fossil goes into the computer as two pairs of X, Y coordinates, representing the point as observed from two slightly different viewpoints. Trigonometric treatment of the data in the computer recreates the three-dimensional image.

For photographic images that are simple shapes, digitizing can be performed manually, but this is so laborious as to outweigh the advantage of having the photograph of a fossil in a machine-recognizable form. Technology has been developed to automate the job of digitizing. A common device for this purpose is illustrated in Figure 2-5. A photograph is placed on a drafting surface and the operator moves a stylus over those areas of the photograph that he wishes to digitize. When a point to be digitized is reached by the stylus, the operator pushes a button or a foot pedal to record automatically the coordinates of that point. A more fully automatic form of digitizing involves *scanning*. In this method, a light-sensitive instrument scans a photograph in a series of traverses. The variations in darkness that compose the photographic image are recorded in some machine-recognizable form. Figure 2-4 shows an original photomicrograph (left page) of a series of primate chromosomes, and the digitized version (right page). The quality of the reproduction depends on the number of traverses made across the original photograph and the sensitivity of the scanning instrument.

The reading of photographic information into a computer does not in itself solve any problems of fossil description, except that it may facilitate the communication of descriptive information. We are still faced with the problem of what to do with a photograph of a fossil: what attributes to choose for a particular paleontologic problem, and how to express them. This emphasizes a basic disadvantage of photographic description. Because it is objective, it includes too much information, the irrelevant as well as the relevant. When we look at the photograph of a fossil, we see almost all its attributes. Some are biologic in origin. Others are geologic (relating to the organism's post-mortem history). Even if we are successful in separating out a set of attributes of particular interest, the set may include more attributes than the human mind can absorb and manipulate. It is common for a student presented with photographs of two fossils to recognize immediately that they are different, but then to be unable to identify their differences and similarities. Description demands simplification: reducing the attributes observed and described to a manageable number and in particular eliminating those likely to be irrelevant or redundant. A certain degree of subjectivity is essential for effective fossil description. The perfect photograph so completely lacks subjectivity that other methods of description are often preferred.

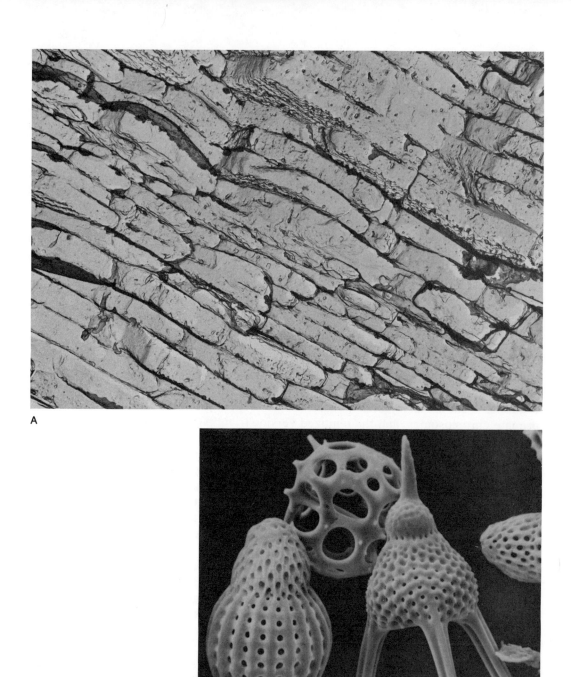

FIGURE 2-3

Electron microscopy of fossil and Recent invertebrate skeletons. A: Shell structure of the Jurassic bivalve *Praemytilus* enlarged 7140 times. B: Radiolarians from the Eocene of Barbados at a magnification of 425. C and D: Surface of the test of the Recent ostracod *Carinocythereis* aff. *carinata* from the Bay of Naples, Italy at magnifications of 110 and 1000. (A from Hudson, 1968; B–D provided by W. W. Hay and P. A. Sandberg.)

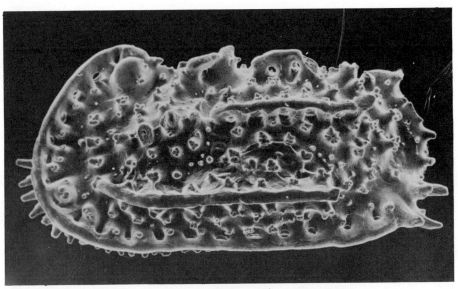

C

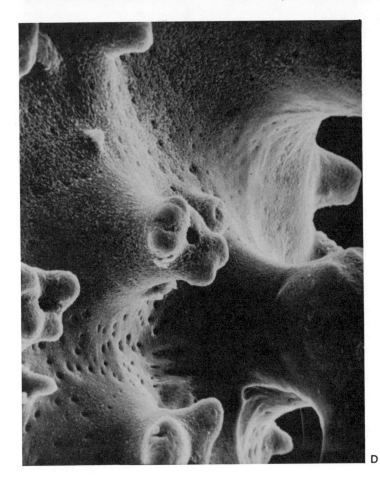

D

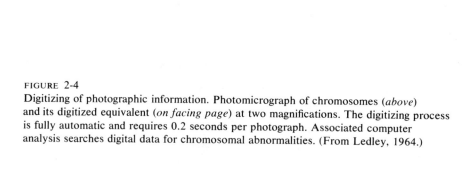

FIGURE 2-4

Digitizing of photographic information. Photomicrograph of chromosomes (*above*)
and its digitized equivalent (*on facing page*) at two magnifications. The digitizing process
is fully automatic and requires 0.2 seconds per photograph. Associated computer
analysis searches digital data for chromosomal abnormalities. (From Ledley, 1964.)

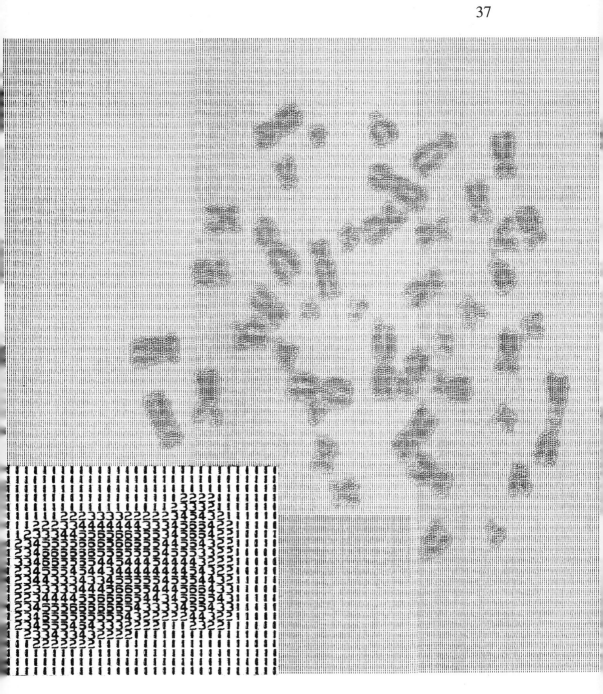

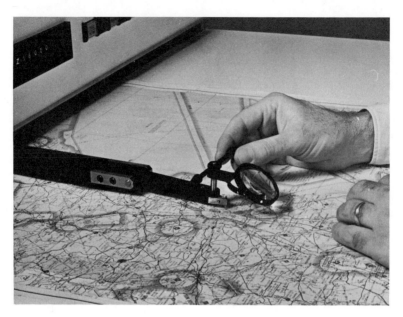

FIGURE 2-5
Automatic digitizing machine. Drawing or photograph to be
digitized is placed on drafting surface. Points located by
the stylus are recorded in X, Y coordinates on magnetic tape.
(Photograph provided by Concord Control Inc.)

The Line Drawing

The most common and often the most effective step toward limiting description is to draw a sketch of the salient aspects of a fossil structure, illustrating certain attributes and ignoring others.

Some discussion is in order concerning the techniques of drawing. Sloppiness is to be avoided not only because it introduces unnecessary "noise," but also because it introduces *uncontrolled* subjectivity. An accurate drawing is a rigorously executed replica. Because a drawing is usually two-dimensional, it is most appropriately made from a two-dimensional projection of the entire object. For this reason, drawings are commonly traced from photographs. As an alternative, a variety of instruments have been developed for projecting photographs or optical images onto a drafting surface. The camera lucida is the simplest such instrument.

The line drawing is more readily converted into machine-recognizable form than is the photograph because it contains less information. A scanning system need only distinguish between black and white to transcribe sketched information. By the same token, computer processing of such information is easier because less information is involved. For some paleontologic purposes a line drawing is too complex and must be further simplified before it can be used.

Descriptive Terminology

By far the most common medium for describing a structure is the word or combination of words. The word is an extremely powerful and economical tool, and in some instances is worth a thousand pictures. All fossil horses may be described as being either one-, two-, three-, or four-toed (with respect to the forefoot). By the use of these simple words, fossil horses can be divided into descriptively valid groups. The significance attached to these words is, of course, highly subjective. By using them we *imply* that a major difference (involving many attributes) exists among groups of horses that can be expressed in terms of the number of toes on the forefoot.

The use of descriptive terminology to describe form and structure has many obvious advantages. Most important, it reduces a great deal of information to a single word or relatively few words. If carefully chosen, the words are self-explanatory or relatively easy to learn. The terminology soon becomes a natural part of the vocabulary of the experienced biologist or paleontologist. Descriptive terms are very readily codified and thus can easily be placed in machine-recognizable form and are readily manipulated by a computer.

As an example of the effective use of descriptive terminology, the formal description (from Sohl, 1960) of a gastropod species follows. Photographs of a specimen of the species are shown in Figure 2-6.

> Shell small, trochiform, phaneromphalous with nacreous inner shell layer; holotype with about $7\frac{1}{4}$ rapidly expanding whorls. Protoconch smooth on early whorl with coarse axial costae appearing at slightly more than one whorl, followed almost immediately by fine spiral lirae; suture impressed. Whorl sides slope less steeply than general slope of spire giving an outline interrupted by overhang of periphery of preceding whorls; periphery subround to subangular; whorl side slopes steeply below periphery to broadly rounded base. Sculpture of axial and spiral elements same size; 8 spiral lirae on upper slope possess subdued tubercles where overridden by somewhat coarser and closer spaced axial cords; base covered by about 10 unequally spaced spirals with poorly defined tubercles and numerous axial lirae. Umbilicus narrow, bordered by a margin bearing low nodes. Aperture incompletely known, subcircular, slightly wider than high and reflexed slightly at junction of inner lip and umbilical margin.

To a person unfamiliar with gastropod morphology and its descriptive terminology, this description may be nearly unintelligible. For the person acquainted with the subject, however, the description should provide a convincing sketch of a group of specimens.

Any system of descriptive terminology, though, can lead to difficulty. Note the word "small" in the first line of the description. This implies a size comparison with other organisms—but what other organisms? Certainly all gastropods are small when compared with elephants, but large when compared with foraminifera or plant spores. In this description, the word "small" means small in comparison with other gastropods of this same

FIGURE 2-6
Cretaceous gastropod *Calliomphalus conanti*. See text for the formal
description of this species. (From Sohl, 1960.)

general type, or at most with all gastropods. By convention, terms such as
small, large, wide, and narrow imply comparison only with closely related
organisms. The dimension that conventionally denotes the general size of
gastropods is the maximum height of the shell. The use of the descriptive
term "small" is, therefore, a shorthand form of expressing a fairly rigorous
description. Once a worker has learned the basic vocabulary, he finds this
mode of description is remarkably efficient.

Notice the term "trochiform" in the first line of the description. It refers
to a general shape common among gastropods, which is illustrated together
with other common shapes in Figure 2-7. The origin of such terms, some
of which are several centuries old, is quite varied. Some terms are derived
from descriptive words: "turbinate" comes from the Latin *turbinatus,* meaning top-shaped. Others are from geometry: conical, biconical, obconical,
and so on. Still others are derived from the names of particular organisms
(genera or species), which display the form well or are common enough to
be familiar.

A common problem encountered in using descriptive terminology is that
of deciding where one category leaves off and another begins. What do we
do, for example, if we are faced with a gastropod shell that is somewhere
between naticiform and turbinate? One way of solving such problems is
simply to add more terms for the intermediate shapes. Thus we have such
terms in gastropod description as high turbinate and low turbinate. Inevitably—however many categories we may have—some fossils will be
found that are best described as being somewhere between two or more of
the categories. A system should have few enough categories to be efficient,
yet enough that the whole body of material may be subdivided meaningfully.

One reason that intermediate forms do not present more problems than
they do, is that form in the animal and plant worlds is not randomly distributed. Certain shapes are much more common than others. If a system
of descriptive terminology accurately reflects the natural clusters of pre-
dominant forms, then relatively few specimens that are between two cate-

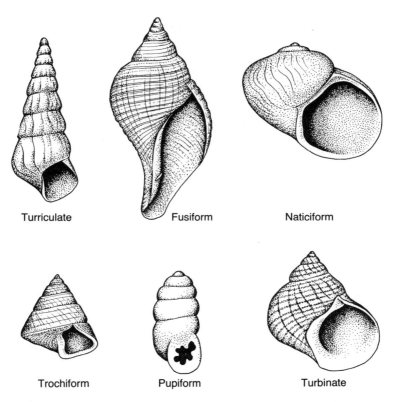

FIGURE 2-7
Some of the terms for shape used in morphologic description of gastropods.

gories will be found. Thus, the establishment of a system of descriptive terminology may be an important scientific contribution in itself, representing a fundamental interpretation of the natural world.

The risk, of course, is that errors, or misinterpretations, will be promulgated. Many completely inappropriate systems of descriptive terminology have become fixed in paleontologic literature. Because they do not represent natural groupings, they have obstructed and delayed development of meaningful interpretations. The risk must be looked upon as one we knowingly take when we move beyond the photograph as a means of description.

Description by Measurement

If precise measurements of fossil form are used rather than descriptive terminology, some problems are avoided but others arise. As long as enough precision is used in measurement, the problem of intermediate forms does not exist. Also, measurements are machine-recognizable data and can be manipulated directly and efficiently. Manipulation of quantitative data enjoys a good reputation. It is generally thought that the quantitative approach in science is the only truly objective approach. In actual fact, measurement is often the most subjective of the descriptive methods.

Consider the measurements indicated in Figure 2-8, in which length and height of a bivalve shell are defined by linear dimensions. Length is defined as the maximum linear dimension in an antero-posterior direction. Height is defined as the maximum dimension perpendicular to the length. These two dimensions yield quite a bit of information. They not only tell us something about the overall size of the shell but also tell us something about shape. The length to height ratio is a measure of shape. As the shell becomes more equidimensional, this ratio approaches one. As the shell becomes more elongate, the ratio increases.

The three bivalve shells in Figure 2-8 have nearly identical length to height ratios, but even a child could see the great difference in shape displayed by the specimens. Here, we see one of the greatest drawbacks of a system of description based solely on measurement. The obvious differences in these shapes are not reflected by their length to height ratios. If we describe shape by giving two dimensions perpendicular to each other, *we are tacitly as-*

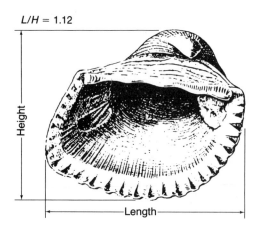

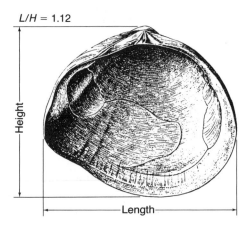

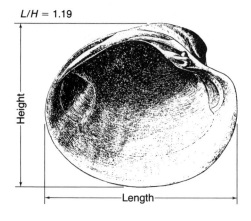

FIGURE 2-8
Interior views of three bivalve shells. These forms all have approximately the same ratio of the two measurements selected as principal dimensions. (From Vokes, 1957.)

suming that the shape is rectangular. If we knew in advance that all bi-valve shells were rectangular, there would be no problem; the length and height measurements or the length to height ratio would be completely descriptive. The important thing to remember here is that *when we choose to define shape by a set of dimensions, we are assuming a model. We are assuming that there is an ideal geometric form, and we can only measure or observe differences in shape with reference to this form.*

Four hypothetical coiled cephalopods are shown in Figure 2-9. For each, the largest diameter, *d,* is indicated, which distinguishes the large forms from the small. The diameter, however, tells us little about shape. The model tacitly assumed is a circle. Any number of quite different spiral cephalopods can be inscribed in the same circle, and thus the diameter gives us little indication of anything except size. For each of the shapes in Figure 2-9, a pair of unequal radii (r_1 and r_2), each originating in the morphologic center of the spiral, gives us more information. Notice that the ratio of the two radii in A is larger than their ratio in D, and that this, indeed, reflects some of the differences between the two shapes. (Note that the sum of the two radii is slightly less than the diameter.) The use of radii in describing coiled shapes can be carried much farther. It has been known for one hundred and fifty years that the form of many spiral invertebrate shells is mathematically rigorous. In the coiled cephalopod, radii measured at equal angular intervals

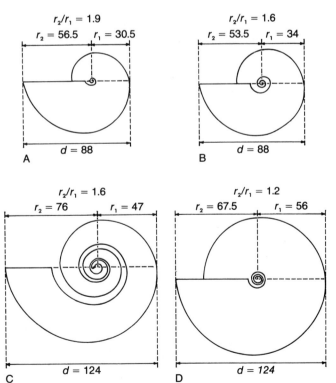

FIGURE 2-9
Four generalized shell forms
common in coiled cephalopods.
(See text for discussion.)

around the shell usually maintain a constant ratio to each other. This is illustrated in Figure 2-10, in which the ratio of radii separated by 180 degrees is always the same. With this in mind, we can look at Figure 2-9 again and describe each of the cephalopod shell forms by the ratio of radii separated by half a revolution. Using this method of description, we can successfully distinguish three cephalopod forms. The differences between B and C are evident in the drawings, but the two forms are indistinguishable from each other by the method of comparing ratios of radii.

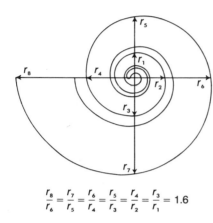

FIGURE 2-10
Generalized cephalopod shell form illustrating the constancy of the ratios of radii.

$$\frac{r_8}{r_6} = \frac{r_7}{r_5} = \frac{r_6}{r_4} = \frac{r_5}{r_3} = \frac{r_4}{r_2} = \frac{r_3}{r_1} = 1.6$$

In description by measurement, the number of measurements and their precision are generally less important than the *choice* of what is to be measured. We could have employed a wide variety of other linear dimensions of the cephalopods to describe their form, but probably no other set would be as economical or direct as the ratios of radii (though it should be kept in mind that more dimensions would be necessary to distinguish B and C in Figure 2-9). At best, measurement involves a selection of information. That is, only a part of the information that would be included in a photograph is used, and thus morphology is generalized and simplified. Whether this is justified in a given inquiry depends entirely upon the problem and upon the scientific questions being asked. There is no unique set of dimensions that most satisfactorily describes a given organism. However, it is fairly common for a basic set of dimensions to be adopted by most paleontologists for a particular fossil group. The dimensions accepted for the cephalopods, for example, are somewhat arbitrary, but quite effective in communicating to the paleontologist an impression of the morphology. The conventional set of measurements often serves as an alternative to descriptive terminology.

Measurements of dimensions can be used for describing variation among specimens. When such measurements of a large number of specimens are available, statistical analyses of variation can be made.

After it has been decided what dimensions will be measured for a given investigation, two problems remain: what method of measurement should be used, and what degree of precision should be sought. All measurements thus far discussed were of linear dimensions: that is, a curved surface projected onto a plane for measurement. Projection makes measurement easier and is generally valid, but depends upon a simplification: a curved surface is assumed to be planar. The same assumption is made when calipers are used to measure the distance between two points on the curved surface of a specimen. Measurements made on photographs produce no serious distortions as long as the plane of the photograph represents the desired projection. Any optical projection of an image (including photography) produces some distortion, which should of course be minimized. What distortion is tolerable depends upon the desired precision and accuracy of the results.

No general rules can be made governing precision in measurement. What precision is appropriate and valuable depends entirely upon the problem being tackled. For example, if we wanted to distinguish between a trilobite and a dinosaur, measurement to the nearest foot would suffice. To measure to the nearest inch or hundredth of an inch would be superfluous. If, however, we wanted to distinguish between two very similar organisms, much greater precision would be necessary. The expert uses only as much precision as his problem demands.

Both elaborate and simple devices for measuring fossils have been invented. Most of them are for specific purposes. An example shown in Figure 2-11 was designed by Alcide d'Orbigny in 1842 for measuring the so-called apical angle of spired gastropods. It is a very simple device employing a protractor and straight edge, but serves the purpose of basic description.

The digitizing method discussed earlier is often applicable to the problem of measuring dimensions. The straight-line distance between two points is easy to calculate, given their coordinates. For a single measurement or a very few measurements on each specimen, the digitizing method is not time saving. However, if many measurements are to be made on each specimen, digitizing and automatic computation are of great benefit. The precision possible by this method depends only on the precision with which a point can be located in a coordinate system.

FIGURE 2-11
Instrument designed by d'Orbigny in 1842 for measuring the apical angle of gastropod shells. (From Thompson, 1942.)

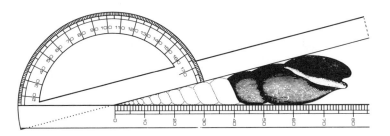

Describing Internal Structures

Most of the descriptive modes discussed thus far are usually applied to the surface of a fossil. Often equally or more important are internal structures such as the fabric of crystals making up a shell or the internal pore pattern in a mammal bone. The problems in describing internal structures are much the same as those for external form. But the techniques are different. We have seen that special types of photography, such as X-ray photography, yield information on internal structures. A specimen that is not amenable to such photographic investigation must usually be cut or sectioned to show its internal structures. The result may be a polished section that can be observed in reflected light or a thin section that can be observed in transmitted light.

The references at the end of this chapter include several basic summaries of the range of techniques available for paleontologic description. It should be kept in mind that new techniques are constantly being developed, many in fields far removed from paleontology, which have direct applicability to problems of describing internal structures in fossils. Historically, some of the greatest advances in paleontology have been made by those who have successfully adapted techniques from other disciplines.

Supplementary Reading

Brown, C. A. (1960) *Palynological Techniques*. Baton Rouge, La., 188 p. (A privately published handbook of techniques designed for work with fossil pollen and spores.)

Camp, C. L., and Hanna, G. D. (1937) *Methods in Paleontology*. Berkeley, University of California Press, 153 p. (A classic treatment of paleontological techniques.)

Kummel, B., and Raup, D. M., eds. (1965) *Handbook of Paleontological Techniques*. San Francisco, W. H. Freeman and Company, 852 p. (A collection of 86 articles written by specialists on various aspects of paleontological techniques, including collecting, preparing and illustrating; it also contains comprehensive bibliographies of techniques.)

McLean, J. D., Jr. (1959–date) *Manual of Micropaleontological Techniques*. Alexandria, Va., McLean Paleontology Laboratory. (A continuing publication in loose-leaf form covering many topics in the preparation and laboratory treatment of microfossils, particularly foraminifera.)

CHAPTER 3

Ontogenetic Variation

This chapter is the first in a series dealing with the problems of describing and interpreting variation in fossils. Differences between fossil specimens result from innumerable biologic and geologic causes. Difference in age of individual organisms at the time of their deaths is one of the most important of the purely biologic factors. Two individuals may have been genetically identical and may have lived in identical environments, yet their fossilized remains may be strikingly different simply because one was older than the other at the time of death. The differences are not limited to size. More often than not, growth is accompanied by changes in form. Growth stages may in fact be so different that fossil specimens in different stages may not be recognizable as members of the same species, particularly with species whose *ontogeny,* or normal life cycle, includes a metamorphosis, as does that of arthropods.

Ontogeny must be understood in order that fossil specimens which are members of the same species can be recognized and in order that the range of morphology displayed by a species can be assessed and interpreted properly. Ontogenetic change in morphology is just as important to the total description of an organism as is the adult form, especially in the study of organic evolution. To consider an evolutionary series as a sequence of adult forms is to oversimplify. Rather, evolution must be looked upon as a sequence of ontogenies.

48

Types of Growth

Organic growth is extremely complicated; it usually involves several types of change, among which are changes in cell size, number of cells, number of cell types, and relative positions of cells. During life, an organism may change in form abruptly (undergo metamorphosis) or it may change gradually.

All organisms that depend on a hard skeleton—for support, for protection, or for muscle attachment—must enlarge the skeletal structure to accommodate growth of the soft body. Postembryonic growth of skeletons is accomplished in four ways.

ACCRETION OF EXISTING PARTS

Most shelled molluscs increase their skeleton size simply by adding new material to the shell throughout life, which has the obvious advantage of permitting continued use of skeletal material deposited at earlier ontogenetic stages. It has the disadvantage, however, that the form of the juvenile shell must be incorporated as part of the adult shell. This type of growth is well illustrated by gastropods (see Figures 2-6 and 2-7). The shell of the coiled gastropod may be looked upon as a hollow tapering tube that is coiled about

FIGURE 3-1
Relation between number of plates and specimen size in the echinoid *Dendraster*. The data come from counts of plates constituting one portion of a plate column in a series of specimens from a single living population. The rate of addition of new plates (as a function of size increase) decreases during ontogeny. (From Durham, 1955. Reprinted by permission of The Regents of the University of California.)

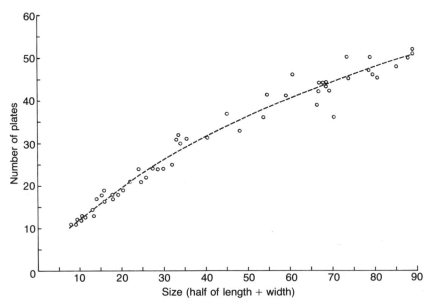

an axis. As the animal inside the shell grows larger, new material is added to the aperture, or opening, of the tube. Growth lines showing increments of growth can often be seen paralleling the outline of the aperture. Most adult gastropods occupy their entire shells and thus the shape of the soft body of a gastropod is, in effect, an internal mold of the shell.

Growth by simple accretion is found in many other animal groups, particularly among animals whose shell is external and serves primarily for protection and muscle attachment.

ADDITION OF NEW PARTS

A common method of skeletal growth for those organisms whose skeletons consist of many parts, either tightly articulated or fitting loosely in the soft tissue, is addition of new skeletal parts. Most echinoids, for example, have a rigid skeleton made up of as many as several thousand tightly articulated calcite plates, arranged in a radial fashion, with columns of plates extending from one "pole" of the crudely spherical skeleton to the other. The continuous addition of new plates largely accounts for growth. The graph in Figure 3-1 relates the number of plates in a portion of one of these columns to the size of the skeleton for a series of individuals of one species. Note, however, that the rate of addition of new plates as a function of change in skeletal size is not constant. An important fact about the growth of the echinoid skeleton is that new plates are always added at the same place, namely at the tops of the plate columns.

Addition of new parts is an integral part of the growth of many other organisms. Figure 3-2 shows several stages in the ontogeny of a trilobite. One of the more obvious ontogenetic changes is the gradual addition of segments to the thorax.

MOLTING

The trilobite shown in Figure 3-2 uses another basic mechanism of growth, the periodic shedding of the entire skeleton and formation of a new one to accommodate the increase in size of the soft parts. Figure 3-3 shows a plot of cephalon length against width in an assemblage of trilobites of one species. As can be seen from the graph, the measurements fall into clusters, each representing a molt stage. Differences between points in a cluster represent minor differences in size and shape among the individuals of that stage. Specimens represented by points intermediate between the clusters are rare.

Skeletal growth by molting has one clear advantage over both growth by accretion and growth by addition of parts. The skeleton of the juvenile individual need not form part of the adult skeleton, and the adult has much more freedom to change shape during growth. At the same time, growth by molting has decided disadvantages in that the organism must pass through rather perilous periods when it lacks a hard skeleton and it must expend considerable metabolic energy in replacing its entire skeleton at intervals.

In terms of numbers of species, skeletal growth by molting is by far the

50

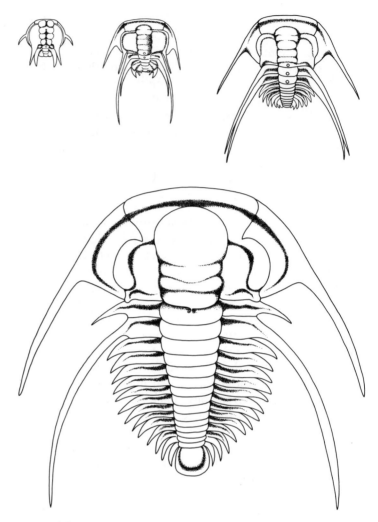

FIGURE 3-2
Four stages in the ontogeny of the trilobite *Paradoxides*. Ontogeny is accompanied by radical changes in the form of existing parts and by the addition of new parts. (From Whittington, 1957.)

most common mode of growth among organisms possessing a skeleton. This stems simply from the fact that arthropods (which make up about three-quarters of all living animal species) employ this mode of growth. In other groups of organisms molting is occasionally encountered, but usually as part of a major metamorphosis following a larval stage.

MODIFICATION

This mode of growth is most common in the formation of certain bones of higher vertebrates. By an elaborate process of replacement and re-formation of skeletal materials, the form and structure of the bone changes

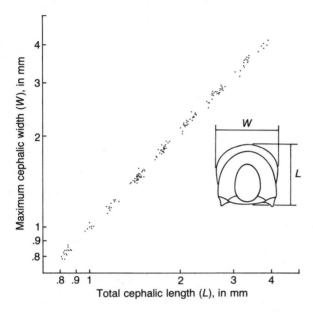

FIGURE 3-3
The ontogeny of the cephalon of the Ordovician
trilobite *Trinodus* as measured by the development
of length and width dimensions. Clusters reflect
molt stages. (From Hunt, 1967.)

as size increases. This mode of growth shares with the molting system the
advantage that skeletal form at one stage of growth is largely independent
of the previous stages but not the disadvantage that the organism is without
a skeleton for certain periods.

It is difficult to classify organisms rigidly according to mode of skeletal
growth, primarily because many organisms employ a combination of the
basic mechanisms just described. Figure 3-4 shows the product of such a
combination, a pair of plate columns from an echinoid. As we have seen,
addition of new plates is a principal means of increasing skeletal size in echinoids. At the same time, however, the existing plates grow by peripheral
accretion. Major periods of accretion are recorded by growth lines that trace
the history of change in size and shape of the plates. Some growth lines are
annual but others are not.

A second example of a combination of the same two modes of growth is
shown in Figure 3-5. Here we have a coiled cephalopod in which the outer
shell grows by *accretion* in much the same manner as in gastropods. But the
two organisms are quite different biologically. The gastropod fills the entire
shell whereas the body of the cephalopod is contained in what is known as
the body chamber, which occupies the last half or three-quarters of a revolution about the coiling axis. The body chamber is separated from the rest
of the shell (phragmocone) by a calcified wall called the septum. During
ontogeny, the animal increases in size and moves outward in the shell so

FIGURE 3-5
X-ray photograph of the living cephalopod *Nautilus* illustrating growth by addition of parts and by accretion. During ontogeny, shell length is increased by accretion and number of chambers is increased by addition of septa. (Photograph by R. M. Eaton.)

FIGURE 3-4
Growth by a combination of addition of new parts and accretion of existing parts. A pair of plate columns from the echinoid *Strongylocentrotus* is shown. Each plate contains a "nest" of growth lines reflecting its accretionary history. New plates are added always at the tops of columns. (From Raup, 1968.)

that it is always in a position near the opening. New partitions or septa are deposited periodically to accommodate the movement of the animal forward in the shell. A succession of septa is thereby deposited, representing an *addition of parts*. The differences in growth between the gastropod and the cephalopod are closely related to physiologic and habitat differences. The cephalopod depends on having a portion of its shell empty in order to make it buoyant for floating or swimming.

Describing Ontogenetic Change

For the biologist, the description of changes undergone by an organism during ontogeny is relatively simple. He can often observe the growing animal directly. Furthermore, he can observe precisely the time at which ontogenetic changes take place. The paleontologist, on the other hand, can only rarely deduce how old an organism whose remains he finds was at the time of its death, and he has no opportunity to observe its growth. Two approaches can be used, however, to obtain a considerable amount of ontogenetic information from the fossil record. First, a "growth series" of specimens may be assembled (as in Figures 3-1, 3-2, and 3-3) and the change in shape with increasing size can be observed or measured. Second, the ontogeny of a single adult specimen may be described from features such as growth lines (like those in Figure 3-4), which really represent temporary cessation, or slowing, of growth.

The choice between these two methods depends largely upon the mode or modes of growth employed by the organism. If growth was by simple accretion with no resorption or modification of previous ontogenetic stages and if growth was recorded by recognizable growth lines, either approach may be used. The same is true if growth was solely by addition of parts. But if molting or modification was a part of the organism's ontogeny, the paleontologist must use a series of specimens to simulate that ontogeny. When either method may be used, the one employing a single adult individual is usually preferable because the difficulties of interpreting the variation between different specimens are avoided.

Figure 3-6 shows graphs of two brachiopod dimensions. The upper graph is based on measurements of approximately seventy-five fossil specimens of a species from a single outcrop and thus may be reasonably assumed to represent growth in a single population. The scatter of points forms a somewhat fan-shaped pattern that extends out from the origin. Notice that the scatter of points is slightly curved (convex toward the "width" axis). This may be interpreted as indicating that the rate of length growth increases relative to the rate of width growth as the animal becomes larger. To make this interpretation, we must assume that size increases as some function of time. The diagram contains no information, however, about the relationship between size increase and absolute time. All we can learn is something about the rate of increase in one dimension *relative* to the rate of increase in the other dimension.

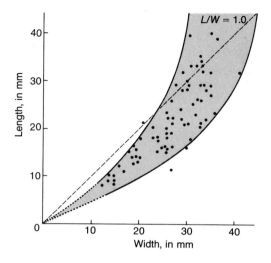

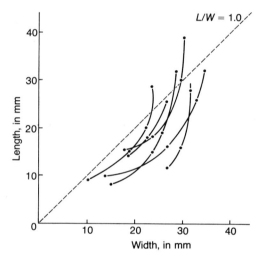

FIGURE 3-6
Ontogenetic change in shape in brachiopods.
Plots are of measurements made on an assemblage
of the Carboniferous brachiopod *Ectochoristites*.
Length refers to the antero-posterior length of the
pedicle valve; width is measured along the hinge
of the pedicle valve. Points connected by curved
lines in the lower graph are based on growth line
measurements of single specimens.
(From Campbell, 1957.)

The curved scatter of points implies a change in ratio between length and width, and therefore a change in shape during growth. A line drawn at 45° is included in the plots in Figure 3-6. All points below this line represent individuals whose width exceeds their length, while points above the line represent individuals whose length exceeds their width. The change in relative growth rates produces a gradual shift from a shell that is wider than long to one that is longer than wide. We could refine this interpretation by fitting a curve to the scatter of data. We would then have a generalized picture of the change in shape during ontogeny for this species with respect to the two dimensions.

What is the explanation of the departure of many of the points from a perfect curve? There are several possible reasons. One is that by describing ontogeny from data drawn from a variety of individuals, genetic differences between the individuals produce variability unrelated to ontogeny. A second possibility is that some of the specimens were damaged by purely geologic causes after death and that the scatter is thus not biologic at all. It is also possible that the method of measurement used was not precise enough and that the specimens themselves fall much closer to the idealized line than would be indicated by the observed scatter.

The lower graph in Figure 3-6 shows several growth curves for the same brachiopod species. The points on each curve refer to measurements of growth lines of a single adult specimen. This plot confirms that the growth curve for the species is curvilinear, and illustrates the importance of being able to trace the ontogeny of a single individual. Both graphs in Figure 3-6 yield essentially the same information—but the lower one gives us a more definitive picture of the growth pattern and of the differences between individuals in the fossil assemblage.

Figure 3-7 shows results plotted in graphic form of ontogenetic reconstruction based on measurements made on a single adult specimen—the coiled cephalopod shown in Figure 3-5. The specimen was measured for a series of radii (see Chapter 2 for discussion of the descriptive significance of radii in coiled cephalopods). Graph A in Figure 3-7 relates the length of a radius to its angular position. Because we know that the outer shell of a

FIGURE 3-7
Relation of length of radius to its angular position during the course of cephalopod ontogeny.
The measurements plotted here were made on the *Nautilus* specimen shown in Figure 3-5.

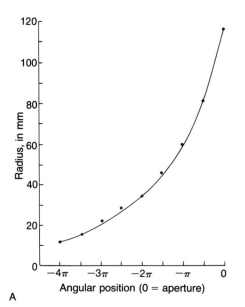

A

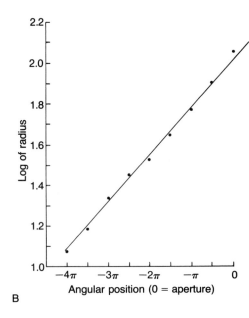

B

cephalopod grows by accretion, each angular increment of shell must represent an increase in age. The curved pattern of points indicates that the rate of radius increase changes relative to the rate of angular movement. (The scatter of points is much less than in the brachiopod graph because only one individual is plotted here.)

Does the curved pattern of points indicate a change of shape? We cannot answer this question directly because the dimensions plotted are not linear as were the brachiopod dimensions. The ontogeny of this cephalopod can be explored somewhat further by plotting the angular position of the radius against the *logarithm* of the radius (graph B in Figure 3-7). Notice that the scatter of points in the new plot falls quite rigorously on the straight line. Thus the angular position of the radius increases with the logarithm of the radius. In other words, successive radii are in geometric progression. We thus have what is known as a logarithmic or equiangular spiral. One of the mathematical characteristics of this spiral is that its shape remains constant. In this sense, we can conclude that the shape of the cephalopod does not change during ontogeny. The shape does change, of course, in that as growth proceeds, there are more and more revolutions about the axis.

In summary it may be said that of the two possible approaches to ontogenetic studies of fossils, the one based on a single specimen is easier to handle and the results generally more accurate. Nevertheless, much the same information can be deduced from a series of specimens if the non-ontogenetic variability inherent in a heterogeneous group of individuals is not excessive.

The examples of ontogenetic analysis shown thus far were done by measuring two morphologic attributes and plotting them against each other to observe their relative rates of development. This can and often is expanded to include more traits. It is possible when working with purely graphical techniques to add a third attribute, thereby seeing the relationship between three growth rates, but to add a fourth or a fifth or a sixth moves beyond the practical limits of graphical analysis. For this, more sophisticated "multivariate" methods of statistical analysis must be employed. Such methods have been used in paleontology with considerable success.

Coordinate transformation is a quite different approach to the description of ontogenetic change, and is particularly suited to describing and interpreting complex changes in a whole organism or an entire structure of an organism. In Figure 3-8, two trilobite growth stages are reproduced as sketches. Their scale has been adjusted so that the sizes of the two are comparable. This makes ontogenetic changes in form more evident. An arbitrary X-Y grid has been superimposed on the first stage. The same grid is indicated on the second stage but is deformed to express the change in shape of the grid necessary to accommodate the ontogenetic change in morphology.

Coordinate transformation was originally proposed by D'Arcy Thompson in his classic work *On Growth and Form*. Relatively little use was made of it, however, because it is very time-consuming and because subjectivity enters into the construction of deformed grids. The high-speed computer has changed things, however. Recently, computers have been combined with

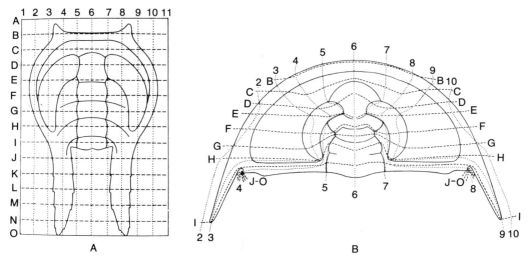

FIGURE 3-8
Trilobite ontogeny interpreted by coordinate transformation. An *X-Y* grid was
superimposed arbitrarily on stage A and the morphologic positions of grid intersections
were noted. After the grid intersections were located at morphologically homologous
points in stage B, the deformed grid was drawn on it. The deformed grid thus indicates
spatially the morphogenetic changes that transform A to B. (From Palmer, 1957.)

methods of "trend surface analysis" (developed in other areas of science),
permitting the paleontologist to create D'Arcy Thompson's deformed grids
with great speed and accuracy and at the same time to reason more rigorously as to the precise mathematical character of the deformation.

Growth Rates

Some plants and animals grow very rapidly; other grow very slowly. About
twenty years are generally required, for example, to complete human growth.
But a monkey completes its growth in about two years. Other organisms,
such as insects, may go through a complete ontogeny in a matter of days or
weeks. Such differences are largely hereditary; that is, they are under genetic
control and have been produced by evolution. In all organisms, however,
there is a certain amount of variability in growth rate that is not under genetic
control but that depends upon the specific conditions under which the organism lives. Factors such as nutrition, crowding, and other conditions imposed by the physical and biologic environment control growth rates to a
certain extent. The geneticist uses the term "norm of reaction" to express the
amount of nongenetic variation possible in a morphologic feature or in the
growth rate of that feature.

In nearly all organisms the rate of growth varies with time; that is, it
changes during ontogeny. Most of this change in growth rate is genetically

controlled although it, too, is subject to a norm of reaction. A typical pattern of growth with respect to time is shown in the first curve of Figure 3-9. Here size in a hypothetical organism is plotted against time. Size increases slowly at first, then more rapidly and finally much more slowly so that in the later stages, growth has nearly stopped. This variation in growth rate is shown more clearly in the second graph in Figure 3-9, where the *rate of growth* (the first derivative of size with respect to time) is plotted against time. In geometric terms, the second curve traces the change in slope of the first curve. The third graph shows the *rate of change* in slope of the first curve and is thus the second derivative of size with respect to time.

Some animals do not follow this general pattern of growth, but it is sufficiently widespread in the plant and animal worlds that we may use it as a general model. One of the principal differences between organisms lies in the form of the upper part of the curve in the first graph. The growth curve for most mammals, for example, actually levels off. The organism may continue to live for a considerable time after a plateau has been reached but growth will have stopped. The period after completion of growth is commonly referred to as *maturity,* or the true adult stage. In most plants, however, and in most invertebrate animals, growth continues throughout the life of the organism, although at a greatly reduced rate, making it difficult to define a true adult stage. At best we can define a stage at which most of the growth will have taken place, but growth will not have ceased.

When the growth of a real organism is measured, the data obtained seldom yield the perfect curve shown in Figure 3-9. This is because the norm of reaction for growth rates is generally quite large. If, during the development of an organism, a sudden improvement in nutrition takes place, the curve may be displaced upward, thus breaking the smooth pattern. Environmental variability often is so high that a clear pattern of growth through time cannot be deduced.

Regardless of environmental change during ontogeny, different parts of an organism typically grow at different rates. We have already seen an example of this in brachiopods (Figure 3-6) in which length and width grew at different rates. The different rates of growth of parts of an animal are well

FIGURE 3-9
Generalized pattern of growth relating the size S of an organism to time t. (From Medawar, 1945.)

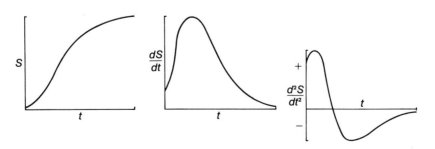

coordinated with respect to each other. If an environmental change causes a decrease in the rate of growth of one part, this is accompanied by a corresponding decrease in rate of growth of another part. The change in the relative size of the human head during ontogeny is well known. The head becomes *relatively* smaller as growth proceeds because it does not grow as rapidly as most of the rest of the body. The proportion of the size of the head to the size of other parts is always the same for a given point in ontogeny although environmental factors may be amplifying or restricting the person's overall growth. From an evolutionary point of view, the advantages of coordinated growth are obvious. The functioning of the entire organism depends to a large extent on maintaining a balance between sizes and proportions of various parts. If growth rates of various parts were under independent control, then normal changes in growth rate caused by the environment would produce an organism that would not be functional.

From the paleontologic viewpoint, the coordination between growth of the various parts is extremely important. It permits us to neglect the absolute time factor and to study the development of an organism and the change in shape of the organism without reference to and without measuring time. Figures 3-6 and 3-7 illustrate this. In both examples it was possible to learn a considerable amount about the ontogeny of the organisms involved, but in neither did we have any knowledge of growth rates with respect to absolute time.

Nearly all paleontologic studies of ontogeny require the measurement of change in one morphologic attribute in relation to change in another. We can define two basically different types of growth: *isometric* growth and *anisometric* growth. If the ratio between the sizes of two parts of an organism does not change during ontogeny, we have isometric growth (Figure 3-10, A and B). If the ratio does change, we have anisometric growth (Figure 3-10, C and D). To say that growth is isometric implies that there is no change in shape during growth; if growth is anisometric, there must be change in shape. Two kinds of isometric growth are shown in Figure 3-10. In the pattern of isometric growth plotted on graph A both parts grow at precisely the same rate. This yields a straight line at 45 degrees to either axis. In the pattern of isometric growth plotted on graph B one part grows more rapidly than the other but the ratio between them is constant, giving a straight line that is separated from one axis by a smaller angle than from the other. In the pattern of anisometric growth plotted on graph C, the X part grows more rapidly than the Y part at first but this relationship is subsequently reversed. The plot of growth is a curve. The pattern of growth plotted on graph D is different from all of the other three patterns in that the plotted line would not pass through the origin if extended. In pattern D as in pattern C, shape changes with growth; that is, growth is anisometric. To summarize, growth is isometric if the plotted line is straight *and* passes through the origin (or would do so if extended). All other conditions produce anisometric growth.

Anisometric growth is by far more common than isometric growth. In other words, *shape change during ontogeny is the rule rather than the exception.* Anisometric growth has been the subject of a tremendous amount of

research in both biology and paleontology and some important generalizations have resulted. Principal among these is that anisometric growth very
often may be approximated by the following exponential relationship between the two parts:

$$Y = bX^a,$$

where a and b are constants and $a \neq 1$. Growth that proceeds according to
this relationship is called **allometry**. (It should be noted that isometric growth
obtains when the constant a is one.) Allometric growth could be the subject

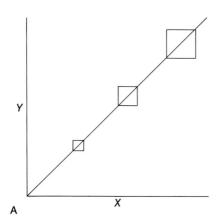

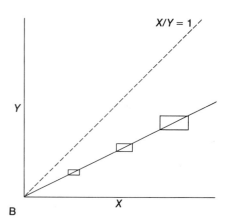

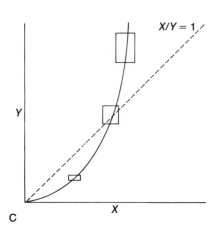

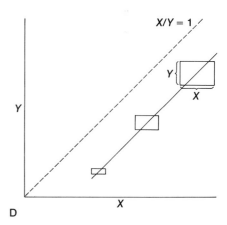

FIGURE 3-10
Typical patterns of growth observed when two morphologic dimensions are plotted against
one another. In each graph, the shape of a hypothetical organism is shown by a square
or rectangle at three ontogenetic stages. In A and B growth is isometric, and there is no
change in shape; in C and D growth is anisometric, and shape is continually changing.

of several chapters. The student should refer to the works of Huxley, Thompson, and Gould listed in the references at the end of this chapter. (Several authors have defined the term allometry more broadly to include all anisometric growth.)

Reasons for Anisometric Growth

So far we have considered anisometric growth and the resulting changes in shape during ontogeny simply as a biologic process and have concentrated on some of the problems of defining and describing it. Of greater interest, however, and of considerably greater importance is the *interpretation of change of shape* during ontogeny. The modern student of evolution subscribes to the idea that changes in form during ontogeny are not the result of whim, are not accidental, but rather have become a part of the hereditary material because of their functional value. We must search, therefore, for functional or adaptive explanations for the changes in shape of a structure or in number of structures that we observe. Why, for example, does the shape of the brachiopod shell change during growth?

Where ontogenetic change is abrupt, as in the metamorphosis of an insect, it can often be related to a sudden change in mode of life. The frog develops through a succession of fairly gradual changes into an adult. The tadpole lives in and depends upon an aquatic environment: it extracts oxygen directly from the water. The adult frog, although partially dependent upon the proximity of water, is essentially a terrestrial organism. Some ontogenetic changes in frog anatomy are produced by addition of new structures and deletion of old; others are brought about by changes in relative rates of growth.

This sort of relation does not apply to the brachiopod. Immature and mature brachiopods occupy nearly the same environment. Both are aquatic. Both use basically the same methods of getting food, and in general have the same life activities. As far as we know the principal change during growth is simply in size. It is fruitful, therefore, to look more carefully into the consequences of size increase.

Consider a leg bone of a terrestrial vertebrate. The bone serves several functions, but one of the most important is to support the body. It must be strong enough to bear the weight of the body. The strength of a bone (or any other supporting structure for that matter) is approximately proportional to its cross-sectional area. Thus, a thick bone will be stronger than a thin bone regardless of its length. Now consider that the bone must grow in order to accommodate an increase in size and mass of the body that is to be supported. Suppose that the growth brings about a doubling in all the linear dimensions of the animal. That is, the length of the body is doubled, the length of the bone is doubled, the diameter of the bone is doubled, and so on. We are thus assuming for the moment completely isometric growth.

The doubling of the linear dimensions produces an eight-fold increase in volume or weight. This stems from the simple geometric fact that as we increase the linear dimension of any structure (without changing shape), the volume increases as the cube of the linear dimension. If the linear dimensions of the bone that supports the weight have doubled, the cross-sectional area of the bone has increased by a factor of only four (area increases as the square of the linear dimension). It is thus inevitable that the volume of the animal supported by the bone increases more rapidly than the cross-sectional area of the bone. Therefore, the weight of the body increases more rapidly than the effective strength of the bones that support it. One way in which the organism can combat this differential is to increase the cross-sectional area of the bone more rapidly than would be the case if all linear dimensions of the bone were to grow in a constant ratio. This, in turn, demands a change in shape of the bone. That is, the bone must become relatively stouter in order to carry the increased weight of the body. This is in fact what occurs in the ontogenetic development of many terrestrial vertebrates.

Species that differ in size often show predictable differences in shape of supporting structures. Galileo was one of the first to recognize this phenomenon, the *principle of similitude,* and he illustrated it by the drawings reproduced here as Figure 3-11. He wrote: "I have sketched a bone whose natural length has been increased three times and whose thickness has been multiplied until, for a correspondingly large animal, it would perform the same function which the small bone performs for its small animal."

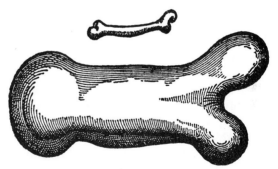

FIGURE 3-11
Galileo's illustration of the relation between size and shape in vertebrate bones. As the size of the bone increases in order to provide support for a vertebrate's increasing mass, the ratio between the bone's diameter and its length must increase if it is to be strong enough. (From Bonner, 1952.)

We can extend the general line of reasoning to include a wide variety of structures. The efficiency of many respiratory devices, for example, depends upon the surface area of the respiratory surface. If the lung of a human adult were simply a scaled-up replica of the child's lung, the efficiency of respiration would be greatly reduced in the adult. The respiratory requirements of the adult increase as the cube of the linear dimensions whereas the sur-

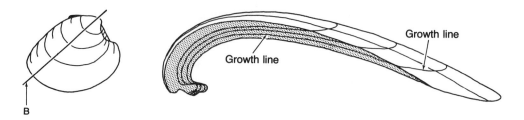

B

FIGURE 3-12
Cross section of a bivalve shell showing the mode of
accretionary growth. The shell is continually thickened so that
its strength keeps pace with the increase in overall size.
(From Pannella and MacClintock, 1968.)

face area of the lung (if it does not change in form) increases only as the
square of the linear dimensions. For this reason, the ontogeny of the human
is accompanied by tremendous elaboration of lung tissues, which enables
the increase in surface area to keep pace with the increase in total body size
(expressed by volume).

A classic example of the relation between surface area and volume is
found among insects. If we could so stimulate growth of a mosquito that it
grew to be 10 feet long but was otherwise a perfect replica of a normal
mosquito it could not survive, because the relation between its surface area
and volume would be unsuitable for temperature regulation and respiration.
In order for a mosquito of such a length to be viable, its morphology would
have to be altered drastically.

Let us return to structural changes more likely to be represented in the
fossil record. Figure 3-12 shows a cross section of the shell of a bivalve.
The bivalve grows by accretion in much the same way that the gastropod or
cephalopod does. In fact, as we shall see in later chapters, the geometry of
the bivalve is not substantially different from that of the other molluscan
groups that we have considered. We may look upon the individual valve of
the bivalve as a hollow tube that tapers from a large open end to a small
"umbo" or "beak." The principal difference is that the shell of the bivalve
tapers much more sharply than does that of the gastropod or cephalopod.

In Figure 3-12 the cross section of the bivalve shell is marked by growth
lines, which can be seen in the shell itself. These growth lines represent
surfaces of accretion. New shell material is added, however, not only at the
leading edge of the shell, but also throughout most of the shell interior. The
primary function of the thickening is to make the preexisting shell thick
enough and therefore strong enough for the adult animal. Gastropods and
cephalopods employ this thickening of the preexisting shell to a much lesser
extent. They are able to "get away" with this only because their shell surface is more highly curved and thus has an inherent strength lacking in the
less curved bivalve shell.

Many other examples of the effect of size on ontogenetic variation could be cited for groups of animals common in the fossil record. Many gaps in our knowledge exist, however. We cannot, for example, offer a convincing explanation for the change in shape of the brachiopod shells discussed earlier. Perhaps the change is size related. There are many fossil groups for which we do not have sufficient knowledge of biologic functions to be able to interpret the observed ontogenetic variation. If such knowledge could be obtained it would be applicable not only to interpreting differences between growth stages of a single species, but also to the understanding and interpretation of differences between species.

In summary, ontogenetic change in shape accompanies two other sorts of change: change in mode of life and change in size. Change in size is usually gradual, as is the change in shape accompanying it. Changes in mode of life may be sudden or gradual. The ontogenetic change in shape observed in the brachiopod may correspond to the gradual cumulative change in size or it may correspond to a rather subtle change in mode of life that has not yet been recognized.

Ontogeny of Colonial Organisms

So far, we have been concerned with organisms that live as autonomous individuals, not intimately affected by or dependent upon others. As we study the ontogeny of colonial organisms several special problems arise. Many of the colonial species in the fossil record belong to the coelenterates and bryozoans.

In colonial corals, the individuals, or corallites, are often bound together by an integrated skeleton. Obvious effects of this mode of life are found when we compare an individual near the center of a colony with one on the margin. The relative lack of crowding on the margin of the colony may produce an ontogeny quite distinct from that of an individual completely surrounded by others. (A comparable effect is sometimes noticed in noncolonial organisms, such as oysters, which in certain environments grow so closely together that the growth of one affects the shape of its neighbors.)

Among colonial corals, the first member of a colony must attach to some solid substrate, such as a pebble, an empty shell of another organism, or even a dead coral. As the colony develops, new individuals attach to older parts of the colony. The ontogeny of a given individual varies considerably with its place in the ontogeny of the colony as a whole.

In some colonial organisms (including corals and bryozoans), there is morphologic and physiologic differentiation between individuals in a colony. That is, certain individuals perform a specialized function, such as reproduction or defense, which benefits the colony as a whole. These individuals may or may not be genetically different. The differences between them may be such that we must consider their ontogenies as individuals as well as the growth of the colony as a unit.

Supplementary Reading

Bonner, J. T. (1952) *Morphogenesis*. Princeton, Princeton University Press, 296 p. (An outstanding biological treatment of growth and form.)

Clark, W. E. Le Gros, ed. (1945) *Essays on Growth and Form Presented to D'Arcy Wentworth Thompson*. Oxford, The Clarendon Press, 408 p. (A collection of essays on various aspects of ontogeny.)

Gould, S. J. (1966) Allometry and size in ontogeny and phylogeny. *Biol. Rev.*, **41**:587–640. (An up-to-date and authoritative review of the problem of allometry. The author uses the term "allometry" to include all anisometric growth.)

Huxley, Sir J. S. (1932) *Problems of Relative Growth*. New York, Dial Press, 276 p. (A classic reference on ontogenetic phenomena.)

Macurda, D. B., ed. (1968) *Paleobiological Aspects of Growth and Development, a Symposium*. Paleont. Soc. Mem. 2, 119 p. (A series of articles on skeletal growth.)

Thompson, D'A. W. (1942) *On Growth and Form*. Cambridge University Press, 1116 p. (A classic work on ontogeny. Essential reading for all interested in the interpretation of morphology.)

CHAPTER 4

The Population as a Unit

All plant and animal species exhibit variation; even genetically identical twins exhibit slight differences, due to differences in environment and to chance. Because of variation, description of a single specimen does not suffice to describe a species.

Populations in Biology

Most species do not have continuous or uniform geographic distribution. Individuals tend to be clustered because of chance, uneven distribution of habitats, gregarious behavior, or geographic barriers. Figure 4-1 shows an example of the discontinuous distribution of a terrestrial plant. For our purposes, we may define *population* as a group of individuals living close enough together that each individual of a particular sex has an approximately equal chance of mating with a certain individual of the other sex. (The terms *local breeding population* and *deme* are often given similar definitions.) The population shares a single *gene pool*. The gene pools of adjacent populations of a species may be partially or completely isolated from each other. When two populations interbreed, there is said to be *gene flow* between them.

Subdivision of species into discrete populations is dramatically illustrated in the West Indies and other island provinces. Restriction to islands inevitably segregates terrestrial species into populations that are genetically separated. Similarly, many marine organisms are limited to shallow-water zones and thus are isolated from their counterparts around other islands. These examples are somewhat extreme, but similar distribution patterns are found throughout the whole biosphere. Furthermore, even if there are no geographic barriers, distance alone tends to reduce gene flow between organisms of a species that live in different parts of the geographic range.

FIGURE 4-1
Discontinuous distribution of the plant *Clematis fremontii.*
Dots (lower left) indicate individual plants. Note that the plants
are clustered at more than one scale of observation (there
are clusters within clusters). (Courtesy of Ralph O. Erickson.)

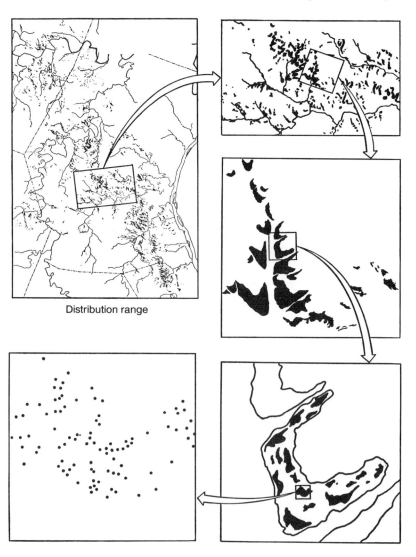

Distribution range

Individual Variation within Populations

ONTOGENETIC DIFFERENCES

In the preceding chapter, we discussed changes that take place during the ontogeny of a single individual. Obviously, a living population at any time contains variation due to age differences. At the very least, individuals will differ in size.

In species whose breeding season is fairly well defined, size variation may depend largely on time of year. Figure 4-2 shows frequency distributions (histograms) of specimen size in a living population of the bivalve *Mytilus edulis* for April and November of the same year. This species spawns during a short period of the year. The histogram for April is based on sampling just after a heavy spatfall and is clearly bimodal: the small, newly settled individuals make up the sharp peak on the left and the rest of the population (consisting of individuals one year old or older) makes up the broad area on the right. The November frequency distribution is also bimodal but both peaks have shifted to the right because most of the individuals in the population have grown. The left-hand peak has become considerably lower because some of the younger individuals have died, and broader because the growth rate varies from individual to individual.

The average age of a population may affect variation in characteristics other than size. In the preceding chapter we discussed a brachiopod species whose shell shape changed systematically through life (Figure 3-6). Any description of shape in a population of such a species is strongly biased by the age distribution of the specimens studied.

FIGURE 4-2
Relationship between age and size in a certain population. Frequency distributions of shell length in a population of *Mytilus edulis* living in Gosford Bay, Scotland. The population was sampled in April, 1961 (shaded diagram), and again in November of the same year (outlined diagram). The peak on the left in each diagram represents the younger bivalves. (From Craig and Hallam, 1963.)

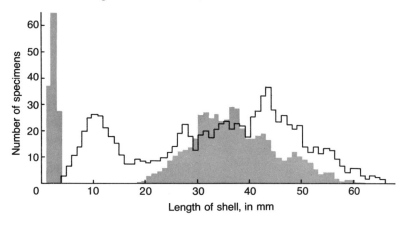

The complex life cycles of certain taxonomic groups produce different types of ontogenetic variation. Most species of foraminifera, for example, have a life cycle in which a generation produced sexually alternates with a generation produced asexually. Consequently, many species are dimorphic. The dimorph resulting from sexual reproduction is usually larger, but grows from a very small initial chamber. The dimorph produced asexually is usually smaller, but grows from a larger initial chamber. The dimorphs of a common Cenozoic genus *Pyrgo* are shown in Figure 4-3. In the cross-sectional view, some chambers are numbered to facilitate comparison. Dimorphism is especially pronounced in large species of foraminifera, in which the sexually produced dimorph may be as much as five times as large as the asexually produced dimorph.

GENETIC DIFFERENCES

In most bisexual organisms, the two sexes differ morphologically because of basic genetic differences. Sexual differences may or may not be reflected by fossilized skeletal morphology. In the example of probable sexual dimorphism in fossil ammonites that is included as Figure 4-4, the two sexes differ in overall size, in several aspects of ornamentation, and in aperture form. The differences are about as great as those observed between many different species and genera, and, indeed, the two sexes have sometimes been classified as separate species. In making a case that particular specimens in the fossil record are sexual dimorphs, the paleontologist must con-

FIGURE 4-3
Pyrgo fischeri (Recent). A: External apertural view.
B: External side view. C: Equatorial section of megalospheric
(asexually produced) test. D: Equatorial section of
microspheric (sexually produced) test. Some chambers in
C and D are numbered. (From Easton, 1960.)

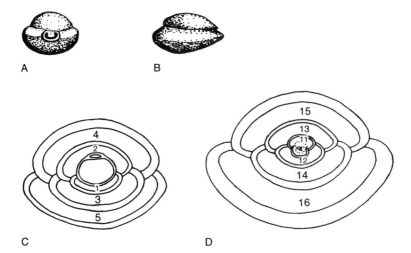

Sexual dimorphism as an expression of genetic variation. Jurassic ammonoid cephalopods from the Middle Jurassic of England show what has been interpreted as sexual dimorphism. Each larger form (the "macroconch") is considered to be the dimorph of the smaller form immediately below it (the "microconch"). In this case, the names predate the interpretation of dimorphism. All these forms were assigned to the same genus but different subgenera (given in parentheses) and species. (From Callomon, 1963.)

Genus *Kosmoceras*

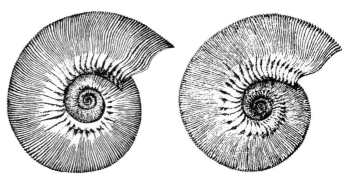

K. (Zugokosmoceras) grossouvrei   *K. (Lobokosmokeras) phaeinum*

K. (Gulielmiceras) aff. *gulielmi*   *K. (Spinikosmokeras) acutistriatum*

sider such factors as mutual occurrence and relative numbers of the two variants in question.

Even individuals of the same sex and growth stage differ, owing to genetic variation. Each individual carries only a small part of the total gene pool. Some genetically controlled attributes of an organism are more variable than others. In human populations, for example, the number of fingers and toes is virtually constant, but other genetic characteristics, such as hair color, body size, and fingerprints, vary widely.

Genetic variation in a population may be continuous or discontinuous. That is, it may be expressed by discrete and notable differences or by a complete spectrum of intermediate forms that overlap each other. Coiling direction in certain molluscs and foraminifera is an example of discontinuous variation. Coiling may be to the left (sinistral) or to the right (dextral). Specimens of a species in which both coiling directions are expressed are usually identical except that one type is a mirror image of the other. Both coiling directions may be present in the same breeding population.

Figure 4-5 shows an example of discontinuous variation from the fossil record. The spacing of tubercles on the pygidium of a Silurian trilobite varies within what is considered to be a single species. The tubercles appear on prominent rings that make up the axial portion of the pygidium. In fossil material studied by Best (1961), the first and third tubercles (counting from the anterior end of the pygidium) are six, seven, or eight rings apart. This variation was interpreted by Best as being genetic (perhaps deriving from only a single gene with two allelic forms). He also reasoned from the ratios of numbers of specimens showing the various spacings that these trilobites constituted a random sample of a single living population.

It is much more common for genetically determined variation to be continuous than discontinuous. Continuous variation in overall size of individuals of the same growth stage and environment is especially common, but may, however, be produced by chance events during growth or undetected environmental differences as well as by variations in genetic material.

How much genetic variation should be expected within a single population? A simple answer to this question would be useful, especially to the paleontologist, who could then expect to find a certain amount of variation within a fossil population; a larger amount could be used as evidence that what was being studied was the remains of more than one population. Unfortunately, there are inherent differences in the size and complexity of gene pools, making rigorous prediction of variation difficult or impossible. The problem is further complicated by the fact that in fossil populations, we can observe a genetic difference only through its morphologic expression.

The amount of variation in a population differs not only from species to species, but from time to time and place to place within species. Some of the controlling factors are:

Selection Pressure.   Genetic variation in a population is always subject to natural selection. Inevitably, some variants are better suited to a particular environment than others, and these individuals are likely to produce more viable offspring: the composition of the gene pool thus reflects the relative suitability of the several variants. Where relatively few of the possible variants are tolerated by an environment, we say that there is high selection pressure.

Mutation Pressure.   The ultimate source of nearly all genetic variation is gene mutation. Gene mutation is only partly understood, but its rate is known to vary from species to species and from population to population. All things being equal, variation in a population will be high if mutation rates are high.

System of Reproduction.   In organisms that always reproduce asexually (by simple fission or budding), there is little mixing of genetic material within populations and the potential for morphologic variation is low. As we have seen, a system of reproduction can combine sexual and asexual processes.

Anterior end
of pygidium

First tubercle

Second tubercle

Third tubercle

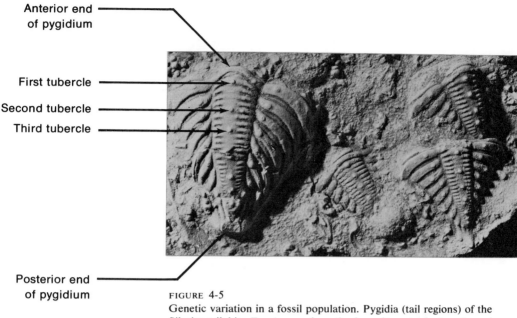

Posterior end
of pygidium

FIGURE 4-5

Genetic variation in a fossil population. Pygidia (tail regions) of the
Silurian trilobite *Encrinurus* show the four alternate modes of spacing
of the first (anterior) three axial tubercles. The second tubercle is
either three or four rings after the first; the third is either three or four
rings after the second. The ratios of numbers of specimens showing
these spacings have been interpreted as indicating that this fossil
assemblage is a random sample of a Mendelian population at equilibrium.
(From Best, 1961.)

This sort of combination is prevalent among plants and leads to clustered
variation, in which many individuals resemble each other closely but differ
markedly from individuals of other clusters.

We have touched on only a few of the factors that affect genetic variation
in populations. The amount of variation in a population cannot be predicted,
but the paleontologist or biologist who works with a group of organisms becomes accustomed to the amount of variation typical of populations within
that group and may become extremely adept at recognizing unusual amounts
of variation.

NONGENETIC DIFFERENCES

In Chapter 3 the "norm of reaction" was discussed. We saw that the
genetic control over the form of an organism is not complete and some variation is caused by the environment and some by chance.

Most nongenetic variation is attributable to ecologic differences. An extreme example of discontinuous nongenetic variation is found among social
insects, such as honeybees and termites, where the distinction between a
worker and a queen may simply result from the fact that one received more
or different food than the other during an early stage of development.

The shells of many invertebrate species that nestle in crevices and cavities in hard substrata exhibit continuous nongenetic variation. The shells of such species tend to conform to the shapes of confining cavities.

Some nongenetic effects of the environment are related to function and may even benefit the organism; others seem quite unrelated to function. An example of the former is the development of calluses on the hands and feet of primates. A callus is an acquired characteristic, being the result of the reaction between the animal's tissue and the physical environment. The callus is often an advantage as protection from abrasion and bruising.

Genetic composition of an organism (or a population of organisms) can rarely be observed directly. Separation of genetic and nongenetic factors is especially difficult because the same morphologic variants may sometimes be under genetic control and sometimes under that of the environment. Among echinoderms, for example, amounts of the isotopes oxygen-16 and oxygen-18 in the skeleton may differ markedly from species to species as an evolutionary phenomenon (that is, under genetic control), but differences of the same magnitude may also be produced by the effects of water temperature.

Fossil Populations

Fossil preservation commonly affects the variation evident within a population. The principal geologic process tending to increase apparent variation is distortion produced by compaction of sediments or deformation of sedimentary rocks. Reduction of variation as a result of fossilization is considerably more common, primarily because fossilization is rarely perfect. It destroys many features of organisms, obscuring differences that may have been very striking in the living population. Thus, few fossil assemblages exhibit as much variation as comparable living populations.

The paleontologist does not expect to find that an organism's soft tissues have been preserved; the absence of information about them is therefore predictable and must simply be tolerated. More serious is the biasing effect of fossilization. The preservation process serves as something of a filter for variation; that is, readily preserved organisms represent a disproportionately large part of the fossil record. Generally, individual animals with robust and solid skeletons are more likely to be preserved than members of the same species that happen to have lighter, more fragile skeletons, which may lead the unsuspecting paleontologist to characterize a species as having had a heavier skeleton than actually typified it. Selective destruction of ontogenetic stages is also common. For example, because of their fragility, few skulls of children are found in the Late Cenozoic fossil record (except at burial sites). The phenomenon of selective preservation is exceedingly complex and not yet well understood, but is potentially one of the most fruitful areas of paleontologic research. With the odds against preservation being high, a subtle morphologic difference may determine whether a particular specimen is preserved.

Sorting or selection during post-mortem transport is also important in influencing the composition of a fossil assemblage. A particularly striking example (with considerable paleontologic implication) is illustrated in Figure 4-6, based on a study by P. Martin-Kaye of bivalve shells collected on Trinidad beaches. The species he studied comprise burrowing individuals that live in nearshore environments. He was surprised to find that in a given area of beach, the left and right valves were not present in the equal proportions that would be expected from the fact that each living organism had one of each. Furthermore, he found that the ratio of left to right valves changed quite systematically along a single beach. Each of the species that Martin-Kaye worked with has virtually indistinguishable right and left valves except that one is the mirror image of the other. Representative specimens from one of the species are illustrated in Figure 4-6. The sorting is readily explained by the fact that wave action is not the only agent transporting the shells. Because waves usually approach beaches obliquely and produce a longshore current, most shells are not deposited on the beach directly opposite their life habitat, but are carried along by the current, finally being destroyed by abrasion or thrown up on the beach at some distance from their habitat. It appears that because left and right valves of the species in Figure 4-6 are asymmetrical mirror images of each other, they were not transported in the same manner by the waves. One valve tended to be sheared toward the beach and preferentially cast up on it.

The effect of differential transport of the Trinidad shells is striking, and its interpretation is reasonably obvious because we know that the ratio of left to right valves in the original population must have been exactly one. If we were to compare the assemblage of shells on the beach with original populations we probably would find other relations between morphology and transport.

Biases in the fossil record produced by differential transport are difficult to evaluate. In some groups of organisms, such as buoyant cephalopods, transportation after death is probably the rule rather than the exception. But we do not know how important post-mortem transport is among various types of plants and animals that normally live on a firm substratum. Recent work in oceanography and submarine geology has demonstrated widespread currents that are capable of transporting some bottom-dwelling organisms considerable distances.

A method commonly used for estimating transport distance for those bottom-dwelling marine species having two shells makes use of percentages of disarticulation. Valves of some brachiopods, for example, are more easily separated after death than others. From structural considerations it is often possible to estimate the types and amounts of agitation or transport that would be necessary to produce disarticulation of particular species. Amount of breakage and wear may be interpreted in the same manner.

Boucot, Brace, and Demar (1958) studied the effects of transport on three brachiopod genera in a large block of Lower Devonian sandstone. They found that two genera were each represented by nearly equal numbers of pedicle and brachial valves (*Leptocoelia* — 879 pedicle valves, 893 brachial

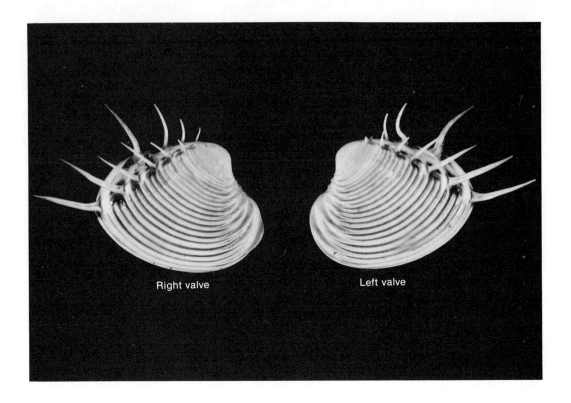

Right valve

Left valve

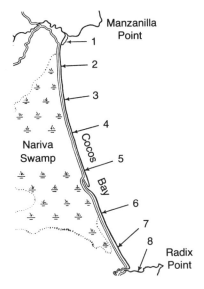

Manzanilla Point

Nariva Swamp

Cocos Bay

Radix Point

Locality	Total Number	Percent right valves
1	45	11
2	154	13
3	11	18
4	51	approx. 50
5	73	approx. 50
6	284	46
7	27	approx. 50
8	450	87

FIGURE 4-6
Effect of selective transport on bivalves. The proportions of right and left valves in beach assemblages collected on the eastern coast of Trinidad varied from place to place, but systematically relative to geography. The two valves of the bivalve *Pitar dione* are identical except that one is the mirror image of the other. (From Martin-Kaye, 1951.)

valves; *Platyorthis*—561 pedicle valves, 548 brachial valves). A third genus, *Leptostrophis,* was represented by 378 pedicle valves but only 35 brachial valves. The authors concluded that members of the first two genera had not been transported far from where they had lived and that the third genus had probably lived in a distant environment and had been transported after death by bottom currents. Of the two genera considered to have been preserved near their life site the percentage of whole *Leptocoelia* fossils was much higher than that of *Platyorthis* fossils. The authors concluded that *Leptocoelia* tended to disarticulate less readily than *Platyorthis.*

Population Dynamics

Population dynamics is the study of changes in population size and composition through time. Birth rate, growth rate, and death rate are the kinds of basic information considered. Reproduction in most animal groups is limited to a certain part of ontogeny, beginning at sexual maturation and ending late in ontogeny or at death. Birth rate (rate of offspring production) may remain constant for a single parent during this time or may change. Growth is usually rapid at first and then slower; it either ceases rather abruptly (as in humans) or its rate decreases continuously until death (as in most invertebrates). Death rate is often high early in life (as in humans); it often increases again later, either abruptly (as in humans) or gradually (as in many invertebrates).

Population biologists commonly make use of what is called a survivorship curve—a graph on which the number of survivors is plotted against their age. The data may represent a single population or several populations of a species. Usually the data are plotted on a logarithmic scale because, as illustrated in Figure 4-7, constant mortality then produces a straight line. (In Figure 4-7, 90 percent of existing individuals die each year.)

The paleontologist seeking to study population dynamics first faces the problem of determining whether a fossil assemblage actually represents a former population. As demonstrated in the previous section, this problem may not be easily solved. Not only do destructive and sorting effects produce change, but it is often difficult to determine whether all of the specimens in a fossil assemblage lived at the same time.

It was once assumed by many paleontologists that a simple size-frequency plot could be used as a means of distinguishing between altered and unaltered fossil assemblages of shells or skeletons. It was assumed that high infant mortality is the rule for most animal species and that accumulation of dead shells or skeletons with little or no selective transport or destruction should produce a fossil assemblage in which small individuals greatly outnumber large individuals. The size-frequency distribution for such an assemblage would resemble the one shown in Figure 4-8, A. A normal, bell-shaped size-frequency distribution (Figure 4-8, B) was then taken as evidence of selective sorting during transport, with the mean size determined by factors such as the velocities of transporting currents.

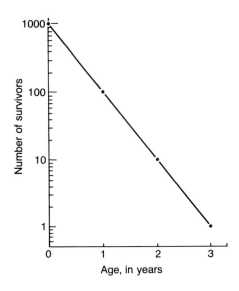

FIGURE 4-7
Survivorship curve showing a constant mortality
of 90 percent per year. The logarithmic scale on the
ordinate yields a straight-line plot.

FIGURE 4-8
Left-skewed (A) and normal (B) size-frequency
curves for a single-species fossil assemblage, once
(but no longer) thought to be evidence of
preservation with little transport or sorting
(A) and of sorting by currents (B).

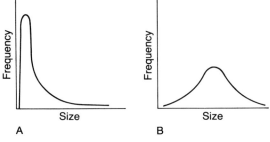

Craig and Hallam (1963) have pointed out, however, that for most bottom-dwelling marine invertebrates infant mortality occurs chiefly in planktonic larval stages, which are not preserved with adults. Hence, post-larval fossil assemblages of these organisms do not reflect high infant mortality. Mortality increases during the post-larval life history of most species. Craig and Hallam have shown that, by considering measured growth rates for *Cardium edule* (the European cockle) and various possible survivorship curves, the paleontologist can mathematically generate hypothetical death assemblages with a variety of size-frequency distributions. In Figure 4-9, *A* shows the normal growth curve for *Cardium, B* shows four possible survivorship curves, and *C* and *D* show the size-frequency distributions generated by combining the growth curve with constant mortality (survivorship curve II) and constantly increasing mortality (survivorship curve III).

In other words, a variety of size-frequency distributions can develop depending on growth rate and survivorship in a particular population or species. The distributions represented in *C* and *D* are, in fact, similar to those of Figure 4-8, indicating that a normal, bell-shaped curve does not

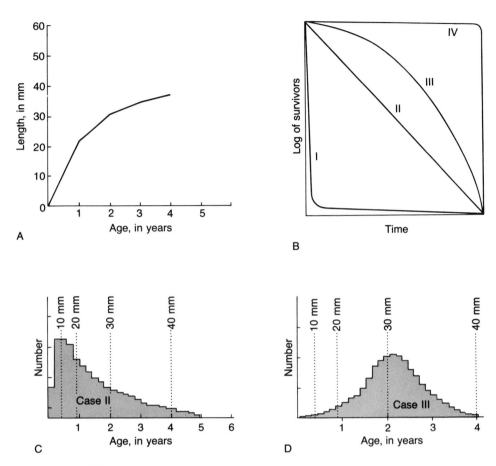

FIGURE 4-9
The production of left-skewed and normal size-frequency curves by combining various death rates with a single growth curve. A: Typical growth curve for *Cardium edule* (European cockle). B: Four possible survivorship curves. C: Size-frequency curve for shells of dead cockles, based on growth curve A and survivorship curve II, which represents a constant death rate. D: Size-frequency curve for shells, based on growth curve A and survivorship curve III, which represents a constantly increasing death rate. (From Craig and Hallam, 1963.)

necessarily give evidence of sorting by currents and that a skewed distribution does not necessarily give evidence of high infant mortality.

It is important to understand that growth and survivorship curves vary from population to population within species and from time to time within populations. The most important factors causing variation in survivorship curves within species are environmental. The many variables in production of a fossil assemblage, however, often make analysis of fossil population dynamics very difficult.

A critical factor in applying population dynamics to fossil taxa is recognition of relative age groups within a species, based on such features as tooth wear (mammals), growth rings (fish scales, echinoderm plates, and mollusc shells), and molt stages (arthropods). Some of these features permit segregation into distinct year classes. Growth rings, for example, commonly represent very slow growth during winter, and molt stages of arthropods are sometimes produced by annual shedding of the exoskeleton. In the absence of distinct growth markers, size-frequency plots may yield distinct age groups (Figure 4-2).

Kurtén (1953) has outlined two approaches by which the population dynamics of fossil assemblages may be studied. The first applies to fossil assemblages composed of the remains of organisms that died of natural causes (*natural mortality*) over a period of time. Kurtén studied remains of Pleistocene European cave bears, most of which died during the annual period of hibernation; thus, their skeletons tended to accumulate in the caves in discrete year groups. The significant point in the *dynamic approach,* which is used here, is that it is unimportant whether a group of preserved skeletons are the remains of animals born during a single year or during a period of several years if growth rate and survivorship were nearly constant within the population through time.

The second approach applies to fossil assemblages formed by sudden death of an entire living population (*mass mortality*). Kurtén calls this the *time-specific approach.* Such catastrophic agents as storms have been observed to bury entire populations in sediment; fossil assemblages formed in such a manner can sometimes be recognized in the fossil record.

It is usually important in analyzing a fossil assemblage to recognize whether the dynamic or time-specific approach is appropriate. We will apply both approaches to data provided by Spjeldnaes (1951) for the Silurian ostracod *Beyrichia jonesi,* following in a general way the procedure of Kurtén (1964).

Figure 4-10, A shows the size-frequency distribution plotted by Spjeldnaes for a collection of 972 valves from the Mulde Marl of Gotland. Ostracods have two valves. Spjeldnaes counted articulated fossils as two valves. Unfortunately, he did not provide information about the number of right and left valves, but he did report that 82 valves were articulated. Like other arthropods, ostracods grow by molting so that their skeletons tend to fall within distinct size groups. The first three molt stages of *Beyrichia jonesi* are very small and are not labelled in Figure 4-10, A because of their poor preservation. It is generally impossible to tell whether a fossil ostracod valve, or test, is a molt or was emptied by the death of the organism. The Mulde Marl assemblage was apparently produced by much more complete preservation than other known fossil assemblages of the same species. Note that each peak in Figure 4-10, A is lower than the one preceding it. The assemblage probably formed from a local population by gradual accumulation both of molts and of tests left behind as animals died and decayed. The dynamic approach is therefore appropriate.

In Table 4-1 the data used for the plot in Figure 4-10, A are presented in the two columns at the left. If we assume nearly complete preservation,

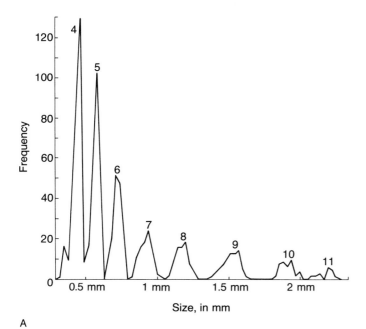

A

FIGURE 4-10
Survivorship in the Silurian ostracod *Beyrichia jonesi* from the
Mulde Marl of Gotland. A: Size-frequency distribution for
a collection of 972 valves. Numbered peaks represent successive
molt stages. B: Survivorship curves based on A. The solid
curve is produced by both dynamic analysis, assuming that all
molts have been preserved, and time-specific analysis. The
dashed curve is produced by dynamic analysis, assuming
that no molts have been preserved. (A from Spjeldnaes, 1951;
B modified from Kurtén, 1964.)

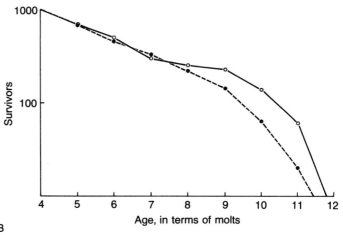

B

TABLE 4-1
Life Tables for the Silurian Ostracod *Beyrichia jonesi*

Approach	1 Age, in Terms of Molt Stages	2 Number of Valves	3 Equivalent Number of Bivalved Tests	4 Number of Molts	5 Number Dying During Age Interval of Molt Stage	6 Number Surviving at Beginning of Interval	7 Number Surviving at Beginning of Interval per 1000 Alive at Beginning of First Interval
Dynamic, assuming all molts preserved	4	309	155	105	50	155	1000
	5	210	105	73	32	105	677
	6	146	73	46	27	73	471
	7	92	46	38	8	46	297
	8	75	38	35	3	38	245
	9	70	35	21	14	35	226
	10	42	21	10	11	21	135
	11	19	10	0	10	10	65
Time-Specific	4	Same as above	Same as above	0	50	155	1000
	5			0	32	105	677
	6			0	27	73	471
	7			0	8	46	297
	8			0	3	38	245
	9			0	14	35	226
	10			0	11	21	135
	11			0	10	10	65
Dynamic, assuming no molts preserved	4	Same as above	Same as above	0	155	483	1000
	5			0	105	328	679
	6			0	73	223	462
	7			0	46	150	311
	8			0	38	104	215
	9			0	35	66	136
	10			0	21	31	64
	11			0	10	10	21

the number of bivalved tests before disarticulation would have been one-half that in column 2; this number is given for each age group in column 3. The first question is what fraction of the tests for each age group are molts and what fraction were emptied by death. Consider age class 4, which contains 155 tests (column 3). We can assume that every animal that survived beyond age interval 4 left behind a test in age class 5, either by molting (if it grew to age class 6) or by dying. The 105 tests in age class 5 therefore represent all preserved individuals that left behind a molt in age class 4. Likewise, 73 animals would have left a molt in age class 5. In a similar manner the rest of column 4 can be filled in. The number that died during each age interval (column 5) is simply the number of tests for that interval (column 3) minus the number of the tests that are molts (column 4). The total number of deaths in the sample is 155, which is the initial number of bivalved tests. This number minus the 50 individuals that died during age interval 4 gives 105 that survived to age interval 5. Similarly, numbers for the rest of column 6 can be calculated. For plotting a survivorship curve it is useful to convert the survivorship numbers to a scale that is based on 1,000 individuals present at the start of the first age interval considered, as shown in Figure 4-7. Survivorship numbers have been converted in this manner to produce column 7. The survivorship curve, based on column 7, is shown as the solid line in Figure 4-10, B.

The preservational circumstances of the Mulde Marl *Beyrichia* assemblage do not suggest mass mortality. For the sake of comparison, however, we will plot a survivorship curve for the data of Figure 4-10, A using the time-specific approach. Here we assume that all the tests were contributed to the fossil record by a sudden killing. None would then be molts. In examining the assemblage from this point of view, we are, in effect, taking a census of a population as it existed at a single moment in time. It follows that all members of a size class were born or hatched at the same time. Refer to the second set of columns in Table 4-1. The first three columns will be the same as before. We will assume a constant death rate for each age and a constant birth rate for the entire population (these are not always reasonable assumptions). There would then have been 155 individuals in each of the preserved age classes when they were at age class 4. The smaller number of tests in each older age class (column 3) is the number that survived to the beginning of the corresponding age interval (column 6). The number dying during an age interval (column 5) is the number surviving to the beginning of that age interval minus the number surviving to the beginning of the next age interval.

It turns out that columns 5 through 7, and the corresponding survivorship curves, are identical for the two approaches we have taken. This result stems from the subtraction of molt stages in the dynamic approach. Thus an assemblage that includes all molted exoskeletons yields the same results as an identical one formed by mass mortality when each is properly analyzed. We would have obtained different results from the dynamic approach had we been considering an animal—say a brachiopod or a bivalve—that did not leave behind molts during growth. We can examine what the difference would be for the data of Figure 4-10, A by treating the ostracod assemblage as if it consisted only of gradually accumulated tests of dead animals (as if no molts were preserved). Then the number that died during each age interval (column 5) would simply be the number of tests in column 3, with no subtraction for molts. The number that survived to the beginning of each age interval (column 6) would be the number that died during that interval plus all individuals that died during later intervals. Column 7 is computed as before. The survivorship curve based on column 6 for this "no molt" approach is shown as a dashed line in Figure 4-10, B. It happens that this curve is nearly the same as the one for the other approaches.

Although the first dynamic approach taken above is probably the right one, judging from the preservational evidence presented by Spjeldnaes, the two other approaches did not yield significantly different results. If the preserved assemblage is as complete as Spjeldnaes believed and as Figure 4-10, A suggests, we have a good idea of the survivorship curve for the species, at least for one environment. In many studies of species that do not molt, the survivorship curve yielded by dynamic analysis will differ significantly from that yielded by time-specific analysis, so that a careful assessment of the mode of formation of the assemblage to be analyzed is essential.

Although rates of growth, and especially rates of mortality, may vary markedly from population to population within species and from time to time within populations, certain fossil species may be found to possess characteristic survivorship curves. It may then be possible to relate such survivorship curves to the species' basic life history, mode of life, and reproductive habits.

Procedures in Describing Variation

We have discussed many of the causes, both biologic and geologic, of variation within and between fossil assemblages. It is important to express this variation in a clear and meaningful form.

VARIATION WITHIN ASSEMBLAGES

One of the simplest and sometimes most effective methods of describing variation is an offshoot of the photographic approach discussed in Chapter 2. Figure 4-11 shows an array of sketches of closely related bivalves collected from rocks of the British Carboniferous. The specimens were chosen not as a random sample but rather to illustrate the principal variants. The result is called a pictograph. The specimens are arranged according to morphologic relations. For example, a general elongation of the shell (increase in length to height ratio) is expressed by the series extending from the top to the bottom of the diagram.

Figure 4-12 shows what are known as variation diagrams, which are based on the kind of analysis and graphic pattern of the pictograph. Circles

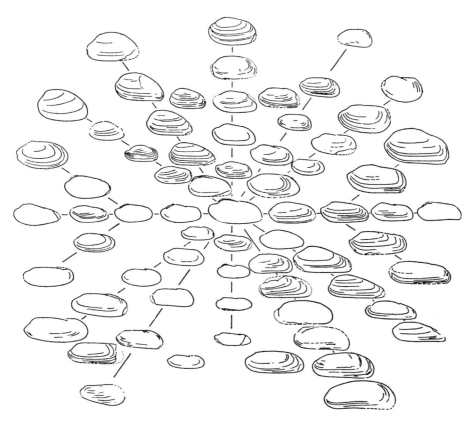

FIGURE 4-11
Pictograph method for expressing morphologic variation.
Sketches of selected individuals of the nonmarine bivalve
Carbonicola (Carboniferous of England) are arranged to show
the principal morphologic departures from the "norm," which
is sketched in the center. Diagrams such as this are used to
evaluate and describe variation in specific assemblages
containing the same variants. (From Eagar, 1947.)

stand for variants in a specific fossil assemblage. The variation diagram, used in conjunction with a carefully constructed pictograph, provides a good first impression of morphologic variation. The method is anything but rigorous, however, because the pictograph is highly subjective and is probably not reproducible by an independent worker. To analyze variation more effectively, we must turn to more formal statistical techniques.

The simplest statistical representation of variation is the frequency distribution, which may be expressed graphically by a histogram or frequency diagram. This representation is an example of ***univariate analysis***. From it we can learn much about variation of a single attribute — the amount of departure from some average value, whether the variation is symmetrical or

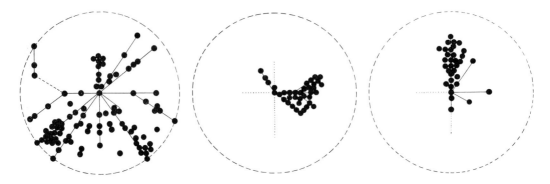

FIGURE 4-12
Variation diagrams—three fossil assemblages of *Carbonicola* analyzed according to
the pattern shown in Figure 4-11. Each circle refers to a single specimen in the assemblage
positioned in the diagram according to its morphologic similarities to other specimens.
The two diagrams to the right are based on assemblages six inches apart in a
stratigraphic section. They were about five feet below the assemblage represented by
the diagram on the left. (From Eagar, 1947.)

skewed, and so on. Many highly sophisticated techniques may be used to
analyze univariate distributions. The paleontologist should become familiar
with these so that he can express and interpret univariate distributions effectively.

As we saw in discussing ontogenetic variation (Chapter 3), many morphologic relations are better expressed if two attributes are handled at a time.
The scatter diagram is the simplest graphic form of such **bivariate analysis.**
The scatter of points may approximate a straight line passing through the
origin (as, for example, in a diagram of perfect isometric growth). The
average relationship between the two variables is most commonly expressed
by the slope of a straight line that represents the line of "best fit" to the
data. The slope of such a line can be calculated in several ways. Very often,
the scatter of points defines a line that, though straight, does not pass through
the origin. To obtain the average relationship between the attributes not only
the slope of a line but also its vertical position must be found.

What does one do when a scatter of points cannot reasonably be approximated by a straight line? We encountered this situation in discussing ontogenetic variation. Allometry is sometimes approached closely enough that a
logarithmic plot renders the distribution nearly linear. Where this is not
the case, more elaborate methods of "curve fitting" must be used. In a
bivariate distribution, the amount and kind of variation can be expressed
only after some line (straight or not) has been fitted to the data. Variation
from the "norm" or average is expressed by departure from this line.

Describing a group of individuals by citing variation in only one or two
attributes greatly oversimplifies the situation. We have seen that some
attributes may be nearly constant in a population, while others vary greatly.

A fuller picture of variation can be drawn only by including many attributes and applying univariate and bivariate techniques to them sequentially or simultaneously. This is **multivariate analysis,** which depends upon a purely numerical analysis to measure the kind and amount of variation in a sample. Because accumulating and processing data for multivariate analysis requires a good deal of time, high-speed computers are of great assistance, not only in accumulating data through various digitizing techniques, but also in statistical data processing. Few areas of science possess more raw data than paleontology and it is thus not surprising that the statistical evaluation of variation by computer has become prominent in paleontologic research.

Even with the most ambitious and sophisticated systems of digitizing and computing, the quantitative description of variation is subject to the problems described in Chapter 2. Since we cannot describe all aspects of variation, nor do we even want to, we must make a selection. The selection requires a subjective appraisal, often in advance, of what characteristics in the population are likely to be significant or interesting, or are likely to be directly applicable to some larger problem. Because variation can be described in an almost infinite number of ways no single way is unique or perfect.

VARIATION BETWEEN ASSEMBLAGES

Many pictorial, graphic, and numerical methods are useful in describing differences between living populations or fossil assemblages. Some are like those used to describe variation within a population or a single fossil assemblage. The primary difference is that *individuals* are compared in describing variation within an assemblage and *groups* of individuals in describing that between assemblages. This difference is sometimes negated by the fact that a particular group of individuals (the population or fossil assemblage) may be characterized by a single individual or by the description of a hypothetical average individual. The validity of characterizing a population by a single individual depends upon the variation shown by the assemblage and upon the differences expected or observed between assemblages. If the differences between assemblages are relatively large, a single specimen from each may suffice to display or describe the difference. In fact, the differences may be so great that any individual, typical or not, drawn from each assemblage may be reasonably used to indicate or describe differences.

A problem in every study is deciding how many morphologic characteristics should be used. Much depends on whether the ultimate purpose is to explore the *differences* or the *similarities* between the assemblages. If we are only concerned with showing that two assemblages differ, a single characteristic may suffice. Figure 4-13 shows frequency distributions of specimen size (measured by maximum width) for two subspecies of brachiopods. The two frequency distributions overlap, but clearly differ. In Figure 4-14, the frequency distributions for the width to length ratio of the same brachiopods are shown. Even without elaborate statistical treatment, it is evident that the two assemblages do not differ significantly in width to length

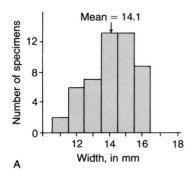

A

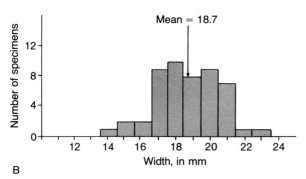

B

FIGURE 4-13
Frequency distributions of shell width
in two subspecies of the Devonian
brachiopod *Pholidostrophia*.
A: *Pholidostrophia gracilis nanus*.
B: *Pholidostrophia gracilis gracilis*.
The mean values are different enough
(in comparison with variation about
the mean) that the difference is not likely
to be due to chance in sampling alone.
(Data from Imbrie, 1956.)

ratio. We cannot argue from this, however, that the two assemblages do not differ because we have already seen that they differ in width (Figure 4-13)! It is common in analyses of this sort for the unsuspecting researcher to find near identity in frequency distributions for some morphologic characteristics and conclude that the samples do not differ, when there may actually be striking differences that are not reflected in the measurements used.

The two examples illustrated in Figures 4-13 and 4-14 are extremes and between them lie an infinite number of intermediates. We must have statistical devices to distinguish the intermediates. Above all, we must be able to determine whether an observed difference between frequency distributions should be looked upon as real (and, therefore, subject to geologic or biologic interpretations) or whether it is simply the result of chance (in sampling). In a univariate analysis, a statistical test may be used to determine whether a difference observed between the mean values of two frequency distributions might have occurred by chance alone. Inevitably, this requires assessment of the difference between the means as well as assessment of the variation within each of the two assemblages. To explain the importance of assessing variation, Figure 4-15 shows two pairs of hypothetical frequency distributions. The differences between the means are identical, but the variation in the individual sample is much greater in B than in A. Obviously, the difference between the means in A is less likely to have occurred by chance than that in B. It is reasonable, therefore, that an expression of variation within the sample enter into the judgment of the significance of the difference between the means. The "answer" given by the statistical test is not a simple "yes" or "no" but an expression of the probability that the observed difference between the means could have occurred by chance.

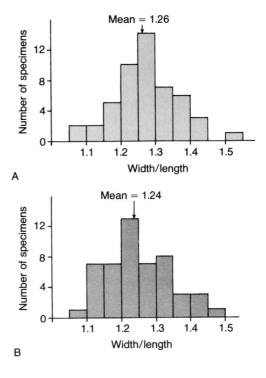

FIGURE 4-14
Frequency distributions of width-to-length
ratio in two subspecies of the Devonian
brachiopod *Pholidostrophia.*
A: *Pholidostrophia gracilis nanus.*
B: *Pholidostrophia gracilis gracilis.*
The slight difference between the mean
values could easily be due to chance errors
in sampling. (Data from Imbrie, 1956.)

An everpresent problem is that of deciding how low the probability must be before we are safe in concluding that the difference between the means is significant. A value of 0.05 is commonly used as the cut-off point. In other words, we must be 95 percent sure that the difference could not have occurred by chance before we accept the hypothesis that the samples are different. At first glance, this may seem to be a rather stringent requirement. There are several good reasons for it, however. Perhaps the most compelling is the possibility for error that exists even with this cut-off point: if a researcher makes twenty decisions and each is based on a 95 percent chance of being correct, then the odds are that one of the decisions will be wrong! Even more important, if a researcher makes a final decision, which is dependent upon a sequence of several decisions made earlier, the probability

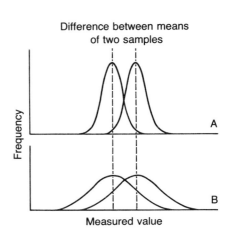

FIGURE 4-15
Pairs of hypothetical frequency distributions
illustrating the importance of variation in
assessing the significance of an observed
difference between mean values. It is much
more likely that the observed difference
occurred by chance in B than in A.

that the final decision is correct will be substantially less than 95 percent. If the calculated probability of chance occurrence is less than 0.05, the difference between the means is said to be "statistically significant" at the 5 percent level. The test of statistical significance in itself gives no specific geologic or biologic information, but only tells us whether the observed differences may be due to chance in sampling. If a difference "passes" this test, we are left with the problem of *interpreting* the difference in terms of causal factors.

The greatest number of quantitative studies of morphologic differences between fossil assemblages are based on bivariate analysis. Figure 4-16 shows a typical bivariate comparison of data from two assemblages. A straight line has been fitted (statistically) to the data for each sample. The two lines differ in slope and vertical position. Thus, the two subspecies differ in shape and in ontogenetic development of shape. But are the differences statistically significant? Tests of significance must be applied. If the difference between the lines turns out to be statistically significant, we may proceed to make a biologic or geologic interpretation.

Note that the data used for Figure 4-16 are the same as those for Figures 4-13 and 4-14. The fact that Figure 4-13 showed a large difference between the samples and that Figure 4-14 did not, should be understandable from the bivariate relationship shown in Figure 4-16.

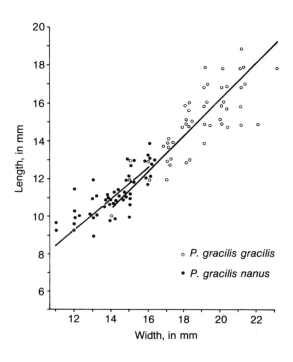

FIGURE 4-16
Bivariate distribution of width and length in two subspecies of the Devonian brachiopod *Pholidostrophia*. The samples are the same as those used for Figures 4-13 and 4-14. (From Imbrie, 1956.)

Number of awns
on lemmas:

    ○ 0–1

    ◔ 2–3

    ◔ 4

Number of awns
on glumes:

    ○ 4–5

    ○̄ 6–7

    ○̄ 8–9

Wedge shape
(numbers give ratio
of length to width):

    ○ broad and short
    (0.5–1.0)

    ◌ intermediate
    (1.1–1.5)

    ◌ narrow and long
    (1.6 or more)

Number of
rudimentary spikelets:

    ○ 3–4

    ○⁄ 2

    ○⁄ 1

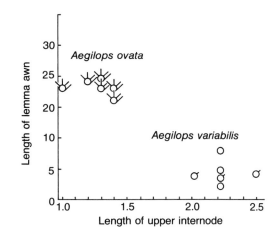

FIGURE 4-17

Glyph diagram used in an analysis of morphologic differences between two species of the wheat genus *Aegilops*. (From Zohary and Feldman, 1962.)

In lieu of formal multivariate analysis, several purely graphic methods have been used to describe variation between assemblages with respect to more than two variables. One of these is known as the "glyph" system and is illustrated in Figure 4-17. The basic framework of the analysis is a bivariate distribution. A sample of specimens from each of the two populations has been measured for several characters and two are chosen for the vertical and horizontal axes. The rest are added to the diagram by an arbitrary system of embellishment of the glyphs, an explanation of which is given beside the diagram. By this method we are able to see differences in several characters simultaneously and thus gain a subjective but quite valid impression of the patterns of morphologic difference between the two assemblages. (A nonpaleontologic example is used for Figure 4-17, because glyph diagrams have not yet been used extensively with fossils.)

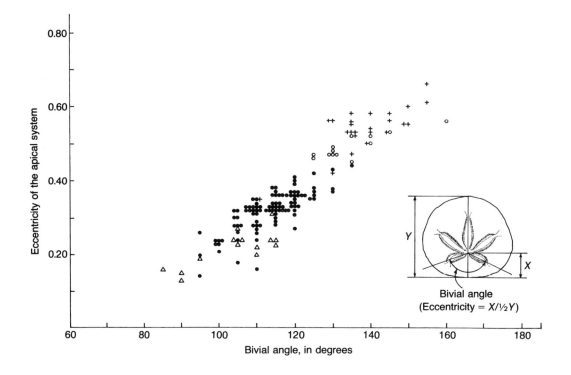

FIGURE 4-18

Plot of bivial angle and eccentricity of the apical system in the sand dollar *Dendraster* illustrating the redundancy of characters. A change in one character is accompanied by a predictable change in the other, regardless of the ontogenetic stage of the specimen. (Data prepared by Raup.)

Although the glyph system is able to handle more characters simultaneously than do normal bivariate or univariate systems, a selection of characters to be compared must still be made. If fully computerized multivariate analysis is employed, the number of characters is limited only by the energy of the researcher and by the computing facilities available.

In the discussion thus far it has been assumed tacitly that the various attributes of a fossil that may be measured and compared are biologically and geologically independent of each other—an assumption that must be explored in greater detail. Many morphologic characteristics are strongly influenced by other characters. The connection between two such characters may be purely mechanical. A structure, for example, must inevitably be larger than the structures contained within it, and if the size of one structure is plotted against the size of a structure within it, the smaller structure will tend to increase in size as the larger structure increases.

More commonly the connection is functional. In a brachiopod, for example, the two valves of the shell must meet along the commissure. Since the outline of the two shells must therefore be the same size and the same shape, neither the high correlation that will certainly be produced by plotting the length of one brachiopod valve against the length of the other nor the measurements of the second valve will add any information to what is known from the measurements of the first valve. The dimensions of the second valve are said to be redundant.

An example of redundant characters from fossil and Recent sand dollars is shown in Figure 4-18. One character is the eccentricity of the apical system: that is, the position of the apical system (at the center of the petaloid area) with respect to the antero-posterior diameter of the skeleton. (Eccentricity increases as the apical system moves from the center to the posterior margin.) The other character is the bivial angle: the angle formed by the two posterior petals radiating out from the apical system.

In Figure 4-18, the scatter diagram relates the two characters for samples of four genetically different sand dollar populations (belonging to different species or subspecies). Because the plot involves an angle and a ratio, specimen size (and, therefore, age) is not explicit in the diagram. The redundancy shown by the scatter is thus independent of ontogenetic stage. The explanation for this redundancy is fairly simple when we realize that the point on the circumference of the sand dollar where a petal ends is more rigidly fixed anatomically than is the position of the apical system. For purposes of comparison, the bivial angle yields no information not already provided by the position of the apical system. The choice of characters to be used in a given study is usually determined by convenience. In this example, the bivial angle would probably be avoided because of the lack of precision inherent in its measurement.

The paleontologist has at his disposal an arsenal of descriptive procedures for comparing fossil assemblages. We have touched on but a few. Because many of the procedures have quite different objectives, the choice of method is extremely important and can only be made after the scientific problem is clearly defined. Thus, it is customary to work from the problem to the technique rather than from the technique to the problem.

Supplementary Reading

Dobzhansky, T. (1951) *Genetics and the Origin of Species,* 3rd Ed. New York, Columbia University Press, 364 p. (A comprehensive treatment of variability and its significance in evolution. The first six chapters are of particular interest.)

Imbrie, J. (1956) Biometrical methods in the study of invertebrate fossils. *Bull. Amer. Mus. Nat. History,* **108**:211–252. (A brief but excellent summary of statistical techniques applicable to the analysis of variation in fossil assemblages.)

Mayr, E. (1942) *Systematics and the Origin of Species.* New York, Columbia University Press, 334 p. (A companion volume to Dobzhansky (1951) emphasizing morphological aspects of variability and evolution.)

Newell, N. D. (1956) Fossil populations. *In* Sylvester-Bradley, P. C., ed. *The Species Concept in Paleontology.* Systematics Association Publication 2, p. 63–82. (A summary and analysis of some of the problems encountered in the biologic treatment of fossil assemblages.)

Rudwick, M. J. S. (1964) The inference of function from structure in fossils. *Brit. Jour. Philos. Sci.,* **15**:27–40. (A thought-provoking analysis of adaptive, nonadaptive, and inadaptive characters and the identification of them in fossils.)

Simpson, G. G., Roe, A., and Lewontin, R. C. (1960) *Quantitative Zoology.* New York, Harcourt, Brace, 440 p. (A standard reference on quantitative techniques applicable to the study of morphology.)

Srb, A. M., Owen, R. D., and Edgar, R. S. (1965) *General Genetics,* 2nd Ed. San Francisco, W. H. Freeman and Company, 557 p. An excellent modern textbook of genetics, covering material essential for any paleontologist.)

CHAPTER 5

The Species as a Unit

The tremendous diversity in form, physiology, and behavior of living organisms is not random. If our imaginations were good enough to conceive of the total spectrum of possible biologic forms, we would find that all of the living and extinct organisms, taken together, are only a small part of the imagined whole. Living species are clustered within this spectrum; there are relatively few basic types, with each type capitalizing on the requirements and the advantages of particular roles in particular habitats. In other words, we find that most species are adapted to one or a few environments.

If biologic diversity is to be maintained, *species units must remain distinct.* In biologic terms this means that hybridization between species must be uncommon. Natural hybridization does, of course, occur between species, particularly in plants. Sometimes it produces new structures or new combinations of structures unknown in either parent species. In the long run, successful hybridization blends parental characteristics and results in the formation of a single species where there were originally two. The fact that high species diversity has been maintained over a long period is the best evidence we have for the relative rarity of hybridization.

Diversity is maintained because species are reproductively isolated from each other (even though they may live in the same geographic area and occupy similar habitats.) Therefore, *the evolution of reproductive isolation* is basic to the origin of species. We must understand how reproductive isolation evolves in order to understand the concept of the biologic species.

The paleontologist usually deals only with the *results* of species evolution rather than the process itself, but knowledge of the process is as important to the paleontologist as an understanding of hydrodynamics is to the sedimentologist.

The Biologic Species Definition

The most widely accepted biologic definition of the species was formulated by Ernst Mayr: *"A species is an array of populations which are actually or potentially interbreeding and which are reproductively isolated from other such arrays under natural conditions."* As an operational definition, this is not directly applicable to the fossil record because hybridization cannot be observed between fossil species. In fact, most working biologists are rarely able to apply the definition *directly* either. Biologists have therefore developed several indirect methods of identifying the boundaries between species. Many of these methods are also applicable to fossils.

The species definition just given has several important elements. The species is referred to as an array of populations, emphasizing the fact that most species are divided geographically into subunits or breeding populations. It is explicit in the definition that such breeding populations are actually or potentially interbreeding with each other. We can define two types of relationships between populations. *Allopatric populations* are those that occupy separate geographic areas; *sympatric populations* are those having the same or overlapping geographic ranges. Allopatric populations may be separated by geographic barriers, making gene flow between them impossible. We refer to such populations as being geographically isolated but by convention do *not* refer to them as being reproductively isolated (even though the geographic barriers may in fact prevent interbreeding). Two populations are said to be reproductively isolated *only* if interbreeding would not occur if they both lived in the same area. Thus, "potentially" in the species definition is particularly critical.

The crux of the species definition is that the array of populations constituting one species is reproductively isolated from other such arrays: two species can live sympatrically without interbreeding. Must the reproductive isolation between species be complete? If not, how much gene flow can be tolerated? If two populations are found to hybridize successfully (that is, to produce fertile offspring), are they necessarily one species? The answers to these questions must remain indefinite, but it is possible to point out—since the issue is the maintenance of distinct species—that an amount of gene flow between species that does not lead to the breakdown of the differences between them is compatible with their being properly described as reproductively isolated.

An important part of the species definition is that populations of different species are reproductively isolated from each other under "natural conditions." There are many recorded examples of species hybridizing readily in captivity or under domestication. This stems from the fact that reproductive isolation often depends on rather minor ecologic or behavioral barriers that tend to break down in captivity. The ability of organisms to interbreed in captivity is generally ruled out as a criterion for concluding that they belong to the same species.

We may consider an example that demonstrates the difficulty in a practical application of the species definition. Assume that we suspect that two allopatric populations belong to the same species. Using the definition, we can test this idea only by transplanting individuals of one population into the environment of the other. But to do this we would inevitably disturb the natural conditions under which one or both populations lived and, thus, might upset certain behavioral or ecologic causes of reproductive isolation.

Speciation

Speciation is the evolutionary process of species formation. There are two types: *phyletic speciation* and *geographic speciation.* The first is the evolution of one species into another through time. Since by this process one species replaces another in an "evolutionary lineage," the total number of living species at any given time is not affected by it. Geographic, or allopatric, speciation, the process by which one species evolves into two or more contemporaneous species, leads to an increase in the total number of living species.

PHYLETIC SPECIATION

Phyletic speciation usually results from the operation of natural selection on the genetic composition of one or more populations. Variation in the gene pool produces variation in the *phenotype* (form, structure, physiology, and behavior) of the organisms making up the population. The action of natural selection upon phenotype variability produces, over time, a change in the genetic composition of the population, and any such change, however slight, is evolution.

Evolutionary change in the genetic composition of a population may represent either progressive adaptation to *constant* environmental conditions, or adjustment to *changing* environmental conditions. With each succeeding generation, the gap in genetic composition with respect to an arbitrary starting generation is widened. As a population changes, it becomes almost inevitable (statistically) that differences develop that would inhibit individuals expressing the changed phenotype from breeding with individuals of the original ancestral phenotype.

If we think of a chronologic series of populations and choose one of these as an arbitrary starting point, then there will be a point at which the accumulated differences are such that the later populations *would* be reproductively isolated from the initial population *if* they were living at the same time. At this point a new species has been formed.

This consideration of the evolution of a single population over time may be extended to a species made up of many populations. If there is gene flow between these populations, then the process of phyletic speciation operates on the entire species. Ironically, if all populations of the species are in genetic contact, the process inevitably leads to *extinction* of the ancestral species, a type of extinction known as **phyletic extinction.**

GEOGRAPHIC SPECIATION

Let us consider populations of a species living at a given time but not in geographic contact with each other. That is, one or more of the populations is geographically isolated from the other populations. Two or more segments of the species thus evolve and undergo phyletic speciation independently. If the geographic barriers between populations remain intact long enough, reproductive isolation may develop. When this happens, geographic speciation has occurred: a single species has become two or more contemporaneous species.

The distinction between phyletic and geographic speciation is to some extent artificial in that both processes depend on natural selection. The critical difference is that phyletic speciation is accomplished in the absence of geographic isolation and geographic speciation requires geographic isolation.

In the geographic speciation process, new species may evolve in both lineages or in only one. In any event, the number of species increases. We might suspect that geographic speciation should inevitably lead to a steady, overall increase in number of species but, in fact, extinction counteracts this increase. Lineages are periodically terminated because of changes in environmental conditions which are so rapid that species cannot adjust to them by phyletic speciation. The number of species living on the earth at any given time is therefore the resultant of the positive process of geographic speciation and the negative process of extinction.

THE SUBSPECIES

In the present context, a subspecies may be looked upon as an incipient species. Subspecies is a term usually applied to geographically isolated populations, or groups of populations that are genetically different, but not sufficiently different to be reproductively isolated. From the paleontologic viewpoint the subspecies can also be used to denote the intermediate stage in the replacement of one species by another through phyletic speciation.

CLINES AND RING SPECIES

Some allopatric populations are separated by environments not hospitable to the organism, such as, for example, bird populations in an archipelago separated by considerable distances of water or lowland terrestrial plant populations separated by mountain ranges. Other allopatric populations are not separated by such pronounced barriers.

The gene flow between populations situated at the extremities of the geographic range of a species may be greatly reduced simply by the distance between them, and a gradient in the genetic composition of the populations may be produced, giving rise to what is termed a *cline.* For certain species, it has been shown that the populations living at the extremities of the geographic range are reproductively isolated from each other even though they are connected by a chain of interbreeding populations.

If such reproductively isolated populations from the extremities of a species' range migrate so that their geographic ranges overlap, what is known as a *ring species* is formed. The populations originally at the extremities live sympatrically and are thus good species, but are connected through a chain of interbreeding populations.

Fortunately for the practicing biologist, clines and ring species are relatively rare. It is nearly impossible for the paleontologist to recognize clines and ring species because the fossil record is not complete enough and time determinations are not accurate enough.

ADDITION OF SPECIES WITHOUT ISOLATION

Although there has been considerable controversy over the question of whether a single species can evolve into two or more contemporaneous species *without* the geographic isolation of two or more of its breeding populations, for our purposes the question may be dismissed once it has been noted that the consensus among biologists and paleontologists is that it is an exceedingly rare phenomenon, if it occurs at all.

Rates of Speciation

Determining the rates at which speciation has taken place is a problem perhaps more suitable for the paleontologist than for the biologist. The biologist sees an array of populations at an instant in time and must rely on informed guesswork to interpret the history of the species that he studies. The paleontologist works with stratigraphic sequences of fossils that represent the chronological order of the populations whose remains are preserved; he thus is in a much stronger position to contribute to our knowledge of how the evolutionary process has carried particular species forward.

As we consider investigating rates of speciation, we must keep in mind the distinction between phyletic speciation and geographic speciation. We are concerned on the one hand with rate of change in a single lineage over time and on the other hand, with the incidence of the physical separation of populations, which is necessary for geographic speciation, as well as the rate of change. The most important factors affecting the rate of phyletic speciation are: (1) generation length, (2) selection pressure, (3) mutation pressure, and (4) population size.

Generation length is the basic unit of time in natural selection. The relative success or failure of variants in a breeding population will be expressed by

the influence they have on the genetic composition of the succeeding generation. Obviously, a species having a short generation length will tend to produce more increments of genetic change per thousand years than a species
whose generation length is long. In a sense, therefore, a short generation
length (which in itself is genetically controlled) may be thought of as an
adaptation that favors rapid evolution. We might therefore expect groups
of organisms whose generation length is short to show greater diversity
than groups whose generation length is longer.

.If a population is living in a very favorable environment, quite a variety
of forms may be equally successful and change in genetic composition from
generation to generation tends to be slow. Where competition between individuals is heightened, however, either with respect to the biologic or physical environment, the "unfit" are more rapidly eliminated from the population and the rate of genetic change from generation to generation increases.
For this reason, what is known as *selection pressure* greatly affects the relative rates of evolution of a single lineage.

The presence of genetic variability in a population is essential for evolutionary change. Without such variability a population either remains un-
changed from generation to generation or becomes extinct, depending on
the harshness of the environment. Because the ultimate source of genetic
variability is gene mutation, the frequency of mutation has a direct bearing
on the rate of evolution. The incidence of mutation is referred to as *mutation
pressure*. All other things being equal, an increase in mutation pressure will
produce an increase in rate of evolution.

Population size is an important factor in determining rate of evolution
because it affects the absolute frequency of new variation produced by mutation and also affects the rate at which a favorable mutation can spread
throughout a population. Although several conflicting processes are involved,
it is generally correct to say that evolutionary rates are higher in small populations than in large populations.

The several factors just discussed affect the rate of change in a single
lineage and therefore affect rates of both geographic and phyletic speciation.
For geographic speciation to occur, however, there must be geographic
isolation between populations. Several physical and biologic factors affect
the incidence of geographic isolation. Organisms living in shallow water on
the edge of a continent may have little potential for geographic isolation
because environments favorable to them may be a continuum. The same
species living on the shores of islands, however, may have great potential
for isolation because environments favorable to them may be separated by
scores or even hundreds of miles. It is thus not surprising that many more
species of shallow-water organisms are found per unit of area in the East
Indies, for example, than are found along the eastern coast of Africa.

A striking example is found by comparing marine invertebrate faunas on
the opposite sides of the Isthmus of Panama. The environments on the two
sides are virtually the same yet few species live on both sides. The Isthmus
has been above sea level for several million years and at the time of its final

emergence it formed a barrier separating populations of marine species. Divergence through geographic speciation since that time has produced almost complete distinction between the faunas on the two sides.

It is sometimes possible to interpret paleogeography from an array of fossil species arranged according to the geographic relations of the places from which they came. By this means, the changing relation of the Isthmus of Panama to sea level during the last 120 million years has been traced with considerable accuracy. There is great potential for relating periods of tectonic activity and periods of rapid geographic speciation.

A geographic feature may provide a barrier for gene flow for one organism, but not for others. This emphasizes an important biologic factor controlling the incidence of geographic speciation: the ***dispersal potential*** of an organism. Some species are severely restricted geographically throughout their life cycle. Distance alone is enought to isolate completely populations of some species. In the gastropod species *Purpura lapillus*, for example, the fertilized eggs remain on the sea bottom throughout their development, and the adult may not move more than a few yards during its entire life. Movement is truly "at a snail's pace," being accelerated only by occasional storms or by attachment to floating driftwood.

Other species of the same genus employ a quite different reproductive system in which the fertilized eggs develop into free-floating larvae, which live as part of the plankton for as long as six weeks. For these species dispersal potential is extremely high and, not surprisingly, the incidence of geographic speciation is lower than that of *Purpura lapillus*.

Because of the sharp differences in dispersal potential among even closely related species, it is important for the paleontologist to have knowledge of dispersal possibilities. Among gastropods and bivalves the question of whether there was a free-floating larval stage may often be revealed by an examination of the larval shell (preserved at the apex of the adult shell) and thus it may be possible for the paleontologist to determine from morphology something about dispersal potential.

Biologic Methods of Species Discrimination

The practical problem of distinguishing species amounts in its most usual form to deciding whether two populations that differ morphologically are different species or whether they are simply minor variants (perhaps subspecies). The biologist's greatest asset in making such a decision is the generally accepted idea that populations cannot live sympatrically unless they belong to different species. This assumes, of course, that the division of a single species into two or more without reproductive isolation is such a rare event that it may be discarded from general consideration. Following this reasoning, if a biologist finds that the geographic ranges of two morphologically different organisms overlap, he may conclude with considerable confidence that they belong to different species.

Some differences can be observed between nearly all allopatric populations. These may be morphologic, physiologic, behavioral, or ecologic. In assessing the probability that two such populations have become reproductively isolated from each other, the paleontologist works largely from comparisons with difference observed between populations known to be distinct species because they occur sympatrically. Distinguishing species on this basis involves varying degrees of uncertainty. It is widely recognized that some species differ very slightly (the so-called "sibling species"), while others are so different that there could be little question that it is correct to designate them as different species. It is generally true that the amount of difference between species is reasonably constant within a single evolutionary group and therefore the specialist usually possesses a backlog of experience that assists him as he interprets species differences among allopatric populations. It is also true that most living species are quite distinct from one another because they have been reproductively isolated for long enough to have developed many and notable differences.

The fact that species discrimination depends largely on the experience of the person making the discrimination has led to an informal but surprisingly valid definition of the species: "A species is a species if a competent specialist says it is."

The geographic distribution of allopatric populations can also be used as an aid in making species discriminations. If populations have been widely separated for a long time, the chances for complete speciation are greater than if the populations are close to one another and have not been geographically isolated for very long. For this reason, biologists often use biogeography as a taxonomic aid. It may be reasonably postulated that widely separated populations belong to different species even though the morphologic differences between them are not great. Although this reasoning may be quite useful if used as supporting evidence, it has led to considerable difficulty in studies in which taxonomists have based species discrimination solely on geographic separation.

In summary, the biologist relies primarily on phenotype differences between populations to define species boundaries. Occasionally he can bring biogeography to bear and occasionally he can gain factual knowledge from finding populations living sympatrically. Above all, biologic species discrimination is an *interpretation* of the evolutionary history of the organism being studied.

At various times the suggestion has been made that we should do away with a system based on a species definition that can be applied objectively only rarely and that the definition should be replaced by one based on morphology, in which boundaries between species could be established by arbitrary rules of statistical discrimination. To this end, various "morphospecies" definitions have been proposed. For example, it has been suggested that two populations differing in one or more morphologic characteristics to a statistically significant degree should be called different species. Such a system has obvious appeal in that it would do away with many of the practical problems.

These proposals have never been widely accepted, primarily because the evolutionary process does not operate in a manner amenable to the use of arbitrary boundaries between species. It has been shown repeatedly that the difference between populations necessary to effect reproductive isolation is only crudely correlated with the absolute amount of difference between the populations. There can be no question but that reproductive isolation is the fundamental condition responsible for the segregation of organisms into discrete adaptive units. Therefore the consensus of taxonomists is that it is better to describe the results of evolution by a theoretically valid method, but one difficult to apply, than by a theoretically invalid method that owes its existence to convenience.

The Species Problem in Paleontology

If we could construct a three-dimensional model of the evolution of life, with time represented by the vertical dimension and evolutionary differences between lineages by distance of horizontal separation, it would take the form of a complex tree, branching toward the top. What the biologist sees in his study of living forms is only the horizontal, two-dimensional upper surface of the tree—the tops of the uppermost branches. In other words, he does not deal with evolutionary lineages (except very short ones, in which little evolution occurs). Biologic species tend to be distinct because most belong to lineages that have been reproductively isolated from other lineages for a considerable time. If a lineage is branching at the present time, the biologist may, as we have seen, have difficulty judging whether the incipient species are reproductively isolated or not.

For the paleontologist, the presence of the time dimension makes it impossible to apply the commonly accepted biologic species definition (based on reproductive isolation) and at the same time recognize species as discrete, nonarbitrary entities. This conflict has given rise to what, in paleontology, is referred to as "the species problem." An analogous problem is posed in biology by the existence of clines and ring species, but these are not common enough to cause major difficulties.

A wide variety of opinions are held as to the ways in which fossil species should be defined and recognized. Some workers believe that the biologic species concept should be abandoned, primarily because there is no way of applying the concept to an evolutionary continuum and because fossils cannot be tested for reproductive isolation. Probably most paleontologists, however, favor applying the biologic species concept, if only in an indirect and imperfect way. A worker can do this by including in a fossil species only specimens that, by his judgment, would have belonged to a single "biologic" species had they lived at the same time. We have emphasized that although the biologist *defines* species on a *genetic* basis, most often he *recognizes* them on the basis of *morphologic* criteria. The paleontologist can do the same, and can use as a reference the degree of morphologic difference between biologic species (preferably species of a living group closely related to the fossil group being studied).

There remains one major problem. Having decided on the amount of morphologic variation within recognized species, where should the boundaries within a lineage be placed? In Figure 5-1, three alternative interpretations of speciation are presented for a simple branching pattern of evolution. Some workers prefer to place species boundaries at branching points and to avoid the situation illustrated in part B of the figure. Because of the added time dimension, some workers use a special term such as "successional species" or "paleospecies" for a species that represents an arbitrarily delimited segment of an evolutionary lineage.

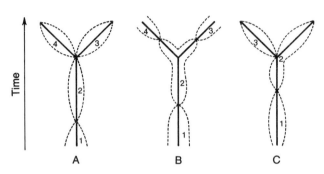

FIGURE 5-1
Three possible ways of dividing a branching
fossil lineage into three or four species.
(Modified from Simpson, 1961.)

It is rather striking that the concept of reproductive isolation was employed in biologic species definitions long before the idea of evolution was accepted in biology. Furthermore, the modern definition of a biologic species ignores evolution altogether. Consequently, Simpson (1961) has proposed a special definition for an **evolutionary species:** "An evolutionary species is a lineage . . . evolving separately from others and with its own unitary evolutionary role and tendencies." The word "role" here means way of life within a particular habitat. This definition is generally consistent with the genetic definition of a biologic species, but does permit a small amount of interbreeding between species (according to Simpson, "as much as does not cause their roles to merge"). Although the criteria used to delimit species boundaries are not described in his formal definition, Simpson suggests that morphologic differences between species should be at least as large as those between living species of the same taxonomic group (or a similar group).

There is no way of eliminating the species problem in paleontology; all general methods of defining and recognizing fossil species must be both subjective and arbitrary. There is, however, one very important factor that eliminates subjectivity from the assignment of many lineage "boundaries." This factor is the widespread presence of gaps in the fossil record. Our preserved sample of the phylogenetic tree of life is so small and so fragmented that often only short segments showing little evolutionary change are found. Successive fossil populations within a hypothetical lineage are represented in Figure 5-2 to illustrate this point. Some preserved segments of the lineage exhibit a small amount of evolutionary change, and others virtually none. The large gaps between preserved segments of the lineage provide convenient locations for species boundaries. Thus, nature has greatly reduced the species problem in paleontology (but at the expense of much valuable information).

Still, it is important to appreciate the uncertainty involved in assignment of many fossils to species. A perusal of several taxonomic articles on local fossil faunas will illustrate this point. Usually a significant percentage of the fossils described in such studies are referred to species equivocally and the species assignments of earlier workers are commonly questioned or

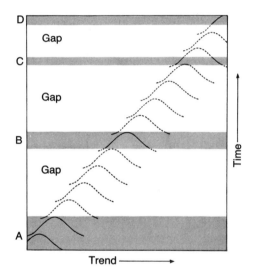

FIGURE 5-2

Diagram showing how gaps in the fossil record can provide arbitrary limits for species discriminations within an evolutionary lineage. The curves are size-frequency plots for an evolving character within a hypothetical lineage. Each curve is for the population that represented the lineage at a particular time. More or fewer curves could have been included. The succession of curves reveals an evolutionary trend. There are major gaps in the fossil record of the lineage, however. Populations are only preserved within four rock units (A, B, C, and D), which together represent less than half the time during which the lineage evolved. Suppose the morphologic differences between the populations in the four rock units are so great that we judge that if the populations in A and those in B had lived at the same time, they would have been unable to interbreed. Similarly, suppose we judge that interbreeding would have been impossible between the populations in B and those in C, and between those in C and those in D. Suppose, on the other hand, we judge that interbreeding would have been possible among the populations within each rock unit. We would conclude that the population group within each rock unit should be assigned to a distinct species. The gaps in the stratigraphic record have made it unnecessary for us to divide continuous lineages subjectively into species. (From Newell, 1956.)

rejected. The paleontologist's problem arises not only because he is attempting to divide up continua (lineages), but also because he works with fossil material lacking many features commonly available to biologists. (Many of these features are morphologic, but others, as we have seen, are behavioral, ecologic, and geographic.)

We have already stated that a single specimen does not suffice to describe a species and have provided examples of variation within and between populations. In the past, even after evolution was widely accepted, many taxonomists assumed that an ideal form existed for each species and that a single specimen, the *type specimen,* or *holotype,* could be chosen to represent this ideal form; the idea of variation within the species was suppressed. With better understanding of the genetics of populations, this "typological approach" has been largely supplanted in the twentieth century. Species morphology is now fully recognized to be best described from a statistical point of view that allows for considerable variation.

Any single worker's *concept* of a living or fossil species is represented by all specimens known to him that he believes belong to the species. For this group of specimens, Simpson (1940b) has proposed the useful word **hypo-digm.** All specimens of the hypodigm, including any formerly designated as type specimens, have equal weight in the species description. A practical problem is that of communicating one's concept of a species to other workers. Obviously, all specimens of most hypodigms cannot be listed in taxonomic articles. They may number in the thousands, or even millions, and many may lack numbers or other means of identification. The problem is partly solved by providing what is called a **synonymy** at the beginning of a species description. The synonymy is a listing of specimens or populations described by earlier workers that are included in the present worker's hypodigm. Many of these specimens or populations may have been assigned other species names by earlier workers, but these names are rejected by the synonymy writer and are listed only for historical reference. Thus the synonymy seldom represents the complete hypodigm, but nonetheless provides a shorthand account of the writer's hypodigm that may be used for purposes of comparison.

Formal Naming and Description of Species

A new species may be erected in biology or paleontology either because previously unnamed specimens have become available or because a previously recognized species is judged actually to be two or more species. In paleontology, a combination of the two reasons is common: a worker may erect a new species based on new fossil material, but include in it specimens or populations formerly assigned to different species. The individual worker's *concept* of a species and his views on the division of genera into species are subjective matters, as we have seen. In contrast, the *naming* of species is governed by objective rules. Were this not the case, nomenclatural chaos would result.

INTERNATIONAL CODE OF ZOOLOGICAL NOMENCLATURE

A wide variety of systems for naming and describing species have been proposed and used for varying lengths of time. Systems have often differed from country to country and from taxon to taxon, leading to considerable confusion. For the last fifty years or so, however, there has been increasing agreement among workers in different countries and among workers studying different plant and animal groups. Most matters concerned with species names are now under the control of international organizations. One of the most important of these is the International Commission on Zoological Nomenclature, which (operating under the continuing International Congress Organization) is responsible for administering and updating the *International Code of Zoological Nomenclature*.

The Code was adopted by the Fifth International Zoological Congress, which met in Berlin in 1901. It is a rather long legal document dealing with procedures to be followed in establishing names of species and other taxonomic groups and with problems posed by names proposed under earlier systems. The Commission acts as a combination court and legislature for treatment of questions of taxonomic procedure.

The International Code applies equally to fossil and living organisms, which is indeed fortunate because relatively few taxonomists actually work with both groups. Without the unifying effect of the International Code, two quite independent and perhaps contradictory systems of nomenclature might develop.

The International Code applies to taxonomic categories from the subspecies to the superfamily. Emphasis in this chapter will be on its application to the species category.

The International Code is almost universally applied to animal taxonomy. A comparable set of procedures for plants is known as the International Rules of Botanical Nomenclature. In addition, one or two sets of international rules are commonly used for special groups like the bacteria. For our purposes, the differences between these codes are minor; in our examples we will concentrate on the Zoological Code.

Regarding the establishment of a new species name, the most important rules of the Zoological Code cover the following topics: choice of the name, publication of the name, description of the new species, and designation of one or more type specimens.

In order for a species to be officially recognized it must be given a name. The choice is limited by certain conditions. Most important is that the name must be in binomial form; that is, it must consist of two words. The official name for the human species is *Homo sapiens: sapiens* is the specific name and *Homo* is the name of the genus to which the species belongs. According to the International Code a specific name like *sapiens* (sometimes called the trivial name) is meaningless unless associated with a genus name. In practice, most newly discovered species can be assigned readily to an existing genus and thus the act of describing a new species involves the invention of only one name. If the new species is apparently distinct from all established genera, a new generic name is assigned at the same time.

Except for the generic assignment just mentioned, the International Code does not insist on the complete taxonomic classification of a new species, in recognition of the fact that the complete classification is often difficult or impossible, particularly if the new species is quite distinct from all other known species.

The most important requirement of the new name is that it not be already in use ("occupied"). This restriction against homonyms refers to the combination of generic and specific names. That is, the name used for the second part of the binomial can be one that is used in other genera, but cannot be one already in use with the genus name accompanying it. By convention repeating species names in closely related genera is also avoided because generic affiliations are often changed as knowledge of the evolutionary relationships between species changes.

The names for species and genera must be Latin words or words that have been latinized. There is considerable latitude in the choice of words to be used as names—latinized place names, names of people, or descriptive words are all used. Certain practical and aesthetic limitations are placed on the selection, however, and full detail of these matters may be found in the International Code.

In order for the name of a new species to be officially recognized it must be published in an approved medium; that is, it must be in print, published in quantity, and circulated through normal bibliographic channels. The precise rules governing the form of publication are complex but their intent is to insure that the announcement of a new name is readily available to taxonomists throughout the world. A new name is not officially recognized if it has been used only in the labeling of a museum specimen or described orally before a scientific meeting.

Species cannot be described anonymously. Thus, the authorship of a new species is an important part of the official procedure. Although most species are described by a single person, more than one person may participate in the official authorship.

The requirement of publication is obviously necessary. The resulting bibliographic problems are, of course, immense because thousands of species are described each year in a great variety of languages and media, from museum monograph series to large international journals. The International Commission has made some attempt to simplify the bibliographic problems by urging that publication be made in one of a relatively few recommended languages. French, German, English, and Russian are generally recommended although this does not have the status of an absolute rule in the International Code.

The International Code does not require that the publication of a new species name be accompanied by a photograph or other illustration of specimens. Illustrating is, however, strongly recommended in appendices to the Code and a considerable amount of convention has developed; for all practical purposes, illustration of specimens may be considered mandatory.

The International Code says surprisingly little about the manner in which a new species should be described. The principal requirement is that the name be accompanied by a "statement that purports to give characters differentiating the taxon." A set of general recommendations published by the International Commission and appended to the International Code expands on this statement and a large body of convention has developed.

The International Code specifies that each newly described species be accompanied by the designation of a type specimen or set of type specimens. These are the only specimens that officially bear the name of the new species. For practical reasons, the Code requires that type specimens be clearly labeled and that suitable measures be taken for their preservation and accessibility to interested scientists, which means that they are usually deposited in a major museum where curatorial facilities are available.

As well as the specifications for naming species just mentioned, the International Code contains a vast array of rules covering various contingencies brought about by historical changes in procedure and by the need to evaluate species that have not been described in adherence to the rules. It often happens that a new name is proposed for a species that has, in fact, already been described and named by someone else. In such a case we apply what is known as the *law of priority*. Except in special circumstances, the name proposed first has precedence over all subsequently proposed names. The International Code contains a set of procedures to be followed for the clarification of such duplications and for the arbitration of any disputes that may arise.

The complications that can and do arise in species nomenclature are many and varied. This discussion presents only an introduction to the general problem of assigning names to species.

THE FORMAT OF A DESCRIPTION

The following format for proposing a new species has been suggested by Mayr, Linsley, and Usinger (1953):

Scientific Name

Taxonomic references and synonymy (if any)

Type specimen (including information about
   where it was found and repository)

Diagnosis

Description

Measurements and other numerical data

Discussion

Range (geographic)

Habitat (ecologic notes) and Horizon (for fossils)

List of material examined

The scientific name is in binomial form, following the rules of the International Commission. "Diagnosis" refers to a listing of characteristics by which the new species can be distinguished from other species. "Description," in this context, means a full assessment of the characteristics without particular reference to similarities and differences with respect to other recognized species. The number of measurements and other numerical data included in a description varies from author to author and from one biologic group to another. At a minimum, the major dimensions of the type specimen or specimens should be included. The "discussion" section may include information about nongenetic variation, ontogenetic stages, the derivation of the name, evolutionary affinities with other species, and, for fossils, state of preservation.

The "geographic range" is usually a list of places at which the new species has been found (in addition to the type locality). With regard to habitat, the paleontologist is most concerned with the geologic setting (rock type, for example). The "list of material examined" usually includes reference to museum repositories of specimens other than those formally designated as types.

To illustrate further the format of species description, three actual examples are given in Boxes 5-A, 5-B, and 5-C, with photographs of type specimens. They illustrate both the consistency of species descriptions and some of the variation among them.

DIAGNOSIS AND DESCRIPTION

The form and content of diagnosis and description has varied considerably from time to time and from place to place. The increased emphasis on intraspecific variation has had substantial and significant effects on the manner in which species are described and diagnosed. Because one of the prime objectives is communication of information, there is an understandable premium attached to consistency. This means, for example, that standardized morphologic terminology is used wherever possible. Chaos would result if each taxonomist invented new terms for describing fossils.

A truly complete diagnosis or description is impossible. As we saw in Chapter 2, an infinite number of attributes can be used to describe a fossil. It is neither possible nor advisable to describe exhaustively either a single specimen or an array of specimens. For diagnosis, enough attributes should be included to distinguish specimens of the new species from those of closely related species. Emphasis is usually placed on attributes that are most noticeable in all states of preservation. If too many are included, the system becomes cumbersome and obscures the most significant attributes. Therefore, if no reference to a particular character is found in a diagnosis, it cannot be assumed that the character is not diagnostic. This is where the actual specimens become most important in serving as a "backup" for the diagnosis.

The description, as distinct from the diagnosis, serves several purposes, not the least important of which is to provide an assessment of attributes

BOX 5-A   *Chonetes pachyactis* Imbrie, new species

The Devonian brachiopod *Chonetes pachyactis* Imbrie (×1). The original species description is reproduced here. (Courtesy of U.S. National Museum.)

DISTINGUISHING CHARACTERS: This species is distinguished by its low length-width ratio; by its strong, wide, rounded ribs; by the moderate and equal depth of its valves; and by its large size. *Chonetes pachyactis* is most closely allied with *C. maclurea* Norwood and Pratten from the St. Laurent limestone of Illinois, but the latter differs in the shallower brachial valve and a greater inflation of the umbo.

DESCRIPTION: Large shell; width of mature individuals commonly 30 mm. Outline transversely semicircular to subquadrilateral. Costellae strong, broad, rounded, about equal to the striae in width and shape, increasing by bifurcation and some implantation; rib count about seven in a distance of 5 mm. at the front margin.

Pedicle Valve: Six to eight spines on each side of the beak along the cardinal margin.

Brachial Valve: Lateral and anterior profiles moderately and evenly concave, about equal in depth to pedicle valve. Postero-lateral regions slightly less concave than the general surface of the valve.

MEASUREMENTS OF HOLOTYPE: Length, 20 mm.; width, 32 mm.; thickness, 5.7 mm.

OCCURRENCE: Norway Point formation (localities 41, 46, 47) of Michigan.

HOLOTYPE: U.S.N.M. No. 124397; Norway Point formation, locality 47.

REMARKS: Analysis of growth stages of this species indicates that its major proportions change regularly with growth, so that larger individuals are relatively wider and thicker than smaller individuals.

BOX 5-B   *Xenoceltites youngi* Kummel & Steele, n. sp.

The Lower Triassic ammonoid *Xenoceltites youngi* Kummel and Steele (×1). The original species description is reproduced here. (From Kummel and Steele, 1962.)

Conch, evolute with compressed, discoidal whorl sections. Venter narrowly rounded, lateral flanks convex, umbilical shoulders rounded but well marked, umbilical wall nearly flat and sloping to umbilical seam at an angle of approximately 45 degrees. Three specimens in the collection are assigned to this species and their measurements are as follows:

	Holotype *MCZ 5266*	*Paratype* *MCZ 5268*	*Paratype* *MCZ 5267*
D	56.4	38.7	28.0
W	13.7?	8.7?	6.1
H	22.9	16.8	12.5
U	18.3	11.5	7.6
W/D	24.3?	22.5?	21.8
H/D	40.6	43.4	44.6
U/D	32.4	29.7	27.1

that may at some future time be critical in diagnosis. If the species is part of a well-known group and is similar in most regards to other species, much of the description may be neglected in deference to the existing descriptions of closely related species. The weight may then be carried by a simple diagnosis. For a species belonging to a relatively unknown group, the description must be more comprehensive in order that relevant comparisons may be made if related species are discovered subsequently.

The flanks of the inner whorls bear blunt indistinct ribs. These blunt ribs are present on the holotype and larger paratype; the smaller paratype, however, has perfectly smooth inner whorls but in every other respect agrees completely with the other paratype and the holotype. On the outer whorls there are coarse growth lines which are periodically bunched to give the appearance of sinuous ribs which are not visible on the internal cast. The coarse growth lines cross the venter and occasionally form a weak constriction over the ventral region.

The suture is only poorly displayed but consists of two minutely denticulated lateral lobes and low, broad saddles.

REMARKS: The assignment of this new species to *Xenoceltites* is based on the presence of the indistinct blunt ribs on the inner whorls, the suture, and the general overall similarity to *Xenoceltites subevolutus* Spath (Frebold, 1930, p. 14, pl. 3, figs. 1–3; Spath, 1934, p. 130, 131, pl. 2, fig. 2; pl. 8, fig. 2; pl. 9, fig. 4; pl. 11, fig. 2). Most of the species assigned to *Xenoceltites* by Spath (1934) have rather marked constrictions on the outer whorls, a morphological feature only weakly developed in *X. youngi*. Peripheral constrictions are particularly marked on *X. spitsbergensis* Spath and *X. gregoryi* Spath, forms which are more evolute than *X. subevolutus*. Spath (1921, 1934) states that he had 100 examples of *Xenoceltites* from Spitsbergen, and in the discussion of each of three species he established (Spath, 1934) he stressed the transitional nature of each to the other species. It is difficult to assess the range of variability in the Spitsbergen species, but from the data available it seems probable that forms like *X. subevolutus* had the peripheral constrictions on the outer whorls greatly reduced, to a degree as found in *X. youngi*.

Xenoceltites is represented by *X. cordilleranus* (Smith) and *X. intermontanus* (Smith) from the *Meekoceras* limestone of southeastern Idaho; *X. hannai, X. douglasensis* and *X. matheri* Mathews (1929) from the *Anasibirites* Zone of Fort Douglas, Utah; *X. robertsoni* and *X. warreni* McLearn (1945) from the *Anasibirites* Zone of British Columbia; *X. russkiensis* Spath, 1934 (for *Ceratites minutus* Diener, 1895, *non* Waagen) from a mid-Scythian horizon near Vladivostok, Siberia; and *X. evolutus* Waagen (1895) from near the junction of the ceratite sandstone and the upper ceratite limestone.

OCCURRENCE: Bed "a," *Meekoceras gracilitatus* Zone, Crittenden Spring, Elko County, Nevada.

REPOSITORY: Holotype MCZ 5266, paratypes MCZ 5267, unfigured paratype MCZ 5268.

If possible, a description should include discussion of ontogenetic development, particularly if the organism's ontogeny is accompanied by a change in form. Also important is an assessment of variation encountered within and between populations of the species. When a new species is recognized on the basis of few specimens or fragments of specimens, this is not possible.

There has been considerable controversy over how much specimen material is necessary to establish a new species. It has been argued that because

114

DIAGNOSIS: A species of *Pterocephalops* having a slightly tapered glabella that bears glabellar furrows and lacks a median glabellar crest.

DESCRIPTION: The cranidium is about as wide as long. The glabella is subrectangular, tapering only slightly toward the front. The greatest glabellar convexity is transverse and anterior to the glabellar midpoint. Two pairs of posteriorly directed glabellar furrows show on exfoliated specimens and less clearly on the exterior of the shell. The occipital furrow is well impressed. The occipital ring is longest at the midpoint and bears an occipital node. The convex brim is sharply downfolded to a level well below that of the glabella. The border is of the same length as the brim and is horizontal except at the margin where it is turned up. Lochman-Balk (1959b, p. 0260) considered this as a wide border furrow and a narrow border; perhaps our view of the structure is incorrect. The fixed cheeks rise abruptly from the dorsal furrows, reaching a level just below the top of the glabella. Palpebral lobes are not preserved. The fixed cheeks are about one-third the width of the glabella. From a point slightly forward of the glabellar midpoint the fixed cheeks

are sharply downfolded anteriorly and posteriorly. The posterior limb is wide at its base and is well marked by the posterior border furrow. The facial sutures diverge only slightly both in front of and behind the eyes.

The free cheek, thorax, and pygidium are not known.

The surface bears scattered punctate tubercles, and the border shows a transverse line of regularly spaced tubercles.

MEASUREMENTS: USNM 158449a: A_1, 8.4; A_2, 6.8; B, 4.7; B_1, 5.7; D, 3.6 (estimated); J_1, *10.1;* J_2, 7.6; J_3, 7.4; K, 4.1; K_1, 3.9; L_1, 3.6. USNM 158449b: A_1, 6.9; A_2, 5.5; B, 3.7; B_1, 4.6; J_1, *9.4;* J_2, *6.8;* K, 3.1; K_1, 2.9; L_1, 3.0. Note that A_2 is the sagittal distance from the rear of the occipital ring to the foot of the steep preglabellar slope; this may not be homologous with the usual occipital intramarginal cranidial length (Shaw, 1957, p. 194–195), but it is the obvious measurement to make on this head.

COMPARISON: This species is very similar to *P. acrophthalma* Rasetti, 1944, but both of the Gorge specimens differ from both of those that were figured by Rasetti in 1, having somewhat more taper to the

recognizing the variation within a species is very important, a single specimen or a fragment should never be used to establish a new species. No unequivocal answers can be given to this question because what is necessary in a particular description depends on the amount of difference between related species. Often a single specimen demonstrates that the organism is different from all other known organisms, making it folly to wait for the accumulation of large numbers of specimens. Many significant discoveries in paleontology are based on single specimens or on fragmentary material. On the other hand, if a new species belongs to a well-known group in which

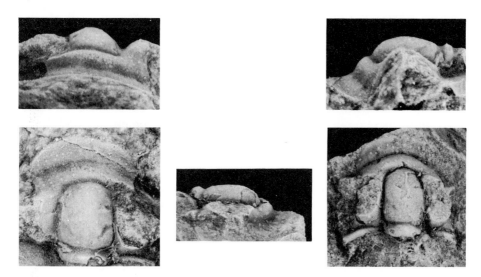

The Upper Cambrian trilobite *Pterocephalops tuberculineata* Clark (×4). The original species description is reproduced here. (From Clark and Shaw, 1968.)

glabella; 2, the rather numerous tubercles as contrasted with the very few visible in Rasetti's illustrations; 3, the presence of two pairs of glabellar furrows, which are absent even from the exfoliated parts of *P. acrophthalma;* and 4, the presence of an occipital node. These minor differences appear to be consistent. The two species are obviously closely similar.

REMARKS: The trivial name, *tuberculineata,* alludes to the line of tubercles on the border.

HOLOTYPE: USNM 158449a (1 cranidium).

PARATYPE: USNM 158449b (1 cranidium).

differences between species tend to be rather subtle, a large amount of material must be accumulated to make the description complete and effective.

Many conventions concerning the amount and quality of description are dictated by practical considerations. A description cannot be so long that it cannot be read and absorbed efficiently. The combined lengths of diagnosis, description, and discussion of a new species usually do not exceed two thousand words. Because automatic methods of storage, processing, and retrieval of data are developing rapidly, many taxonomists are looking to these new facilities to extend the practical limits of description. It would be a relatively easy matter in many instances to codify the description of a new species so that it could be stored in compact form and compared with others at will. Several proposals have been made recently and pilot studies have been carried out. It is tempting indeed to contemplate that the information contained in a two thousand word description of a fossil species might be reduced to an assessment of 30–40 morphologic characters and might then occupy only a few inches on a magnetic tape. It might even be possible to do away with the diagnosis as we know it because descriptions of attributes of several species could be compared automatically and the differences noted by the machine would represent the diagnosis.

An attribute of an organism used in diagnosis is referred to as a ***taxonomic character.*** How does a taxonomic character differ from other characters? The answer is inevitably circular because a taxonomic character is one shown to be useful in taxonomic discrimination. It is a character useful in showing *differences* between taxonomic groups.

To be useful in discriminating between species, a character must satisfy several conditions. Because species discrimination is ultimately a question of assessing genetic differences between populations, the taxonomic character must be one that shows a minimum of nongenetic variation. To be effective, it must also be a fairly obvious attribute, especially in fossil material in which shortcomings of preservation often yield an incomplete picture of the total organism. Ideally, a taxonomic character should be present and recognizable throughout an organism's ontogeny, and not just during certain growth stages.

Many morphologic characters satisfy the conditions just mentioned, but do not constitute effective taxonomic characters. These are most usually characters that, although genetically controlled, are subject to great genetic variation within populations. Hair color in the human species is an example.

How effective a particular taxonomic character is may depend on which taxonomic category is under consideration. Many characters are too variable to be useful in distinguishing species but may be extremely important in describing subspecies. Other characters show no variation from species to species but are valuable in distinguishing genera and families.

The biologist often chooses taxonomic characters based on soft-part anatomy or behavioral traits that rarely leave clues in the fossil record. Problems thus arise in the taxonomy of groups of organisms with both fossil and living representatives. Very often, the paleontologist restudies the living species and develops a set of secondary taxonomic characters based on readily preservable parts of the organism. Certain living species, however, cannot be distinguished by preservable characters. The usual result is that several fossil species are combined under a single name although they would be separated into distinct species were complete knowledge of their morphology available.

In summary, taxonomic characters are chosen primarily for their usefulness in taxonomic discrimination. Their choice is based in large measure on hindsight. Taxonomic characters as a group are probably the most significant characters biologically, but the system contains no assurance of this.

TYPE SPECIMENS

The selection of type specimens presents several problems. If a single specimen is designated, it is called the *holotype.* If several specimens serve this purpose they are called *syntypes.* Both alternatives are officially accept-able although the International Commission strongly urges the use of a holotype rather than a series of syntypes because it is always possible that the series will be judged by later workers to contain representatives of more than one species.

Type specimens were once widely used to define species and as special standards for comparison. We now know, owing in part to the adoption of the population approach in taxonomy, that many of these type specimens are not the most legitimate representatives of their species. Even in a large sample a single specimen may not exist that is average for all observable morphologic characters. Furthermore, the mean value for any variable fea-ture tells nothing of the range of variation.

In modern biology and paleontology the type specimen no longer serves to define a species (unless no other specimens are available). It is relegated, instead, to the position of "name-bearer." When a species is named, the name is formally attached only to the one or more specimens that are desig-nated as type specimens. In practice, fossil type specimens actually tend to be somewhat unusual representatives of the hypodigm. The most common bias is toward large size and good preservation.

Several other kinds of type specimen figure prominently in various phases of taxonomic work. A *paratype* is a specimen other than the holotype that is formally designated by the author of a species as having been used in the description of the species. The designation of a single holotype and a series of paratypes thus contains some of the advantages of both the holotype sys-tem and the syntype system. The holotype remains as the name bearer but the paratypes, which may be numerous, serve to express more fully the author's concept of the species.

A *lectotype* is a specimen originally designated as a syntype, but subse-quently singled out as the definitive type specimen for a species. This dupli-cates what the author would have accomplished if a holotype and a set of paratypes had been designated originally.

It often happens that type specimens are destroyed or lost. The Inter-national Commission has established procedures for designating new type specimens to replace those that are lost. The *neotype* is such a replacement.

For species described many years ago, it is often important to redescribe and re-illustrate specimens, especially if the original description was written under a different set of rules. *Plesiotypes* are type specimens used for re-descriptions of existing species.

A great many other kinds of types have been proposed and are occasion-ally used. For example, a *topotype* is a specimen that is not part of the origi-nal type material, but that has been collected at the type locality.

Changing Species Names

A worker may change the name of a taxon either because he has found that its use violates a rule of nomenclature or because he judges that the taxon has been improperly classified. We will restrict our discussion of name changes to species names.

Homonyms are identical species names that denote different species groups. There are two varieties. *Primary homonyms* are identical names that were erected for different taxa (with different holotypes) belonging to the same genus. The author of the later-named homonym was in error, not knowing that the species name had been occupied when he used it. Once such an error is discovered, only the first published, or *senior homonym* can be retained. The Code states that the *junior homonym* must be permanently rejected. The difference between primary and *secondary homonyms* is that the latter originate by transfer of one species to a new genus that contains a species with the same specific name. The author of neither species is in error according to the Code, because the same specific name can be used for species belonging to different genera. The problem arises when a later worker has decided that the two species belong to the same genus. The worker who discovers secondary homonyms should formally reject the junior name. A rejected primary or secondary homonym must be replaced by the oldest available name, or if no previously published name is available, by a new name.

Synonyms are two different names applied to the same taxon. There are two varieties. *Objective synonyms* are different names that are based on the same type specimen or specimens. Here there is no question of taxonomic opinion; the senior (first published) synonym must be retained, and the junior synonym must be permanently rejected. *Subjective synonyms* are names that were established for different type specimens that are later judged by a worker to belong to one species. This worker must then apply the senior synonym to both of the type specimens and all other specimens belonging to the hypodigm in which he places them. Another worker, however, may judge that the type specimens belong to separate species; he will not consider the names to be synonyms and will retain both. In other words, while a junior objective synonym is eliminated automatically by the Code, a junior subjective synonym remains available as a name, its use depending entirely upon taxonomic opinion.

Rejection of names on the basis of priority is sometimes unfortunate because it eliminates familiar names. The formal change in generic name of the familiar Eocene "dawn horse" from *Eohippus* to *Hyracotherium* was unpleasant to many workers. The Commission on Zoological Nomenclature is empowered by the Code to exercise its plenary powers in order to suspend the rule of priority at special request. Many familiar names found to be junior homonyms or synonyms have been retained by this procedure.

Perspective

Taxonomic procedures form an extremely complex and important subject for the working paleontologist. In actual fact, procedural problems are considerably more complex than outlined here. The student is often impressed and discouraged by this complexity and by the many seemingly archaic and inefficient practices. The possibility of working with the biology and paleontology of organisms from outer space has given impetus to many reform movements. The contention has become common that the exploration of space should provide an opportunity to correct the errors of the past and to start with a new, modern system.

It cannot be denied that the current procedures could be improved upon. To date, the greatest deterrent to sweeping reform has been the obvious cost. With 1,500,000 living species and 130,000 fossil species the problem of effecting change has been overwhelming. The mere size of the problem has been a deterrent to progress, but at the same time has been a stabilizing influence and has effectively prevented the confusion that would undoubtedly result from partial reforms. In the next generation, we can probably look forward to gradual improvement and streamlining, but it is unlikely that the system will be altered significantly.

Supplementary Reading

Blackwelder, R. E. (1967) *Taxonomy: A Text and Reference Book.* New York, Wiley, 698 p. (A comprehensive sourcebook of taxonomic theory and practice.)

Mayr, E. (1963) *Animal Species and Evolution.* Cambridge, Harvard University Press, 797 p.

Mayr, E. (1969) *Principles of Systematic Zoology.* New York, McGraw-Hill, 428 p. (An excellent and well-organized treatment of taxonomy.)

Savory, T. (1962) *Naming the Living World.* London, English University Press, 128 p. (A brief summary of taxonomic nomenclature.)

Schenk, E. T., and McMasters, J. H. (1956) *Procedure in Taxonomy*, 3rd Ed. Revised by A. M. Keen and S. W. Muller. Stanford, Calif., Stanford University Press, 119 p. (A detailed discussion, with examples, of the rules of taxonomic nomenclature.)

Simpson, G. G. (1961) *Principles of Animal Taxonomy*. New York, Columbia University Press, 247 p. (Emphasizes taxonomic theory.)

Stoll, N. R., et al., eds. (1961) *International Code of Zoological Nomenclature*. London, International Trust for Zoological Nomenclature, 176 p. (The official version of the International Code; contains a glossary and extensive appendices.)

Sylvester-Bradley, P. C., ed. (1956) *The Species Concept in Paleontology*. Systematics Association Publication 2, 145 p. (One of several symposium volumes devoted to problems of defining species in the fossil record.)

CHAPTER 6

Grouping of Species into Higher Categories

The naming of species has usually gone hand in hand with arranging them in some kind of classification. Species having characteristics in common have been grouped together and thus distinguished from other such groups, which assists us in communicating information about them. One might say, for example, "I saw a bird eating a worm." There are about 8,600 species of birds living today and about 25,000 species of worms and the statement just given would be meaningless to all but the most specialized audience if species names were substituted for "bird" and "worm." By using the word "bird" we describe the predator as being one of a clearly defined group of similar organisms and by so doing exclude all but about $\frac{1}{10}$ of one percent of the animals living today. Furthermore, the purpose in making the statement might have been such that specific identification would be quite irrelevant or the species might have been very difficult to identify.

Perhaps the most practical reason for grouping species into *higher categories* is that each person cannot learn and remember the names and distinguishing characteristics of a million and a half different organisms. In the example of the bird and the worm, a good ornithologist might be able to recognize the species of the bird (particularly if he were familiar with the fauna of the region), but it is unlikely that he could identify the worm. In this sense, the grouping of species compensates for the lack of storage space in the human memory. The usual procedure for identifying a specimen is to determine its phylum first, and gradually proceed from the general to the specific.

How is the classification of plants and animals best accomplished? A single group of species can be subdivided in many different ways. The species could, for example, be arranged alphabetically in 26 categories, each determined by the initial letter of the species name. Somewhat more logically, they might be grouped according to year of discovery. Neither of these systems, however, would be of any assistance to the person who observes the bird eating the worm, because the organisms have no visible characteristics that serve as an aid in relating them to these systems.

Most classifications are based upon readily observable characteristics, either of appearance, habitat, behavior, or geographic or stratigraphic occurrence. Many of the earlier classifications of vertebrates were based on habitat, with aquatic vertebrates (fish) being distinguished from land-dwelling vertebrates (reptiles, birds, and mammals) and so on. The fact that there were exceptions to these classifications (such as aquatic mammals) did not negate the usefulness of the classifications for most purposes.

Most successful classifications have been based on morphology. We use morphology here in a rather broad sense to include not only external form, but also internal anatomy and even some details of physiology, biochemistry, and behavior.

Most classifications used today had their beginnings long before organic evolution was widely accepted. Pre-Darwinian classifiers assumed for the most part that the species they were dealing with were created spontaneously and independently. Nevertheless, some early classifications, particularly at the higher levels, have changed little as the understanding of evolution has become widespread.

Classification versus Evolution

If classification is to serve primarily for communication and identification, utility is the principal criterion for choosing one system over another. With the rise of interest in evolution there has inevitably been a move toward using classification to express evolutionary relations also.

In most plant and animal groups, increased emphasis on evolution has not brought about very dramatic changes in the actual form and content of classification, primarily because morphologic similarity is highly correlated with evolutionary proximity. Thus a classification intended to reflect evolution is largely dependent upon the same kinds of similarities and differences as one designed simply for identification.

Classifications are constantly undergoing change, partly as a result of our increase in descriptive knowledge. New higher categories are often needed to express the distinction between newly discovered species and ones that have been known for a long time. Classifications also change from time to time because of increase in theoretical knowledge of evolutionary mechanisms. Considering this we may note that classifications at a given time are an index of evolutionary thought. Furthermore, classifications by different but equally well-qualified scientists commonly differ at any one time simply because interpretations of evolution differ.

Taxonomic Categories

Linnaeus in 1759 used only six taxonomic categories: kingdom, class, order, genus, species, and variety. Of these, kingdom, class, order, and genus qualify as higher categories in the present context. Since the **kingdom** as a taxonomic category was used by Linnaeus only to separate plants from animals, in his work as a zoologist he used only three higher categories: class, order, and genus.

Since the number of species known in Linnaeus' time was relatively small, his categories were quite adequate to divide the total array into manageable groups. The tremendous increase in the number of known species has demanded the addition of categories intermediate between kingdom and species. The principal additions are **phylum** and **family,** with the most commonly used hierarchy being: kingdom, phylum, class, order, family, and genus. For some groups, further categories such as subphylum, subclass, superorder, suborder, superfamily, subfamily, and subgenus have been added.

How are each of these categories defined? In simplest terms each category may be defined as a subdivision of the next higher category.

The International Code of Zoological Nomenclature covers many procedures used in dealing with higher taxonomic categories. The Code itself includes rules for formation of categories up to the superfamily level. The recommendations of the International Commission attached to the Code, however, cover the entire range of higher categories (by implication if not by explicit rules). The rules are approximately parallel to those for species. We have already seen, for example, that a new species must be assigned at least to the next higher category (the genus). It is generally assumed that all genera can be assigned to families, orders, and so on, although allowance is given for the possibility that the evolutionary affinities of a genus may be unknown. In such a case, the genus may be defined in isolation (*"incertae sedis"*). Several rules governing the formation of names of higher categories make it possible to recognize the taxonomic rank on sight from the spelling of the ending of the name.

The type concept also extends to the definition of higher categories. When a new genus is proposed, a **type species** designation must accompany the original description. The type species thus becomes the name bearer for the genus. Similarly, a family must have a **type genus,** and so on.

When a newly discovered species is named and described, every attempt is made to find a plausible assignment in existing higher categories. If this proves to be impossible, new categories are sometimes proposed; not only might a phylum name be proposed, but also new names for a class, order, family, and so on. This illustrates an important feature of the conventional system of classification: a higher category may be **monotypic.** A phylum need have only one class; a class may have no more than one order, and so on.

Classification as an Interpretation of Evolution

Most classifications assume that evolution follows a well-defined model. In Chapter 4, we distinguished two kinds of speciation: phyletic and geographic. The first is the gradual evolution of one species into another along a single lineage, while the second involves the branching of one evolutionary lineage into two, which is often expressed graphically by a phylogenetic tree. Quite different phylogenetic trees are shown in Figures 6-1 and 6-2. In Figure 6-1, a few closely related species evolved over a short span of time: each segment of the "tree" is part of an evolving lineage.

At best, the phylogenetic tree is a simplification of nature. At any point in time, a species is usually represented by many geographically distinct populations. Elements of a lineage may diverge with the onset of geographic isolation and later converge if the geographic barriers are removed. The "internal" character of an evolving lineage is shown diagrammatically in Figure 6-3.

Figure 6-2 shows the evolution of a large group of species over a long span of time. The smallest taxonomic unit is the order. Thus, a single line in Figure 6-2 represents a complex of many evolving lineages, each of which contains evolving populations.

What coordinate system is used in a phylogenetic tree? Many different systems are used and it is important to understand the differences. In the examples shown in Figures 6-1 and 6-2, the vertical dimension is time; the horizontal dimension shows the relationships between lineages. In phylogenetic trees such as these, the top of a line indicates the extinction of the lineage unless it reaches to the top of the diagram.

Figure 6-4 shows a phylogenetic tree that is essentially dimensionless. The end of a line does not imply the extinction of a lineage although it may coincide with it nor does the distance between branches indicate geographic or evolutionary distance. Time plays a role in the diagram only in that the chronological sequence of evolutionary events is established.

In Figure 6-5, the vertical and horizontal dimensions are time and geographic area. It was possible to construct the phylogeny in this way because the evolving lineages rarely overlap geographically.

Figure 6-6 shows the classification of thirty-one fossil brachiopod genera into eleven subfamilies (names ending in "-inae") and four families (names ending in "-idae"). If we assume that this classification is a valid interpretation of evolution, we may look upon it as a phylogenetic tree (without geographic or time dimensions). Genera considered to be most closely related are linked in the diagram as members of a subfamily; closely related subfamilies are linked as members of a family. The thirty-one genera, as a group, constitute a superfamily.

Assignment of early forms to higher categories often depends upon our knowledge of later evolutionary history. In the Solnhofen Limestone, three Jurassic specimens have been found that are believed to represent the first true birds. They are reptilian in most morphologic features (they have teeth and long tails, and their general anatomy is reptilian) but have what appear to

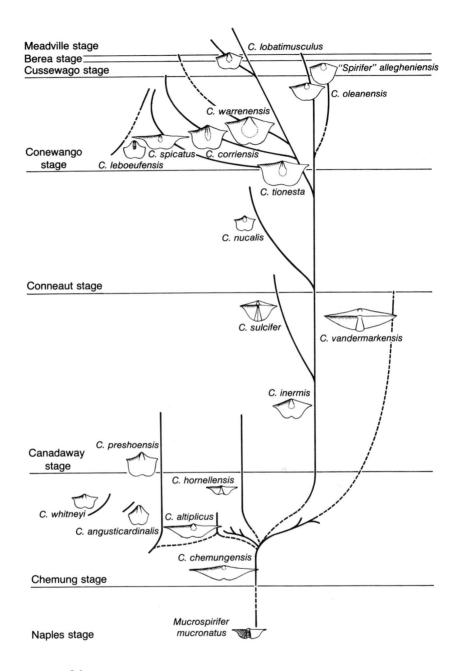

Meadville stage
Berea stage
Cussewago stage

C. lobatimusculus

"Spirifer" *allegheniensis*

C. oleanensis

C. warrenensis

Conewango
stage   *C. leboeufensis*

C. spicatus—*C. corriensis*

C. tionesta

C. nucalis

Conneaut stage

C. sulcifer

C. vandermarkensis

C. inermis

C. preshoensis

Canadaway
stage

C. hornellensis

C. whitneyi

C. altiplicus

C. angusticardinalis

C. chemungensis

Chemung stage

Naples stage

*Mucrospirifer
mucronatus*

FIGURE 6-1
Phylogeny of the species of the brachiopod *Cyrtospirifer* (Upper
Devonian and Lower Mississippian, New York and Pennsylvania).
(After Greiner, 1957.)

126

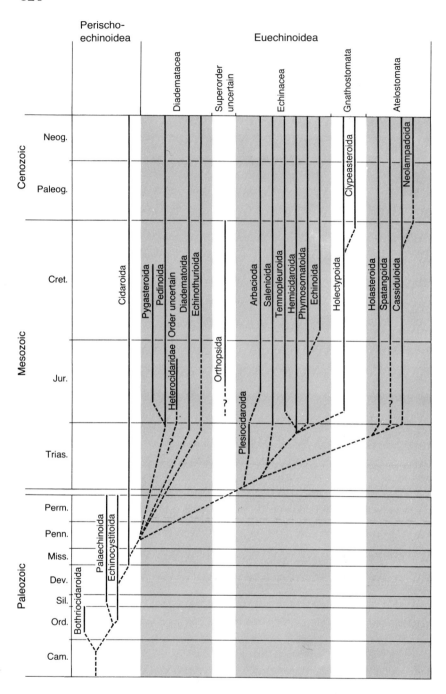

FIGURE 6-2
Phylogeny of the class Echinoidea, based on ranges in the stratigraphic record of
specimens of the various orders and on inferred evolutionary relationships. Gap above
Permian indicates a change in the vertical scale. (From Durham, 1966.)

FIGURE 6-3
The anatomy of an evolutionary lineage.
A: Ontogenies of individuals in a population.
Horizontal lines represent mating.
B: Evolving array of populations making up
the lineage.
C: Relationship between several lineages.
(From Simpson, 1953.)

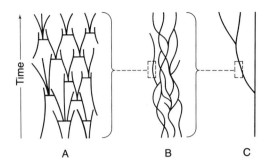

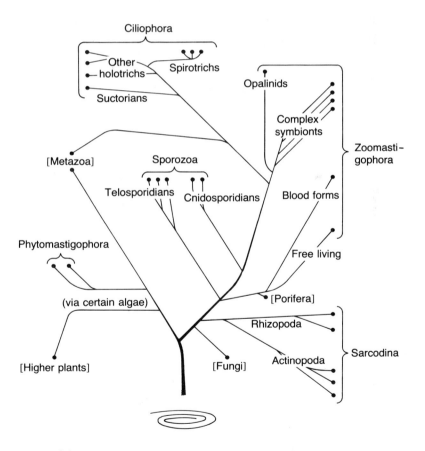

FIGURE 6-4
An inferred phylogeny for the Protozoa. Presumed evolutionary descendants
of the Protozoa are indicated in brackets. (From Corliss, 1959.)

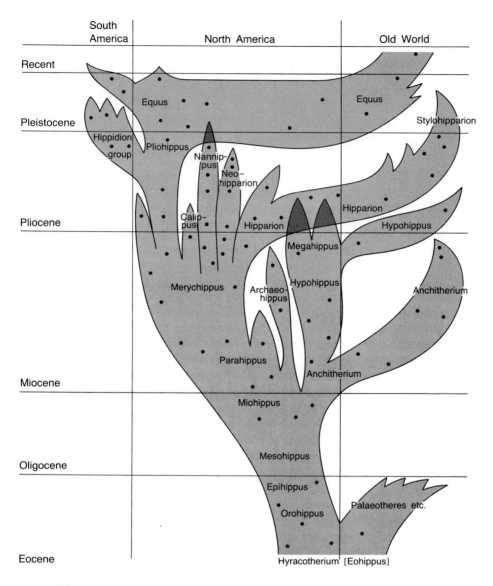

FIGURE 6-5
Inferred phylogeny of the horses. (From Simpson, 1951.)

be functional wings with feathers. Knowing that these modifications of reptilian anatomy led in later Mesozoic times to a great evolutionary radiation (the class Aves), we attach special evolutionary significance to the Soln-hofen specimens. We identify and classify them as bird (class Aves) even though they were more closely related to Mesozoic reptiles than to any modern birds. It is doubtful whether the Solnhofen bird would be considered as anything more than an aberrant reptile had it not been the form known to have given rise to such a wide diversity of organisms having features quite different from reptiles.

At this point, we might wonder whether higher categories have an objective basis. In a sense they do and in a sense they don't. We define the biologic species by using the concept of reproductive isolation and look upon reproductive isolation as a basic biologic phenomenon that makes it possible for evolving lineages to remain genetically distinct. In this sense, Simpson's definition of the **evolutionary species** has some objective basis even though

FIGURE 6-6
Classification of the fossil brachiopod superfamily Chonetacea, arranged in the form of a dendrogram. (From Rowell, 1967; after Muir-Wood, 1965.)

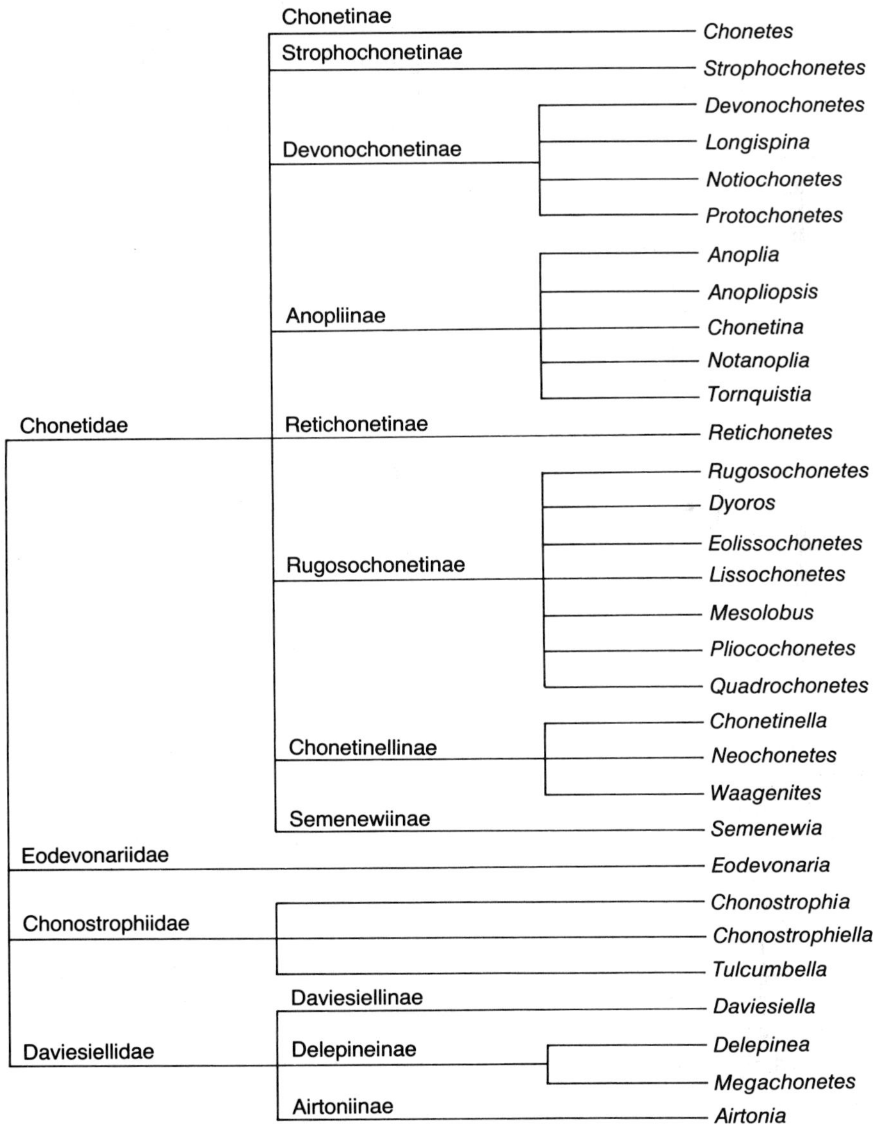

division of lineages into species is arbitrary. The family, on the other hand, has no comparable objective basis. Higher categories are used to express evolutionary relationships but we do not have, in the present state of our understanding, specific rules that can be applied. In other words, there are no thresholds in the transition from genus to family comparable to the threshold between the subspecies and the species (at a single moment in time).

Some objectivity does enter into the development of higher category classification. It is implicit, for example, in the general classification model that all species of a genus have evolved from one ancestral lineage. This follows from the fact that at the *supraspecific* level we assume that phylogeny is a process that expresses itself by continuously branching. If two species can be shown to have a common ancestor that is not also the ancestor of a third species, then the first two species can objectively be placed in a higher category different from that of the third. This does not lead, however, to an objective basis for deciding where the family level leaves off and the order level begins because the difference between the levels is one only of degree. It is not surprising that classifications developed by different taxonomists often differ in assessment of higher categories. If classifications are based on the same understanding of phylogeny, however, the classifications should be correlatable. Rarely does the biologist or paleontologist have a truly accurate knowledge of phylogeny; classification is thus an expression of knowledge at a given time.

The approaches of paleontologists and biologists toward questions of higher categories differ somewhat because the paleontologist is able to observe the time dimension directly whereas the biologist must always infer the chronological development of lineages. The paleontologist is sometimes able to say that one species evolved from another because he finds the two in stratigraphic succession.

Criteria for Definition of Higher Categories

If classification is to reflect phylogeny, we must assess the "phylogenetic distance," or amount of genetic difference, between species. Genetic difference cannot be observed directly in fossils any more than it can in most living forms. We can, however, use morphologic difference as a rough measure of genetic difference (just as we did when dealing with species). The two species in a group that have the largest number of morphologic characters in common are most likely to be descended from a common ancestor and thus are most likely to qualify as members of a single higher category.

A large number of species, each defined by many taxonomic characters, may be divided up into many quite different systems of groups. How do we choose between several alternative classifications? Which alternative is most likely to express phylogeny? It sometimes seems that there are as many different answers to such questions as there are taxonomists. It has even been stated that taxonomy at this level is an art rather than a science and that the methods defy a clear and logical explanation. Let us explore some commonly used methods.

Very often, a few characters are singled out (on *a priori* grounds) as being of greater evolutionary significance than the rest and division is based on these characters in preference to others. This **weighting of characters** takes several forms. In the example of the Solnhofen bird discussed earlier, the presence of feathers was used as a criterion for assignment to the class Aves. This was based partly on the knowledge that feathers were of vital importance in enabling the organism to occupy a different set of habitats from those of its ancestors and therefore, to undergo tremendous evolutionary expansion.

In other cases, choice of characters to be weighted is based on a more nearly objective judgment of the effect of weighting on the resulting classification. Table 6-1 shows purely hypothetical data that we will use to illustrate this type of weighting: ten imaginary species are defined by ten morphologic characters; each character is expressed in a given species as one or the other of two possible states, indicated by plus and minus. The two states might be "large" and "small," or "red" and "white," or "having teeth" and "toothless."

TABLE 6-1
Coded Morphology of Ten Hypothetical Species

		Characters								
	1	2	3	4	5	6	7	8	9	10
A	−	+	−	+	−	+	+	−	+	−
B	−	−	−	−	+	+	+	−	−	+
C	+	+	+	+	−	−	−	+	−	+
D	+	+	−	−	−	−	−	+	+	+
E	+	−	+	−	+	+	−	+	−	−
F	+	−	−	+	+	−	−	+	−	+
G	−	−	+	−	+	+	+	−	+	−
H	−	+	−	+	−	+	+	−	+	+
I	+	−	+	−	+	−	−	+	+	+
J	−	+	−	−	−	+	+	−	−	+

(Species, A–J)

Let us pose the problem of dividing the ten species into two groups (without specifying the relative sizes of the groups). Let us assume that the two groups occupy different branches of a phylogenetic tree and that we wish to use the morphologic information to determine them. There are 637 possible divisions of ten species into two groups. Our job is to select one having a high probability of being correct.

Some of the possible divisions are more plausible (morphologically) than others. Table 6-2 shows three of the alternatives. The first subdivision produces two groups that are each heterogeneous in all morphologic characters.

TABLE 6-2

Three Possible Classifications of Ten Species into Two Genera

Classification 1

Characters

	1	2	3	4	5	6	7	8	9	10	
A	−	+	−	+	−	+	+	−	+	−	
B	−	−	−	−	+	+	+	−	−	+	Genus 1
C	+	+	+	+	−	−	−	+	−	+	
D	+	+	−	−	−	−	−	+	+	+	
E	+	−	+	−	+	+	−	+	−	−	
F	+	−	−	+	+	−	−	+	−	+	
G	−	−	+	−	+	+	+	−	+	−	Genus 2
H	−	+	−	+	−	+	+	−	+	+	
I	+	−	+	−	+	−	−	+	+	+	
J	−	+	−	−	−	+	+	−	−	+	

Species (label along left side)

Classification 2

Characters

	1	2	3	4	5	6	7	8	9	10	
A	−	+	−	+	−	+	+	−	+	−	
C	+	+	+	+	−	−	−	+	−	+	Genus 1
F	+	−	−	+	+	−	−	+	−	+	
H	−	+	−	+	−	+	+	−	+	+	
B	−	−	−	−	+	+	+	−	−	+	
D	+	+	−	−	−	−	−	+	+	+	
E	+	−	+	−	+	+	−	+	−	−	
G	−	−	+	−	+	+	+	−	+	−	Genus 2
I	+	−	−	−	+	−	−	+	+	+	
J	−	+	+	−	−	+	+	−	−	+	

Species (label along left side)

Classification 3

Characters

	1	2	3	4	5	6	7	8	9	10	
A	−	+	−	+	−	+	+	−	+	−	
B	−	−	−	−	+	+	+	−	−	+	
G	−	−	+	−	+	+	+	−	+	−	Genus 1
H	−	+	−	+	−	+	+	−	+	+	
J	−	+	−	−	−	+	+	−	−	+	
C	+	+	+	+	−	−	−	+	−	+	
D	+	+	−	−	−	−	−	+	+	+	
E	+	−	+	−	+	+	−	+	−	−	Genus 2
F	+	−	−	+	+	−	−	+	−	+	
I	+	−	+	−	+	−	−	+	+	+	

Species (label along left side)

The second produces groups that are homogeneous for one of the characters (4) but heterogeneous for all others. The third produces groups homogeneous for three of the ten characters (1, 7, and 8). The last alternative may be said to be "supported" by 30 percent of the characters. Note also that an additional character (6) almost supports the classification: one group is completely homogeneous for character 6 (all plusses) and the other group is nearly homogeneous (all but one species are minus). The three characters infallibly supporting the grouping might be called "excellent" taxonomic characters because they each suggest the same classification. Character 6 might be called a "usable" taxonomic character because it suggests a classification very similar to that supported by the other three.

If we were to expand this hypothetical taxonomic problem to include many more species (perhaps several hundred) we would increase the complexity of the problem to a point at which there would be an almost infinite number of possible classifications. We could not begin to consider all of them. On the basis of the preliminary study of the ten species we might postulate, however, that characters 1, 7, and 8 would be significant in dividing the larger group of species. Character 6 could be added tentatively because of the 90 percent agreement between its distribution and that of the selected three. Characters 1, 7, 8, and 6 could then be tested with the larger group of species. If their separate use produced the same (or nearly the same) divisions of the larger group and if the divisions were geographically, ecologically, or stratigraphically reasonable, we might conclude that the four characters are critical in determining a major evolutionary division. We might forego further testing and agree to give disportionate importance to these characters in all subsequent taxonomic decisions. By so doing, we would have *weighted* the taxonomic characters.

In the hypothetical example, the classification would be completed by further subdivision of each group. We might call the original group of ten species a family. The initial division would provide two genera, a subdivision would produce subgenera, and so on. The characters used to determine genera could not, of course, be used in the determination of subgenera because the genera are homogeneous for those characters. Thus, the process of weighting characters excludes them from use at lower taxonomic levels.

Most weighting of characters is probably done in a manner similar to that used in the hypothetical example. The presence or absence of feathers, for example, is weighted heavily in vertebrate classifications because feathers are always present in a group (the birds) that is united by other characters as well; feathers are absent in other vertebrates. The presence or absence of feathers is thus analogous to characters 1, 7, and 8 in our example. The absence of teeth is almost universal in birds but *Archaeopteryx* had teeth. The presence or absence of teeth, as a character, is thus analogous to 6 in our example. It is a usable taxonomic character.

The taxonomist rarely follows a clear routine and he may not be aware of the logical steps he follows. It is not surprising that the classification process has been called intuitive or an "art."

Probably the greatest pitfall in the history of higher-category taxonomy has been the natural tendency to search for a single definitive character for grouping species. Many taxonomic problems with invertebrates have arisen in attempts to establish orders or subclasses. Attempts to group brachiopod families into orders solely on the basis of shell microstructure or morphology of the lophophore support have failed. Likewise, attempts to classify trilobites into orders strictly on the basis of facial suture configuration have proved unsuccessful. In the Bivalvia, two schemes, each based on a single morphologic character, were proposed. Many biologists favored use of gill type in classification of families into orders whereas other biologists and most paleontologists favored use of dentition (configuration of the hinge teeth of shells). Two contradictory classifications developed. It is now well known that the eulamellibranch type of gill evolved both in burrowing clam groups with heterodont dentition and in oysters, which lack hinge teeth and were derived from scallops. Certain types of dentition also arose independently in two or more taxonomic groups. Modern workers no longer attempt to use a single character in grouping bivalve families into orders. Most use several characters, including dentition and gill type, which they weight in a variety of ways.

Few modern workers employ the single-character approach in higher-category taxonomy. Still, for a few groups it has proven to be generally adequate. As we have seen, feathers in birds represent a definitive taxonomic character. Similarly, pelvic structure alone is sufficient to divide dinosaur families into two major orders (Figure 6-7). In a sense, however, the union of the two dinosaur orders in a single subclass is artificial; they are no more closely related to each other than either is to crocodiles, which constitute a distinct order that arose from the same ancestral group.

Numerical Taxonomy

It is tempting to wonder whether classification could be accomplished by machine. Given the information in Table 6-1, could a computer be programmed to investigate all 637 alternative divisions and choose the one supported by the largest number of characters? A modern digital computer could probably do this job in less than a second! Furthermore, it could be programmed to apply a variety of criteria in selection of the "best" classification. Selecting the classification supported by the largest number of characters is only one of many possible criteria.

Numerical taxonomy is the science of classifying organisms by purely mechanical or mathematical means. If we were to devise a numerical method for assessing the 637 possible classifications of the ten hypothetical species in Table 6-1, we would be doing numerical taxonomy. Numerical taxonomy is almost as old as taxonomy itself but nearly all major developments have come since the middle 1950's because of advances in computer technology.

So far, classifications developed by numerical methods have been based entirely on observable characteristics of the organisms. Such factors as

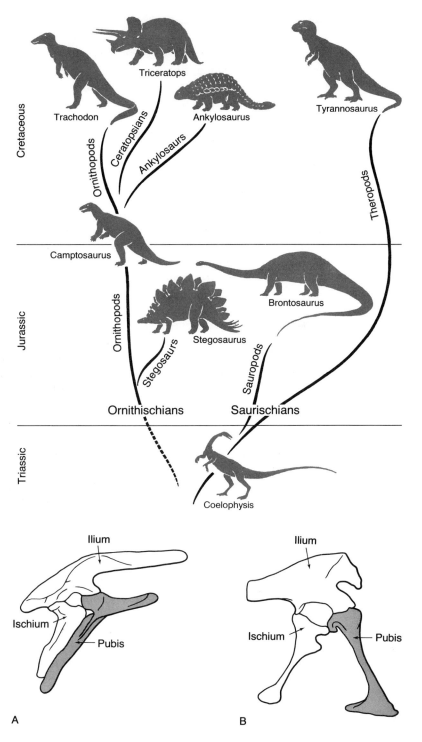

Cretaceous

Trachodon
Triceratops
Ankylosaurus
Tyrannosaurus

Ornithopods
Ceratopsians
Ankylosaurs
Theropods

Camptosaurus

Jurassic

Ornithopods

Stegosaurus
Brontosaurus

Stegosaurs
Sauropods

Ornithischians
Saurischians

Triassic

Coelophysis

Ilium

Ilium

Ischium

Pubis

Ischium

Pubis

A

B

FIGURE 6-7
The two dinosaur orders, which share a common ancestry but are characterized by distinct pelvic structures. A: Ornithischian ("bird-hipped") structure. B: Saurischian ("lizard-hipped") structure. (From Colbert, 1955.)

biogeography, stratigraphic distribution, and ecology have not been used. The resulting classifications are thus called *phenetic classifications* because they are based on the phenotype (defined broadly to include physiology, biochemistry, behavior, as well as what we conventionally mean by morphology). There has been considerable controversy over whether such classifications are "natural" in the sense of reflecting phylogeny. The answer to this depends partly upon the importance one attaches to nonphenetic information in classification. The fact is that most conventional taxonomies are based largely or completely on morphology and thus numerical taxonomy does not necessarily represent a substantial departure.

Most methods of numerical taxonomy are designed to operate at any taxonomic level: the basic units to be classified may be individual organisms, species, or even genera or higher groupings. The unit to be classified is called the *operational taxonomic unit,* or OTU.

The usual procedure in numerical taxonomy includes the following four steps: (1) selection of the OTU's to be classified, (2) selection of a group of phenetic characters (usually 50–100) to describe the OTU's (3) comparison of each OTU with every other OTU and, (4) determination of groups or clusters of OTU's on the basis of the computed similarities.

The data in Table 6-1 may be used to illustrate the procedure even though the methods discussed earlier in connection with this table have not been widely used in numerical taxonomy. The ten hypothetical species are the OTU's and each is described by ten characters.

A wide variety of techniques have been used for assessment of similarity among OTU's. Many of these are complex statistical methods that are beyond the scope of this volume. We can illustrate the procedure, however, by using the simplest possible measure of similarity: the percentage of characters for which two OTU's coincide, or match. This measure of similarity thus varies from 0 (no matches) to 100 (perfect correspondence in all characters). Table 6-3 shows similarity values assessed on this basis.

TABLE 6-3

Similarity Matrix for Ten Hypothetical Species
(Based on Data from Table 6-1)

	A	B	C	D	E	F	G	H	I	J
A		50	30	40	20	20	60	90	10	70
B			20	30	50	50	70	60	40	80
C				70	50	70	10	40	60	30
D					40	60	20	50	70	50
E						60	60	10	70	30
F							20	30	70	30
G								50	50	50
H									20	80
I										20
J										

We see from Table 6-3 that some pairs of OTU's show a much higher degree of similarity than others. In fact, there are two quite clearly defined "clusters" of similarity values. In Table 6-4, the similarity matrix has been rearranged to emphasize these clusters (ABGHJ and CDEFI). Similarity is generally high when members of a cluster are compared but low when members of different clusters are compared.

TABLE 6-4

Similarity Matrix from Table 6-3 Rearranged to Show
Clusters of Species

	A	B	G	H	J	C	D	E	F	I
A		50	60	90	70	30	40	20	20	10
B			70	60	80	20	30	50	50	40
G				50	50	10	20	60	20	50
H					80	40	50	10	30	20
J						30	50	30	30	20
C							70	50	70	60
D								40	60	70
E									60	70
F										70
I										

Further inspection of the original similarity data makes possible subdivision of the clusters and establishment of a fuller picture of the phenetic distribution of the OTU's. The full analysis often takes the form shown in Figure 6-8, known as a *phenogram* or *dendrogram,* which has become a standard format for expressing the results of numerical taxonomic studies. The scale on the left indicates the similarity between OTU's or groups of OTU's connected by horizontal lines. (The similarity between clusters of OTU's is usually calculated by some method based on averaging of similarity values.) Note that the two groupings listed as the third alternative in Table 6-2 are seen in the dendrogram as being separated at the lowest point on the similarity scale.

The clustering technique just used is feasible where the number of OTU's is small. For larger problems, several other methods (amenable to computerization) are used. These vary in the criteria used for establishing clusters and thus dendrograms constructed from the same similarity data may differ. It is important to understand the clustering method used in a given analysis (as well as the technique used to measure similarity).

What relationship, if any, is there between a dendrogram and a phylogenetic tree, or between a phenetic classification and one developed by standard taxonomic means? Many proponents of numerical taxonomy claim that the phenetic classification is as real as a conventional one. They look upon the clusters in a phenogram as denoting higher taxa of rank proportional to their

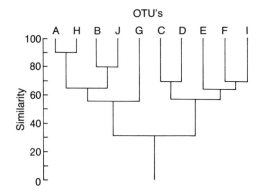

FIGURE 6-8
Dendrogram based on numerical taxonomic data given in Table 6-4.

separation on the similarity scale. In Figure 6-8, separations falling between 60 and 70 might denote *genus rank* differences and those less than 60 might be *subfamily rank,* and so on. The boundaries between ranks are arbitrary — just as they are in conventional taxonomy.

An effective way to test numerical taxonomy is to compare the results with conventional taxonomy for groups of organisms that are well known. Where such tests have been run, the results are encouraging for numerical taxonomy. In a wide variety of instances, the two approaches are in substantial agreement. Many conventional taxonomists who have claimed that classification is an art rather than a science have been forced to examine and explore their methodology. Weighting of characters, in particular, is at issue. Most of the techniques of numerical taxonomy provide no weighting of characters, yet the results are often comparable to classifications in which weighting is an important methodological element. This may mean that weighting serves only as a useful time-saving device and is unnecessary where the work is being done by computer. On the other hand, it may be that weighting is biologically important and that numerical taxonomists have been successful in spite of not weighting because the clusters they normally deal with are sufficiently well defined that a weak method still comes fairly close to the truth.

Figure 6-9 shows a dendrogram, which was produced by numerical taxonomic methods, of nearly the same set of brachiopod genera that were classified by conventional means in Figure 6-6. Although there are obvious differences between the two classifications, there is surprisingly good agreement in essential features. For example, the genus *Eodevonaria* in Figure 6-6 is separated from all other genera at the family level; in Figure 6-9, it is distinct from the other genera at nearly the lowest similarity (phenetic) level. The genera *Dyoros* and *Eolissochonetes* are the most closely linked in Figure 6-9 and are members of the same subfamily in Figure 6-6. The same sort of agreement is found when the genera *Chonostrophia, Chonostrophiella,* and *Tulcumbella* are considered.

Modern numerical taxonomy is still in its infancy. To date, most applications have been concerned with living species. It is inevitable, however, that numerical taxonomy will become more important in both biology and paleontology. The methods and techniques will change but the basic purpose of making classification rapid and rigorous will remain.

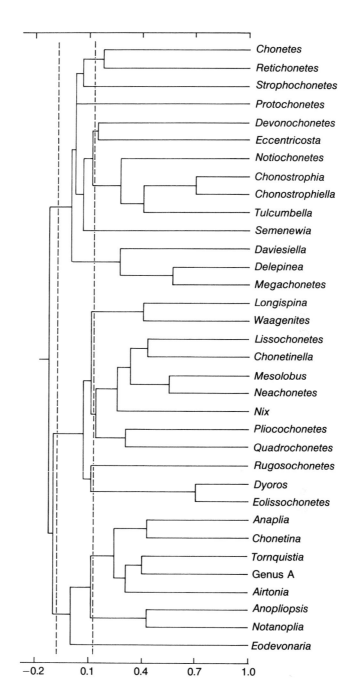

Chonetes
Retichonetes
Strophochonetes
Protochonetes
Devonochonetes
Eccentricosta
Notiochonetes
Chonostrophia
Chonostrophiella
Tulcumbella
Semenewia
Daviesiella
Delepinea
Megachonetes
Longispina
Waagenites
Lissochonetes
Chonetinella
Mesolobus
Neachonetes
Nix
Pliocochonetes
Quadrochonetes
Rugosochonetes
Dyoros
Eolissochonetes
Anaplia
Chonetina
Tornquistia
Genus A
Airtonia
Anopliopsis
Notanoplia
Eodevonaria

−0.2   0.1   0.4   0.7   1.0

FIGURE 6-9
Dendogram produced by numerical taxonomic
analysis of chonetacean brachiopods (for
comparison with the classification developed
by nonnumerical methods shown in Figure
6-6). (From Rowell, 1967.)

Supplementary Reading

Heywood, V. H., and McNeil, J., eds. (1964) *Phenetic and Phylogenetic Classification*. Systematics Association Publication 6, 164 p. (A collection of articles on the relative merits of orthodox taxonomy and numerical taxonomy.)

Kaesler, R. L. (1967) Numerical taxonomy in invertebrate paleontology. *In* Teichert, C., and Yochelson, E. L., eds. *Essays in Paleontology and Stratigraphy*. Lawrence, University Press of Kansas, p. 63–81.

Simpson, G. G. (1953) *The Major Features of Evolution*. New York, Columbia University Press, 434 p. (Chapter XI is a discussion of problems of the evolution of higher categories.)

Simpson, G. G. (1961) *Principles of Animal Taxonomy*. New York, Columbia University Press, 247 p.

Sokal, R. R., and Sneath, P. H. A. (1963) *Principles of Numerical Taxonomy*. San Francisco, W. H. Freeman and Company, 359 p. (A general textbook on numerical methods in taxonomy.)

CHAPTER 7

Identification of Fossils

To identify a fossil is to assign it to a taxon of some pre-existing classification. Actually, no worker who specializes in a particular taxonomic group is likely to do this without formulating his own concept of the selected taxon, which may differ from the concepts of all other workers. Thus, for the specialist identification and classification are not clearly separable. This chapter, then, describes the approach of the nonspecialist who is willing to accept the classifications of earlier workers.

Clearly, some classifications are more widely accepted than others. Because of the species problem in paleontology, identification of a species by a nonspecialist is a risky proposition.

The precision of fossil identification required varies, however, and commonly a species designation is unnecessary. Paleoecologic analysis, given a well-understood stratigraphic framework, seldom requires species identification. It may, for example, be possible to establish a marine origin for sediments simply by noting the presence of fossil remains of Cephalopoda (a class) or Echinodermata (a phylum). Similarly, in stratigraphy or geologic mapping it may be possible to distinguish between nearly identical local rock units of differing age by fossil identification of phylum, class, or order.

Normal Procedures

In establishing a classification, lower taxonomic categories are generally delineated first and then assigned to higher categories. The procedure is usually reversed in identification. By first recognizing a phylum or class we quickly eliminate most of the vast number of recognized species.

Thus, the initial step is phylum identification, and it is important to have a good working knowledge of the taxonomic characters most useful in distinguishing phyla. The necessary information is little enough to be easily grasped. There are approximately twelve important phyla in the fossil record and each is defined by an average of six to ten characters.

Identification of the class of a fossil specimen is considerably more difficult. About thirty-two classes are generally important paleontologically. Few people can correctly identify representatives of all classes at a glance but the distinguishing characteristics of most classes are well enough summarized in recent monographs to permit identification by nonspecialists.

With each successively lower taxonomic category, it is necessary to dig more deeply into the specialized literature. The nonspecialist may therefore find it expedient to send his fossils to a recognized specialist for identification.

We illustrate the general procedures by tracing the steps that might be followed in identifying the fossil shown in Figure 7-1. The specimen was collected from Lower Carboniferous rocks near Moscow, U.S.S.R. It is somewhat broken and therefore photographs of a similar living organism are also shown (Figure 7-2).

The fossil is covered with polygonal, plate-like elements arranged radially. Columns of plates extend from one "pole" of the roughly spherical skeleton to the other. Most plates have one or more prominent knob-like structures called tubercles, which, by analogy to the living relatives, are points of attachment for movable spines (see Figure 7-2). This combination of characters suggest that the specimen belongs in the Phylum Echinodermata and within it, in the Class Echinoidea. No other classes of this phylum or any other phyla share these characters. We thus eliminate at sight all but one class, thereby reducing the identification problem to a choice among about 8,000 species (assuming that the specimen does not represent a previously undiscovered species). More than 120,000 fossil species and more than a million living species have thus been eliminated from consideration.

The characters used to identify the *class* are not the only ones that could have been used: they are simply the most obvious and best preserved in this specimen.

About seventy-five *families* are generally recognized in the class Echinoidea. Approximately two-thirds show a distinct bilateral symmetry superimposed on the radial symmetry. This has led to a somewhat informal differentiation of echinoids into two groups: the regulars (radial) and the irregulars (bilateral). The fossil illustrated lacks bilateral symmetry so the identification is narrowed to about twenty-five families.

A more detailed look shows that the plates making up the prominent columns are of two distinctive types. One type (ambulacral) has small holes or

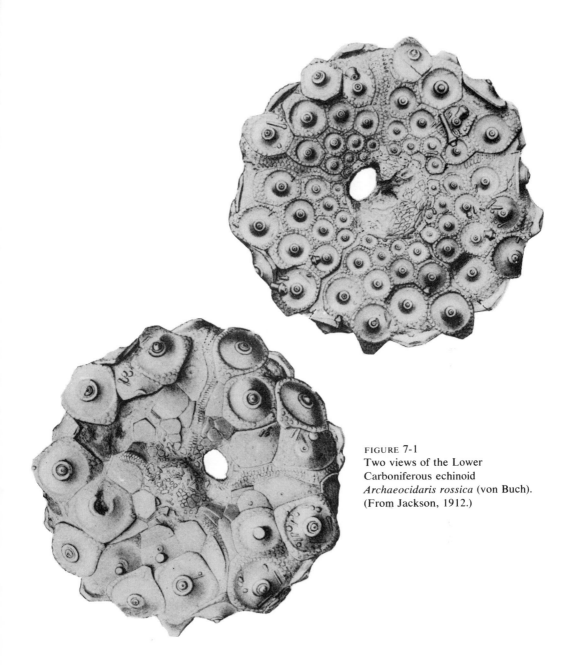

FIGURE 7-1
Two views of the Lower
Carboniferous echinoid
Archaeocidaris rossica (von Buch).
(From Jackson, 1912.)

pores (two per plate), the other (interambulacral) does not. Furthermore, the ambulacral plates are much smaller than the interambulacral plates. (Because of post-mortem sliding of plates, some of the ambulacral columns in the fossil specimen cannot be seen.) In the living species there are five double columns of interambulacral plates. In the fossil specimen, however, there are more than ten columns of interambulacrals. Only six echinoid families contain species with more than ten interambulacral plate columns. All are found in Paleozoic rocks. Of the six, only one (Archaeocidaridae) has the type of tubercles seen on the interambulacral plates in Figure 7-1.

FIGURE 7-2
Recent echinoid *Eucidaris tribuloides* (Lamarck), illustrating the morphology
of cidaroids. (Photographs by R. M. Eaton.)

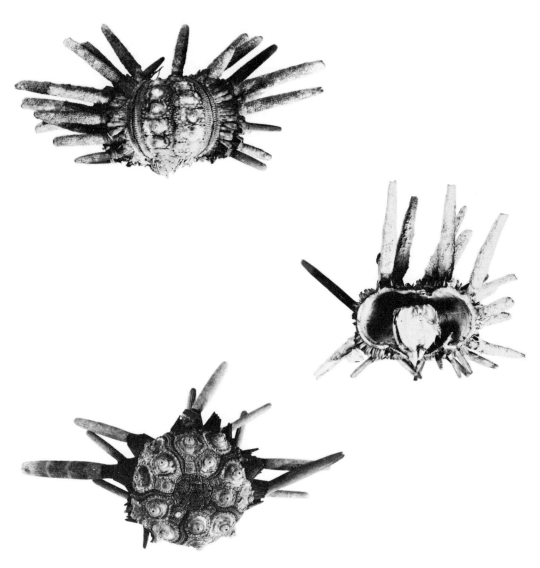

It is important to note that the tuberculation of the living species shown
in Figure 7-2 does not differ (at the level of the present discussion) from that
of the fossil. But the living species does not belong to the Archaeocidaridae.
We eliminated its family (Cidaridae) because of the number of interambulacral plate columns. Tuberculation made it possible to narrow the choice to
a single family *only* after all but six families had been eliminated on other
grounds.

It should be reemphasized that the characters used here are not the only ones that could have been used. Our analysis has been limited to as few characters as possible and to those most easily diagnosed. It is normally prudent to use more characters so that the results from using one can be checked by the others.

To summarize, identification of the family to which the fossil specimen belongs was accomplished by the following steps: (1) choice of phylum and class, (2) recognition of the specimen as being a regular echinoid, (3) elimination of all but six families of regular echinoids, and (4) elimination of all but one of the six families. Quite different routes could have been followed. We ignored the subclass and order level and used instead the informal distinction between regular and irregular. This distinction is not an official taxonomic one because the two groups are not considered to be real evolutionary groupings. Nevertheless, it is a convenient and practical aid to identification.

We do not carry the echinoid identification beyond the family level. Identification of genus and species is done by the same procedures although more characters are used. As the species identification is approached, it becomes more critical to be able to compare the specimen being identified with photographs or with actual specimens known to be members of the various species to which the specimen being identified might possibly belong.

Keys

Many attempts have been made to make the process of identification more systematic. Foremost among the tools devised for this purpose is the *key*. On page 147 is reproduced a key commonly used in echinoid identification. A key consists of a series of paired statements concerning particular morphologic attributes. From each pair the investigator chooses the one that most closely expresses the morphology of the specimen. His choice then leads him to another pair of statements, and so on. The path leads ultimately to the name of a taxonomic group. This may be a species (as in the key on page 147), or a family, depending upon the taxonomic level of the key. Each key assumes that the investigator starts with the knowledge that his specimen belongs to a certain higher category. The key helps him to identify lower taxa. Thus, each key has an upper and a lower taxonomic limit.

The individual statements in the key may describe single characters or combinations. The characters chosen must be selected with care. Obviously the investigator can be "thrown off the track" or stopped at any point if he finds it impossible to choose between the alternatives. For a key to be workable and broadly applicable, the characters must be ones that are commonly preserved and can be diagnosed unequivocally. The choice depends largely on the material that the key is intended for. Botanists, for example, often establish quite different keys for the same group of plants depending on whether the key is to be used with or without foliage or whether flowers are to be used. Living forest trees can usually be identified by what is known as a twig key, which may be based entirely upon such characters as bark, general form of the tree, and morphology of branches and twigs. Such a key can be applied at any season of the year in the study of deciduous trees. Completely parallel keys have been developed that depend only on leaf form and the characteristics of flowers and fruit. Similarly, the key developed by a biologist may be quite different from one made by a paleontologist. The differences emphasize the fact that taxonomic characters commonly used in species discrimination are chosen from a larger group of characters that could have been used. It is important that the student not assume that the characters commonly used in species discrimination are the only characters distinguishing species. All too often a biologist or paleontologist finds it impossible to distinguish between two species by using a certain key and concludes that there are no differences between the two species although they are actually distinguishable by many characteristics in the formal descriptions of the species or the type specimens.

Figure 7-3 shows the structure of the echinoid key in the form of a dendrogram. Each bifurcation corresponds to a numbered pair of statements. What relation is there between this type of dendrogram and one designed to reflect phylogeny? In other words, what relation is there between identification and phylogeny? It is possible to construct a key based literally on phylogeny, but because the primary purpose of a key is to aid identification, many of the most efficient keys do not follow phylogeny rigorously. For example, as we proceeded in the identification of a specimen earlier in this chapter we eliminated about two-thirds of the echinoid families because the specimens lacked bilateral symmetry. This symmetry in echinoids has evolved independently several times and a phylogentic tree thus does not contain a single bifurcation separating radial and bilateral echinoids. Nevertheless, the distinction is relevant to almost every echinoid species and is extremely useful in a key.

Keys are used much more widely for some taxonomic groups than for others, partly because of tradition but also because of differences in identification problems in different organisms. To construct a workable key, the

FIGURE 7-3
Dendrogram for the echinoid genus
Histocidaris constructed from the key
reproduced on the facing page. Numbers
correspond to the numbered pairs
of statements in the key.

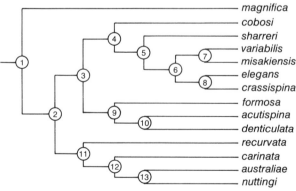

Genus *Histocidaris*

Key to the Species of the Genus *Histocidaris*

1. Primary spines perfectly smooth, at most with some longi-
   tudinal ridges without serrations. Ambital primaries
   downwards curved .... *H. magnifica.*
   Primary spines with more or less fine serrations or
   thorns; ambital spines not downwards curved .... 2.
2. Primary spines uniformly serrate .... 3.
   Primary spines of aboral side with scattered larger
   thorns, mainly in the basal part .... 11.
3. Serrations of primary spines very fine, microscopical .... 4.
   Serrations of primary spines coarser, distinctly visible
   to the naked eye .... 9.
4. Marginal series of ambulacral tubercles very irregular .... *H. cobosi.*
   Marginal series of ambulacral tubercles regular or, at
   most, slightly irregular in larger specimens .... 5.
5. Interporiferous zone with a well-marked, deeply sunk
   middle line .... *H. sharreri.*
   Interporiferous zone without a well-marked, deeply sunk
   middle line .... 6.
6. Valves of large tridentate pedicellariae more or less
   shaped, the whole inside concave .... 7.
   Valves or large tridentate pedicellariae more or less
   slender, not spoon-shaped, usually with a small concavity
   above the apophysis .... 8.
7. Primary spines cyclindrical .... *H. variabilis.*
   Primary spines fusiform .... *H. misakiensis.*
8. Primary spines slender, cyclindrical .... *H. elegans.*
   Primary spines rather thick, fusiform .... *H. crassispina.*
9. Ambital primaries upwards curved, somewhat flattened
   towards the end, the serrations mainly arranged so as to
   form a pair of lateral keels .... *H. formosa.*
   Ambital primaries not upwards curved, not flattened in
   the outer part .... 10.
10. Apical system approximately 50 percent horizontal diameter ..... *H. acutispina.*
    Apical system approximately 36 percent horizontal diameter ... *H. denticulata.*
11. Ambital spines somewhat flattened and upwards
    curved in the outer part .... *H. recurvata.*
    Ambital spines not flattened or upwards curved in the
    outer part .... 12.
12. Valves of large tridentate pedicellariae very broad .... *H. carinata.*
    Valves of large tridentate pedicellariae narrow, slender .... 13.
13. Primary spines scarcely exceeding horizontal diameter .... *H. australiae.*
    Primary spines approximately $2\frac{1}{2}$ horizontal diameter .... *H. nuttingi.*

taxonomy of a group must be reasonably well known. The group must lend itself to key construction; that is, its classification must be based on relatively discrete characters, preferably ones that can be expressed as the presence or absence of a morphologic feature.

Automatic Methods of Identification

We begin the identification procedure with the highest taxa for convenience and practicality. The alternative approach is frighteningly arduous because it entails systematically comparing the unknown specimen with specimens of each species in turn until a strong similarity is found. If time were not a factor we might choose a quite different system because the normal procedure is not without pitfalls. Consider, for example, the problems that arise if an error is made at the higher taxonomic levels. If in the echinoid identification we had chosen the incorrect family, it could lead only to the erroneous establishment of a new species.

The computer has freed science from some procedures that were originally selected for economy of time. With a computer, millions of logical decisions can be made in the time required for the human mind to make a single decision. It is technologically possible to design an analysis that will determine the lowest taxon of a totally unclassified specimen—that is, a program to compare the morphologic characteristics of an unknown specimen with the characteristics of all recognized species within a very large taxonomic group.

We saw in Chapter 6 some of the basic methods developed for codifying the morphologic description of a species and noted that a definitive species description may be recorded on one or two punched cards or on a short segment of magnetic tape. Once the distinguishing characteristics of many species are thus recorded it is simple to program a computer to compare those of an unknown specimen with them.

To date, computerized identification of fossils has been infrequently attempted because the technology and methodology needed are very new (barely a decade old). It is virtually inevitable, however, that these methods will be used more and more. Another important factor delaying computer identification is the difficulty of putting information about morphology into machine-recognizable form. As long as this difficulty exists, the "human element" will remain important in taxonomic identification. A major problem in this context is to select those features most amenable to computerization without jeopardizing the scientific validity of the result.

Presentation of Results

Whether a species identification is for publication or only a museum label, it must be presented in usable form. Certain conventional formats have been developed.

By convention, the generic and specific names are italicized when printed or underlined when written or typed. The generic name is capitalized and the specific name is not. Immediately following the species name, the name of the author of the species is given. These conventions are arbitrary and are used to avoid confusion in communication.

A list of fossils found in a particular Tertiary assemblage follows this paragraph. Notice that the names of some authors of species are enclosed within parentheses, which indicates that the genus has been changed since the species was erected. Notice also that a generic name may be abbreviated to its initial letter in a list of two or more congeneric species. This is acceptable as long as it is unambiguous. In the text of a paleontologic paper generic names are often abbreviated and the names of authors of species deleted.

Typical List of Species Found in a Fossil Assemblage

Rhabdocidaris sp. cf. *R. zitteli* de Loriol
Porocidaris schmidelii (Münster)
Pedinopsis ? *melo*, n. sp.
Porosoma lamberti Checchia-Rispoli
Ambipleurus rotundatus, n. sp.
A. douvillei (Lambert)
Echinolampas fraasi de Loriol
Plesiolampas curriae, n. sp.
P. auraduensis, n. sp.
Conoclypus delanouei de Loriol
Echinocyamus polymorpha (Duncan and Sladen)
Brightonia macfadyeni, n. sp.
Leviechinus gregoryi (Currie)
Pharaonaster sp. cf. *P. ammon* (Desor)
Opissaster farquharsoni Currie
O. auraduensis, n. sp.
O. somaliensis Currie
O. derasmoi Checchia-Rispoli
O. derasmoi var. *angulatus*, n. var.
Hemiaster (*Trachyaster*) sp.
Schizaster africanus de Loriol
S. (*Paraster*) *hunti*, n. sp.
S. (*Paraster*) sp. cf. *S.* (*P.*) *meslei* Peron and Gauthier
S. (*Paraster*) *karkarensis*, n. sp.
S. (*Paraster*) *beloutchistanensis* (D'Archiac)
S. (*Paraster*) *duroensis*, n. sp.
Linthia somaliensis Currie
L. cavernosa de Loriol
Lutetiaster maccagnoi Checchia-Rispoli
Arcaechinus auraduensis, n. sp.
Migliorinia migiurtina Checchia-Rispoli
Eupatagus cairensis de Loriol
E. dainelli (Checchia-Rispoli)
E. fecundus (Checchia-Rispoli)
E. sp. cf. *E. cordiformis* Duncan and Sladen
Brissopsis sp. cf. *B. raulini* Cotteau

In one entry in the list, the subgeneric taxonomic rank is given (*Paraster*). By convention, names of subgenera are enclosed in parentheses and follow the generic name. An "sp.," rather than a species name, following a generic or subgeneric name indicates that the species could not be identified with confidence. In a few cases, "sp." is followed by "cf." and a species name, indicating a questionable or doubtful species identification. "n. sp." following a species name means that the author of the list is naming the species for the first time.

In dealing with groups of organisms whose species and generic affiliations are well known and for which the classification and nomenclature are relatively stable, the simple combination of genus and species names and species authorship is an unambigous identification. For many organisms, however, identifications like those shown in the list could lead to considerable confusion. To exemplify this, let us consider the echinoid shown in Figure 7-1.

This echinoid has been identified as *Archaeocidaris rossica* (von Buch). Following this paragraph is a list of the sort known as a *synonymy* of the species. A synonymy is a brief history of the taxonomic treatment of a species, with bibliographic citations to important works.

Archaeocidaris rossica (Buch)

(?) *Cidaris deucalionis* Eichwald, 1841, p. 88. [Description is unrecognizable so the name cannot hold.]

Cidaris rossicus Buch, 1842, p. 323.

Cidarites rossicus Murchison, Verneuil, and Keyserling, 1845, p. 17, Plate 1, figs. 2a–2e.

Palaeocidaris rossica L. Agassiz and Desor, 1846–'47, p. 367.

Echinocrinus rossica d'Orbigny, 1850, p. 154.

Palaeocidaris (*Echinocrinus*) *rossica* Vogt, 1854, p. 314.

Eocidaris rossica Desor, 1858, p. 156, Plate 21, figs. 3–6.

Echinocrinus deucalionis Eichwald, 1860, p. 652.

Eocidaris rossicus Geinitz, 1866, p. 61.

Archaeocidaris rossicus Trautschold, 1868, Plate 9, figs. 1–10b; 1879, p. 6, Plate 2, figs. 1a–1f, 1h, 1i, 1k, 1l; Quenstedt, 1875, p. 373, Plate 75, fig. 12; Klem, 1904, p. 55.

Archaeocidaris rossica Lovén, 1874, p. 43; Tornquist, 1896, text-fig. p. 27, Plate 4, figs. 1–5, 7, 8.

Archaeocidaris rossica var. *schellwieni* Tornquist, 1897, p. 781, Plate 22, fig. 12.

Cidarotropus rossica Lambert and Thiéry, 1910, p. 125.

This particular species was apparently first described by Eichwald in 1841 under the name of *Cidaris deucalionis* but the name is disallowed by the author of the synonymy (Jackson, 1912) because Eichwald's description was too vague. The next entry is to von Buch's description of the species as *Cidaris rossicus*. As the first valid description of the species, the name *rossicus* has priority over all names subsequently applied to the species (although the spelling has been altered to conform grammatically to a change in generic affiliation).

The third entry in the synonymy records the assignment of the species to the genus *Cidarites* (meaning "fossil *Cidaris*") by Murchison, Verneuil, and Keyserling. Several subsequent entries record similar shifts in generic affiliation, most reflecting changes or differences of opinion regarding the evolutionary relationships of the species. One entry in the synonymy stands out from the others: *Echinocrinus deucalionis*. This is credited to Eichwald (1860) who evidently recognized as valid his 1841 publication of the name *Cidaris deucalionis*. The use of the genus name *Echinocrinus* raises another nomenclatural problem. This name was proposed (quite validly) in 1841 by L. Agassiz. *Archaeocidaris* was proposed independently for the same group of echinoids three years later by McCoy (1844). Technically, the name *Echinocrinus* is the correct name because it was proposed first. A special exception was made in 1955, however, by the International Commission on Zoological Nomenclature partly because *Echinocrinus* had rarely been used by echinoid specialists and partly because it was misleading in being very similar to generic names common in nonechinoid echinoderms (particularly crinoids).

Synonymies often contain names of species that are completely unrelated nomenclaturally to the species in question. This occurs when the synonymy writer feels that two or more species previously considered distinct are actually one. The name proposed first is then used for the species, unless an official exception is made. Because a synonymy is in part an historical record and in part an interpretation of a taxonomic situation, it is common that synonymies written by different specialists for a single species name do not agree.

Bibliographic Sources

Paleontologic information (particularly taxonomic information) has been published in a vast literature extending back well into the eighteenth century. It is published in all major languages and a wide variety of publication media, from regularly scheduled periodicals to occasional monographs of museums, governments, and even private individuals. The paleontologist is thus more dependent on bibliographic aids than is, for example, the nuclear physicist or the electrical engineer.

Identification is greatly aided by definitive monographs on either a specific taxonomic group or fossils found in a particular part of the geologic column. If the monograph has been well prepared, it includes reference to all important literature. The reader need then consult other bibliographic sources only for articles that have been published since the publication of the monograph. In using a given monograph, the reader must understand to what taxonomic categories the writer's definitive summary reaches. The writers of the *Treatise on Invertebrate Paleontology,* for example, attempt to be comprehensive in listing genera, but do not try to include all known species.

When no up-to-date summary treatment is available, the paleontologist must turn to published bibliographies, such as the *Zoological Record* published by the Zoological Society of London. Each volume is a reasonably comprehensive survey of the zoological and paleozoological literature published during the preceding year. The *Zoological Record* is divided taxonomically into eighteen sections and within each is a list of titles, authors, and bibliographic references and a comprehensive subject matter index. The index includes a comprehensive list of all taxonomic names used in the papers cited. The *Zoological Record* is a valuable aid in the development of a synonymy because it permits tracing the bibliographic citations to a genus or species year by year.

The *Zoological Record* is not complete, however. No such bibliography could be and still be issued within a reasonable time after the publication of the literature on which it is based. Therefore the *Zoological Record* must usually be supplemented by other bibliographies such as *Biological Abstracts* and *Bibliography of North American Geology*.

PART 2

THE USES OF
PALEONTOLOGIC DATA

CHAPTER 8

Adaptation and Functional Morphology

We have discussed evolution from several viewpoints and in several contexts. A central idea has been the assumption that species (or populations) evolve by natural selection and thereby adapt. Adaptation makes it possible for organisms to cope with changing environmental conditions, invade new environments, and function more efficiently in a given environment.

As a result of adaptation having taken place in countless independent evolutionary lineages (over hundreds of millions of years), we now have a truly staggering array of forms. In part, this array must reflect diversity in possible ways of life. We cannot argue, however, that there are as many ways of life as species. Many organisms that have evolved independently now live in very similar ways. Some such organisms display *convergence:* the independent evolution of similar morphologies that function similarly. Other organisms adapt to the same environment in quite different ways. Thus, the number of morphologic types produced by adaptive evolution is considerably larger than the number of general ways of life.

In this chapter we will take a more detailed look at the results of adaptation and attempt to measure the success of organisms in coping with and exploiting their environments. We will assume that morphology is primarily adaptive; that is, that the observed morphologic features are functional in terms of life activities. It is theoretically possible, though unlikely, for nonfunctional morphology to evolve. The consensus among modern paleontologists and biologists is, however, that it is relatively rare and requires special genetic or ecologic circumstances.

The Nonrandomness of Adaptation

If we were able to describe the spectrum of biologically feasible morphologies and compare it with the forms that have *actually* evolved, we would find that adaptation is not random. That is, all possible forms have not been used. Those that have been used have evolved in markedly unequal numbers. In a sense, this is analogous to the array of forms *actually* produced by the automotive engineer as compared with the spectrum of forms that *could* be produced. If we could somehow express all possible automobile designs and then plot the frequency of actual types, we would find that the distribution of actual cars (both living and extinct) would be an extremely spotty one and would concentrate on a relatively few designs found to be efficient.

The nonrandomness of adaptation has attracted the attention of many evolutionary biologists. The problem has been stated by Dobzhansky (1951), as follows:

> Every organism may be conceived as possessing a certain combination of organs or traits, and of genes which condition the development of these traits. Different organisms possess some genes in combination with others and some genes which are different. The number of conceivable combinations of genes present in different organisms is, of course, immense. The actually existing combinations amount to only an infinitesimal fraction of the potentially possible, or at least conceivable ones. All these combinations may be thought of as forming a multi-dimensional space within which every existing or possible organism may be said to have its place.
>
> The existing and the possible combinations may now be graded with respect to their fitness to survive in the environments that exist in the world. Some of the conceivable combinations, indeed a vast majority of them are discordant and unfit for survival in any environment. Others are suitable for occupation of certain habitats and ecological niches.

Another evolutionary biologist, Sewell Wright, suggested a graphical method for expressing these principles. This is illustrated in Figure 8-1, whose vertical and horizontal coordinates represent genetically controlled characters. The area bounded by the two axes thus includes all possible combinations of expressions of the two characters. For example, the horizontal coordinate might be the number of ribs on the shell of a bivalve and the vertical coordinate might be the relative depth of these ribs.

In any environment, some morphologic combinations will be better adapted than others. Variation in adaptiveness is shown schematically by the contoured surface in Figure 8-1. The topographic highs are called "adaptive peaks" and are separated from each other by "adaptive valleys." The diagram contains several adaptive peaks indicating that there are several adaptive combinations of the two morphologic attributes (though presumably the nature of the adaptive value is different for each). In between the adaptive peaks are forms not well adapted to any environment available to the group of organisms. Evolving lineages tend to climb adaptive peaks and avoid, or traverse quickly, adaptive valleys. We should actually be able to

FIGURE 8-1
Sewell Wright's graphical interpretation of adaptive peaks and
valleys among organisms sharing two genetically controlled
characters. Peaks are indicated by plus signs. Horizontal and
vertical coordinates are the two characters.
(From Dobzhansky, 1951.)

measure a morphologic attribute on a variety of organisms, ribbing in bivalves, for example, and find that the expressions of the attribute are clustered on the adaptive peaks.

Theoretical Morphology — The Coiled Shell

Enough is known about the morphology of a few groups of fossil organisms
that the spectrum of possible forms in these groups may be defined, and the
sort of analysis suggested by Figure 8-1 may then be performed. This entails, of course, defining and describing the forms that have not evolved as
well as the forms that have. Analysis of this type is referred to as *theoretical
morphology.* Perhaps the most fully developed morphologic model for use in
such analysis is of the external form of the shell of the coiled invertebrate.
This model is used here as the principal example of theoretical morphology
and the analysis of adaptive peaks among fossils.
   The shells of many invertebrates are spirally coiled. Examples may be
found in many diverse groups, from single-celled foraminifera to highly
complex organisms such as molluscs and brachiopods. Spiral coiling has
developed independently in several distinct evolutionary lines, and its functional significance varies.

The basic geometry of the coiled shell may be illustrated by the common gastropod. The shell is a hollow, tapered tube open at its larger end (the aperture). New shell material is added at the aperture so that the tube becomes longer as the animal grows. During growth, more material is added on one side of the aperture than on the other. The effect is to produce a spiral form such that the tube appears to revolve about a fixed axis, which is known as the coiling axis.

Several features of the spiral are remarkably constant during growth and form the basis of an efficient system of description. Four parameters, which are illustrated in the diagram that is included as Figure 8-2, are most useful: (1) The shape of the tube in cross-section, which is usually referred to as the *shape of the generating curve.* In Figure 8-2 the generating curve is circular. (Technically, the generating curve is defined as the shape of the intersection of the expanding tube with a plane that contains the coiling axis. In most gastropods this is coincident with the shape of the aperture.) (2) The rate of expansion of the generating curve with respect to revolution about the axis. This is often referred to as the *rate of whorl expansion,* and is the ratio between the same linear dimension (such as the diameter) on two generating curves separated by a full revolution. In Figure 8-2, the whorl expansion rate has a value of 2, meaning that any linear dimension of the generating curve is doubled for each revolution about the axis. (3) The *position and orientation of the generating curve* with respect to the axis. In Figure 8-2 the circular generating curve is separated from the axis by a distance equal to half its own diameter. If the generating curve were noncircular, its orientation with respect to the coiling axis would also be critical. (4) The movement of the generating curve along the axis, which is known as *whorl translation.* Translation is most conveniently expressed by the ratio of movement along the axis to movement away from the axis during any interval of revolution about the axis. The reference point for determining this ratio is the geometric center of the generating curve. In some forms the rate of translation is zero and the tube revolves about the axis in a single plane and produces what is known as a planispiral shell.

The shape of the generating curve, the whorl expansion rate, the position of the generating curve relative to the axis, and the rate of translation are generally (though not always) constant during growth. When different species are compared, however, marked differences become evident. Some have nearly circular generating curves and others have extremely complicated shapes (defying simple mathematical description). The variation in generating curve shape in gastropods is by no means random, however. If we were to survey a large number of gastropod forms, fossil and living, we would find that certain generating curve shapes appear over and over again while others are only rarely found and still others seem never to have developed.

Gastropods also show considerable variation in whorl expansion rate: from only a little more than 1.0 (the theoretical minimum) to 4 or 5. Similarly, the generating curve varies from being in contact with the axis to being separated from it by a so-called *umbilicus,* that is, a roughly cone-shaped depression in the base of the shell.

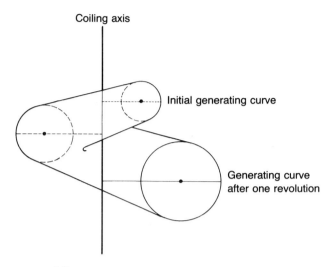

Coiling axis

Initial generating curve

Generating curve
after one revolution

FIGURE 8-2
Schematic diagram of part of a gastropod shell.
(From Raup, 1966.)

One of the most striking variations is in translation, which may vary from zero (planispiral) to extremely high values (in the high spired snails). Among living gastropods, there are no truly planispiral forms but among early Paleozoic gastropods, the planispiral form is fairly common.

The four geometric parameters of gastropod form involve much, but not all, of coiled shell morphology. Many parameters could be added to the list. An obvious one, for example, is the direction of coiling. Some gastropods coil in the left-handed fashion (sinistral) and some in the right-handed fashion (dextral). Curiously, the vast majority of gastropods are dextral.

In order to describe gastropod form fully we also require some means of describing departures from the model just presented. An example is illustrated in Figure 8-3 in which the rate of translation decreases during ontogeny so that the resulting shell has a concave lateral profile (instead of the normal straight-sided spire). Such ontogenetic departures from normal geometry are usually consistent within species and are genetically controlled; presumably they are adaptive.

FIGURE 8-3
A gastropod in which the rate of whorl translation decreases during ontogeny.
(From Raup, 1966.)

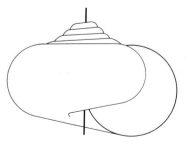

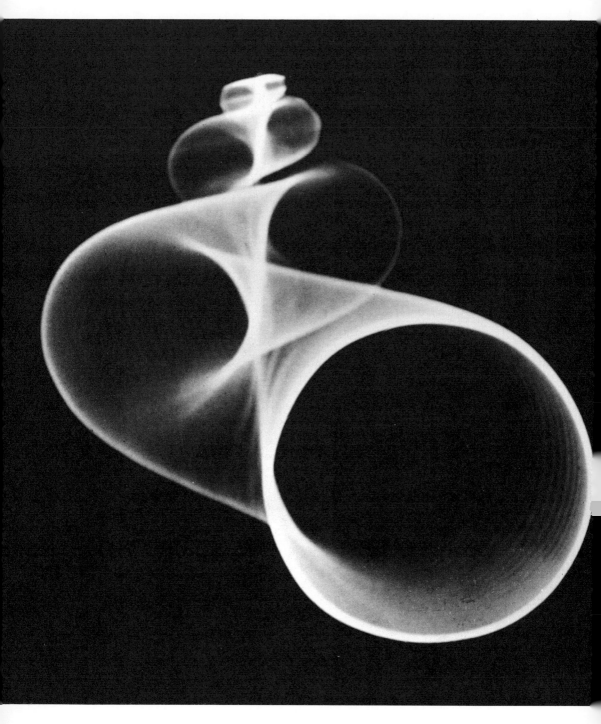

FIGURE 8-4
Analog-computer simulation of a common gastropod shell form.
(From Raup and Michelson, 1965.)

In addition to the morphologic attributes just mentioned a wide variety of minor features are not included in the general coiling model; for example, impressions of muscles on the interior of the shell, color patterns, operculum shape, the many nonspiral details of ornamentation, and so on.

A useful test of the model is to use it to replicate the form of a shell by computer. If we specify the shape of the generating curve, its position, its rates of expansion and translation, we should be able to calculate the shape of the surface that would be produced by a prescribed amount of growth. A computer simulation of a common gastropod type is shown in Figure 8-4. The generating curve is circular and in contact with the axis. It has an expansion rate of 2.0. The translation rate is fairly high. These characteristics were converted into an electrical circuit for an analog computer. The generating curve was allowed to "grow" on an oscilloscope screen. The simulation in Figure 8-4 is a photograph of the trace of the oscilloscope output and is a generalized replica of the shell surface. As a test of the model it convinces us that the four basic parameters do not overlook or misrepresent important aspects of shell growth. Much more important, the computer simulation provides the opportunity of constructing pictures of gastropod forms that are possible but that have not been produced in nature.

Figure 8-5 shows an array of twenty-five forms produced by the same analog computer circuit. For simplicity, only rate of whorl translation and rate of whorl expansion are varied. Thus we see the array of possible forms in a very limited part of the spectrum of forms available for gastropod evolution. Some of the forms are readily recognizable as gastropod types; others (particularly those in the lower left-hand corner of the array) are rare or nonexistent. The geometric area covered by Figure 8-5 thus contains one or more adaptive peaks (concentrated in the upper part of the array) and adaptive valleys (particularly in the lower left-hand corner of the array).

The coiling model can be extended to include coiled forms found in many animal groups. The coiled cephalopods differ little in basic form from many gastropods. Most have planispiral shells (zero translation). The generating curves are elliptical or circular and are symmetrical about the plane of coiling. In many forms, successive revolutions overlap one another so that the entire generating curve is not visible or even formed. If the rate of whorl expansion were larger or if the generating curve were farther from the axis, the whorls would not overlap and the generating curve would be entirely exposed. Most of the cephalopods could be placed geometrically in the extreme upper right-hand corner of the simulated array (Figure 8-5).

Cephalopods and gastropods are, of course, quite different animals and their shells have different functions. The cephalopod shell is partitioned by internal septa into a series of chambers used to hold gas for buoyancy. These chambers do not, however, affect the external form or the overall geometry of the outer shell. It is thus reasonable from a purely descriptive viewpoint to consider cephalopods as variants on the basic model used for gastropod description.

The rate of whorl expansion is generally much higher in bivalves than in gastropods and coiled cephalopods. Whorl expansion rates in gastropods rarely exceed 5, whereas in bivalves they are often as high as one million. The high whorl expansion rate in bivalves is accompanied by some whorl translation, which is minimal among forms such as the scallops but always exists to some degree and produces an asymmetrical shell form. The array of computer simulations in Figure 8-5 contains some types (particularly in the lower right-hand corner) that approach bivalve morphology but most bivalve forms lie outside this simulated array.

The comparison just made between bivalves and other coiled organisms is geometric rather than physiological. The total skeleton of a bivalve of course comprises two articulated coiled shells (one dextral, one sinistral). As we shall see, the geometric differences between bivalves and other groups of coiled organisms may be related to shell function.

Let us consider yet another coiled group: the brachiopods. The individual brachiopod shell is planispiral. Like that of the bivalve, the total brachiopod skeleton is made up of two articulated, coiled shells. Consistently high whorl expansion rates mean that it is virtually impossible to confuse the geometry of the brachiopod shell with that of planispiral cephalopods. The principal source of variation among brachiopods is the shape of the generating curve. In the other groups that we have considered, a reasonable simulation of the shell can be constructed by assuming that the leading, or growing, edge of the shell is equivalent to the shape of a cross-section of the expanding shell (the cross-section in a plane that includes the coiling axis). In brachiopods the leading edge of the shell is often nonplanar, and thus is not coincident with the theoretical generating curve. This is most evident in forms whose shell has a strongly developed fold and sulcus on the anterior margin. A nonplanar margin is produced when some parts of the shell are "ahead" of others during growth. This does not mean that the shell lacks a generating curve in a geometric sense but only that the growth at the leading edge departs in time from the simple geometric model. To handle the array of brachiopod forms we must therefore introduce an additional parameter reflecting nonuniform growth rates.

Many foraminifera have a spiral form that appears to conform to the model established for brachiopods and molluscs. The shell has chambers that have a much greater effect on external morphology than do those of the cephalopods and therefore the geometry does not appear as regular. Where measurements have been made, however, it appears that the basic geometry is the same. A great deal of work remains to be done in this area to confirm and develop a model applicable to foraminifera.

Figure 8-6 shows another array of simulated shell forms produced by the analog computer. The simulated shell in the center is a rather generalized form with a circular generating curve, which is in contact with the coiling axis, zero translation, and a relatively low rate of whorl expansion. The morphologic effect of changing each of these parameters is shown by series of simulated forms leading away from the central form. An increase only in whorl expansion rate produces a brachiopod-like form. Movement of the generating curve away from the axis produces forms most common among coiled cephalopods. An increase in translation accompanied by maintenance of a relatively low whorl expansion rate produces gastropod forms.

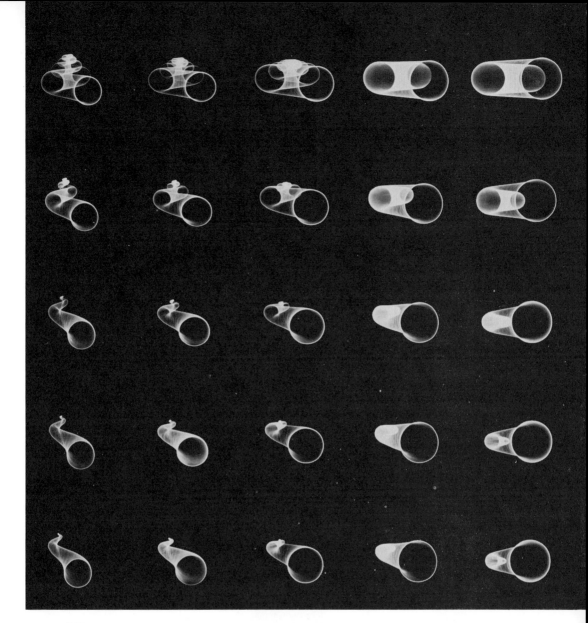

FIGURE 8-5
Array of 25 analog-computer simulations of shell forms. Whorl
translation increases toward the left and whorl expansion increases
downward. (From Raup and Michelson, 1965.)

In Figure 8-7 these relationships are formalized for forms with a circular
generating curve by expressing variation in a three-dimensional block diagram. Translation increases from zero on the right to 4 on the left. Whorl
expansion rate increases from 1 (the theoretical minimum) at the top to
1,000,000 at the bottom. The distance between the generating curve and the
axis (relative to its own diameter) increases from front to back. We may look
upon this block, therefore, as containing *all possible geometric forms within
the restrictions of the model.*

Shaded areas in the block (Figure 8-7) indicate regions occupied by most species in four taxonomic groups. The gastropods typically occupy the region of low whorl expansion rate but show quite a range in translation rate and in distance between the generating curve and the axis. Bivalves generally have low translation and high expansion rate. Brachiopods occupy much the same range in expansion rate but are geometrically separated from the bivalves by their lack of translation. The coiled cephalopods are mostly planispiral, but their whorl expansion rates are consistently lower than those of the brachiopods and thus no overlap between the two regions occurs.

FIGURE 8-6
Series of analog-computer simulations showing the effects of change in geometric parameters. (From Raup, 1966.)

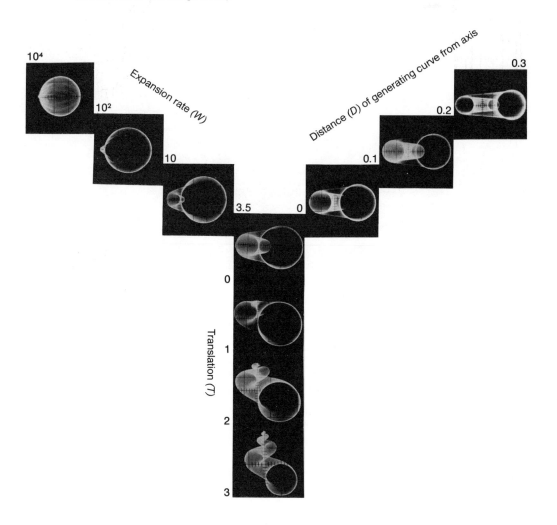

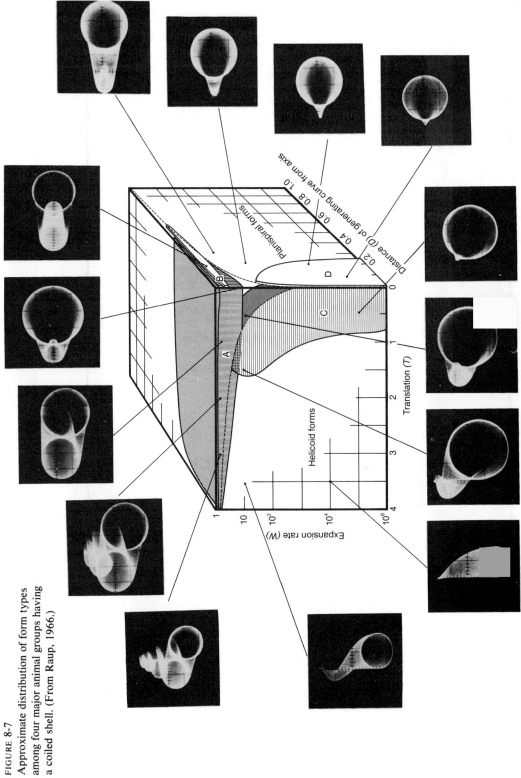

FIGURE 8-7
Approximate distribution of form types among four major animal groups having a coiled shell. (From Raup, 1966.)

The three-dimensional block in Figure 8-7 indicates that the four evolutionary groups occupy virtually nonoverlapping regions of the geometric space available to them. This is not surprising because the four groups have quite different functional and environmental problems and inevitably approach different adaptive peaks.

Perhaps the most striking element in Figure 8-7 is the fact that large regions in the block are virtually empty. There are several possible explanations. The empty spaces may represent adaptive valleys. Alternatively, they may represent forms that are biologically impossible. A third possibility is that there has been insufficient time for the evolutionary development of various forms that would exploit the entire geometric range. Of the three alternatives, the first, that empty spaces represent adaptive valleys, seems the most reasonable.

Methods of Functional Morphologic Analysis

There is no single method of inferring function from morphologic features. Many approaches are possible. In part, the approach depends on the information available. For a fossil organism, partial preservation and only spotty knowledge of life habits and habitats impose limits on the method of analysis. For a living organism, soft anatomy, life habits, and habitat preferences can usually be observed more directly and in greater detail, permitting more accurate interpretation.

The most common approach to the interpretation of fossil structures is through comparison with living species. Consider first the case in which fossil and Recent taxa that bear similar structures are closely related and the structures are judged to have had a common origin. The fossil structure may then, by *homology* (having the same origin), be judged to serve the same function as does the Recent structure. In other instances, the similarity of structures may be more superficial, having arisen independently in separate taxonomic groups. The fossil structure may still be interpreted as having served the same function as the Recent structure, but by *homoplasy* (having the same form). Once it is established that homoplastic structures serve the same, or similar, functions, they are known as *analogous* structures. Homology and homoplasy represent pathways of comparison, rather than methods of inference of function.

In recent years, it has been recognized that most interpretations of function are ultimately based on mechanical analysis, even when applied to fossil taxa by homology or homoplasy. Because most fossils represent skeletal remains and most skeletons are rigid supportive structures, fossil structures generally served mechanical functions in life. Often we can observe the mechanical function of a structure in Recent taxa and, by homology, apply our observations to closely related fossil taxa. For example, we can observe antler fighting among male members of the deer family in autumn. Inasmuch as antlers are restricted to males, are grown and shed annually, and reach their full development for the autumn mating season, we can conclude that their primary function is to serve as weapons in intra-species combat for females. (We cannot, however, rule out secondary functions.) By homology we can interpret fossil deer antlers of similar construction as having served the same primary function.

The function of fossil structures is seldom as obvious as that of deer antlers. Even if Recent analogues or homologues of a structure are available it is often necessary to deduce their function or functions before tackling the fossil structure. A common procedure is as follows: first, postulate hypothetical functions for the similar structure in the Recent group. Next, test each hypothetical function in terms of whether the function would be useful to the living organism in light of known life habits and habitats and whether the function is mechanically *feasible* in terms of the morphology, life habits, and habitat preferences. The most reasonable function is chosen on this basis. The conclusion can then be applied to the problematic fossil structure by homology or homoplasy.

Many variations are possible within this basic framework. If the fossil group is closely related to the living group and the structures are considered to be homologous, data from both the living and fossil groups may be used together to test the hypothesis. If the fossil and living structures are not homologous, it may be necessary to justify application of conclusions derived from the Recent group. This can be done, for example by considering evidence from fossil or lithologic associations to determine the likelihood that the fossil taxa did, indeed, have life habits and habitat preferences similar to those of the Recent taxa.

When information from living organisms is inadequate or absent, a variant on the procedure described above, called the **paradigm** approach, is particularly useful. The use of paradigms was formalized by Rudwick (1964). In Rudwick's scheme, one or more functions for a given structure are postulated, but they are used to define abstract mechanical models called paradigms — one paradigm for each possible function. Each paradigm represents what the structure *should* look like in order to perform the function best. The result is a set of purely hypothetical structures. (It is sometimes helpful to construct three-dimensional replicas or computer simulations of these hypothetical structures.) The paradigm that most closely fits the actual structure is the one whose associated function is chosen as the most probable for the real structure.

An important requirement that must be observed in using the paradigm approach is that the hypothetical ideal structures must be so formulated as to be consistent with the genetics and physiology of the organism. For example, a cephalopod may be noticed to function like a submarine, but shipbuilding and cephalopod growth are not analogous processes and the differences between them rule out the submarine's being a paradigm. Furthermore, the cephalopod's skeletal composition, basic organ system, and other characters limit — even exclude — the possibility of making a comparison with the submarine.

In any approach to functional morphology, it is important to consider what might be called **multiple-effect factors,** of which there are three basic types: (1) a structure may perform more than one function, (2) a structure may be affected by more than one gene, and (3) a gene may affect more than one structure. The third factor is what geneticists call "pleiotropy."

Because of multiple-effect factors, many structures are subjected to selection pressure in more than one direction. The situation is comparable to a mechanical system in which two or more forces pull on an object in different directions. The direction and magnitude of each force may be represented by a vector and the direction and magnitude of the resultant force (which, if strong enough, will move the object) are, in effect, a compromise of the component forces.

Likewise, the direction of evolution of most morphologic features is a compromise. For example, most external mollusc shells serve a protective function in addition to supporting muscular systems. Extremely thick shells would be useful in making molluscs invulnerable to many types of predation. But high mobility — most common among molluscan species having thin, light-weight shells — permits an organism to escape from its predators. Shell thickness in most molluscs thus represents a compromise between opposing selection pressures. We must also bear in mind that a fossil species may represent a stage of evolution in which the final compromise has not yet been attained.

The examples of functional analysis that follow illustrate the variety of existing methods and approaches. Some rely heavily on neontologic information; others are relatively independent of living organisms.

Functional Morphology of the Coiled Shell

For our first example of actual functional analysis, we will return to the coiled shell. An important question stimulated by the block diagram in Figure 8-7 is: why do the evolutionary groups shown occupy more-or-less nonoverlapping regions? There are many answers, involving detail quite beyond the scope of this volume. We can, however, touch on some of the salient factors. Notice in Figure 8-7 that the two front faces of the block contain a curved dashed line. This is a calculated line above which whorls overlap, on which they are barely in contact, and below which the shell is an open spiral. Two hypothetical forms from opposite sides of this line are illustrated in Figure 8-8. The shell of the one with a whorl expansion rate of 10 has a nonoverlapping spiral; the one with a rate of 3.5 has an overlapping spiral. The two dashed lines on the block (Figure 8-7) actually represent the emergence of a curved surface that slopes toward the lower, right-hand, front corner.

An important consequence of this relationship is that a shell having no overlap between succeeding whorls has the inner margin of the generating curve exposed. This margin of the generating curve forms the hinge line in two-valved shells (brachiopods and bivalves). Teeth and sockets for articulation of the two shells usually develop along the hinge line and grow larger as the inner margin of the generating curve grows. Hinge-line development would be exceedingly difficult in a form in which successive whorls overlap:

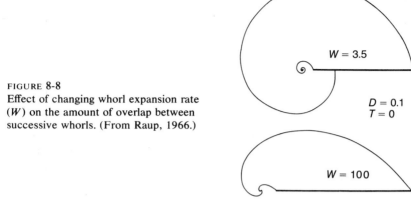

FIGURE 8-8
Effect of changing whorl expansion rate
(W) on the amount of overlap between
successive whorls. (From Raup, 1966.)

the hinge line would fall on the outer margin of the previous whorl and there would be no space for the development of a ligament area or interarea. It is probably for this reason that two-valved shells of bivalves and brachiopods are found below the dashed line in Figure 8-7.

We must also explain the relative lack of univalved shells (gastropods and cephalopods) in the lower part of the block. The univalved shell functions in part for internal support of the soft-part anatomy and in part for protection of the animal. If the aperture is very large, the shell's protective function is defeated: furthermore, a shell with a large aperature has relatively less interior surface for support and attachment of the soft body. The inappropriateness of a very large aperture for a gastropod can be imagined if we picture a gastropod occupying the single valve of a brachiopod.

Some univalved molluscs, such as the limpet, do indeed have apertures that are extremely large in relations to the organism's total volume but they live attached by a muscular foot to a firm substratum. The large aperture provides the necessary area for strong attachment of the foot to the substratum, and as long as the animal remains attached, the shell gives it enough protection.

Turning to a somewhat more detailed analysis of adaptive peaks, let us examine the geometry used by the coiled cephalopods represented in the upper left-hand corner of the right-hand face in Figure 8-7. Figure 8-9 shows an array of hypothetical shell forms in this region. Expansion rates range from 1–4 and distances between the generating curve and the axis range from 0–0.6. The line separating overlapping and nonoverlapping forms is also shown. This is the region occupied now and in the geologic past by most coiled cephalopods. Figure 8-10 is drawn on the same format and shows a contoured surface that expresses varying frequency of geometric types in 405 ammonoid genera of the late Paleozoic and Mesozoic. Representative species of each of the 405 genera were measured with respect to expansion rate and the distance of the generating curve from the axis. A point was plotted for each species and the density of these points was contoured to produce Figure 8-10.

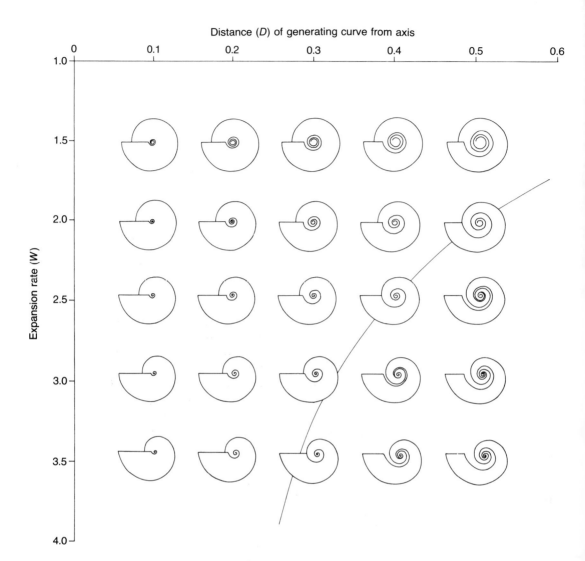

Distance (*D*) of generating curve from axis

Expansion rate (*W*)

FIGURE 8-9
Array of hypothetical planispiral shells showing changes in form resulting from change
in whorl expansion rate and the position of the generating curve. (From Raup, 1967.)

Geometry in the region of $W = 2$ and $D = 0.35$ is the most common for the
ammonoid sample. In a sense, this has been the most "popular" geometry.
The high in the density surface thus represents a generalized adaptive peak
for the entire group.

Functional morphology in coiled cephalopods is not as simple as suggested by Figure 8-10. Within the class Ammonoidea, there are smaller
groups that have evolved to fit subsidiary adaptive peaks. An example is
shown in Figure 8-11. The format is the same as that of the previous illustration and the outline of the total ammonoid density distribution is shown as

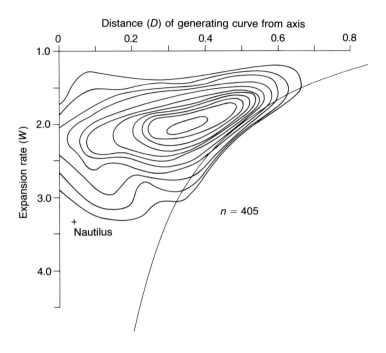

FIGURE 8-10
Contoured density of natural occurrences of
planispiral ammonoids. Ninety percent of the sample
lies inside the outermost contour; inner contours
measure increase in density of genera per unit area
on the plot. (From Raup, 1967.)

a dashed line. The contours refer to the distribution of geometries among a
subsample of 35 genera of the Paleozoic order Goniatitina. The high in
density is much farther to the left than for the total ammonoid sample. This
may represent a different adaptive peak from that of the main sample (representing a different environmental preference or a different mode of life), or
it may represent a stage in the evolution of cephalopods that came before
the attainment of the principal adaptive peak (at $W = 2, D = 0.35$). It is clear,
however, that we should expect to find many subsidiary adaptive peaks
reflecting slight differences in habits or habitats. For example, some cephalopods may have adapted to shallow-water environments, other to deep-
water environments. Some may have been fast swimmers and others slow
swimmers. Some may have been bottom-dwellers, and so on.

Our brief analysis of cephalopod coiling geometry has been limited to the
ammonoids. The complete picture would also include the other large group
of coiled cephalopods, the nautiloids. A contoured density distribution for
nautiloids would probably be somewhat different from that for ammonoids

but would occupy the same general part of the coiling spectrum. It is in-teresting that the sole surviving genus of nautiloids (the genus *Nautilus*) falls outside the ammonoid region shown in Figure 8-10. *Nautilus* has an expansion rate of about 3.5, considerably higher than that typical for am-monoids. We can only speculate as to whether this is due to chance or re-flects the fact that *Nautilus* (and perhaps all nautiloids) evolved to meet the requirements of a different adaptive peak.

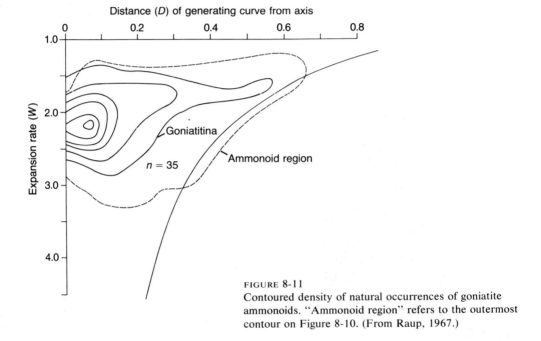

FIGURE 8-11
Contoured density of natural occurrences of goniatite ammonoids. "Ammonoid region" refers to the outermost contour on Figure 8-10. (From Raup, 1967.)

The Cephalopod Suture Problem

As an illustration of the paradigm approach, let us consider the long-debated question of the function of complex suture patterns in ammonoids. We can-not hope to solve this problem conclusively here but can at least shed some light on it.

First, let us review some basic facts about ammonoid morphology, with reference to Figures 8-12 through 8-15:

1. The typical ammonoid shell is thinner, and therefore weaker, than the typical nautiloid shell.
2. The partitions (*septa*) that compartmentalize the ammonoid shell are characteristically folded (fluted) where they join the outer shell (Fig-ure 8-12). The septa, like the coiled shell, were secreted by the fleshy mantle, which lined the living chamber in life.

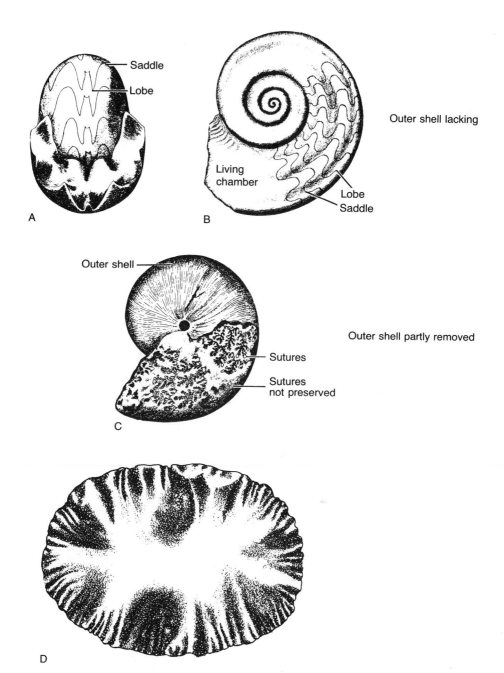

Saddle

Lobe

A

Outer shell lacking

Living
chamber

Lobe

Saddle

B

Outer shell

Outer shell partly removed

Sutures

Sutures
not preserved

C

D

FIGURE 8-12
Septal morphology of ammonoids. A: *Goniatites kentuckiensis* Miller, showing
goniatite sutures and the surface of one septum. From the Mississippian of Kentucky.
B: *Xenaspis skinneri* Miller and Furnish, with ceratite sutures. From the Permian of Texas
and Mexico. C: *Phylloceras (Hypophylloceras) onoense* Stanton, showing ammonite
sutures where the outer shell is peeled away. From the Lower Cretaceous of California.
D: Septal surface of *Baculites,* a common straight-shelled Cretaceous genus.
(From Fenton and Fenton, 1958.)

3. The pattern of the *suture* (the line marking the juncture between the septum and outer shell) typically increases in complexity during ontogeny (Figure 8-13).
4. The suture pattern varies greatly from species to species and is an important character in ammonoid classification.
5. The *siphuncle,* which is a flesh-filled tube that passes back from the living chamber to all older chambers, forms the only connection between chambers.
6. Thickness of the outer shell wall increased less rapidly than whorl diameter during ontogeny (Figure 8-14).
7. Ammonoid septa are commonly convex toward the aperture; nautiloid septa are usually concave toward the aperture.
8. Septa commonly become crowded together as the ammonoid shell approaches its maximum size.

We can derive additional relevant information from the living *Nautilus* (Figure 8-15), although there are certain important differences between it and the ammonoids. Since we know that ammonoids evolved from nautiloids (page 306), we will consider septa and associated structures of ammonoids and *Nautilus* to be homologous. *Nautilus* has simpler septal sutures than most ammonoids, and as we have seen, has a higher whorl expansion rate (W). It possesses a large lateral muscle scar on either side of the

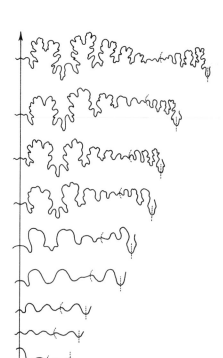

FIGURE 8-13
Ontogenetic increase in suture complexity of the Jurassic ammonite *Creniceras renggeri* (Oppel). Suture lines are represented for the right half of the shell. The arrow points toward the aperture. (From Palframan, 1966.)

FIGURE 8-14
Graphs showing increase in shell thickness
relative to diameter during ontogeny of
the ammonites *Promicroceras marstonense*
(left) and *Dactylioceras commune* (right).
(From Trueman, 1941.)

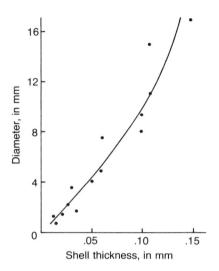

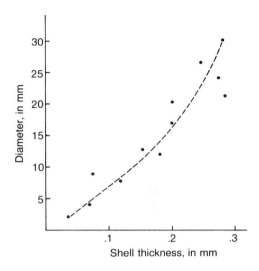

living chamber, just anterior to the last septum, where the body is attached
to the shell. We must certainly assume that ammonoids also differed from
Nautilus in some details of soft anatomy, although we have no way of determining such differences. We can be certain that most ammonoids, like
Nautilus and other cephalopods, were predaceous jet-propulsion swimmers.
Typical cephalopod features associated with predaceous swimming are a
horny **beak**, a file-like **radula** for rasping pieces of food seized by the beak
(beaks and radulas have, in fact, recently been discovered in association
with ammonoid fossils), **tentacles, eyes,** and **buoyant gas** filling the sealed
chambers of the shell.

Denton and Gilpin-Brown (1961, 1966, 1967) have recently provided
extremely significant studies of the physiology of buoyancy in living cephalopods. These workers studied *Nautilus* and two quite different groups of
the order Sepioida (*Sepia* and *Spirula*). All three have chambered shells.
The shell of *Nautilus* is external, like those of the ammonoids; the shells of
the other groups are internal. In *Nautilus, Sepia,* and *Spirula,* a whole-
animal specific gravity nearly equal to that of sea water is maintained regardless of how deep they are in the water, so that the animals have nearly
neutral buoyancy. Their sealed chambers contain a gas mixture similar to
air; the last-formed chambers also contain varying amounts of watery fluid.
The shell of *Nautilus,* shown in Figure 8-15, illustrates the condition. The
amount of liquid is largest in the last chamber and resembles sea water in

composition but contains a lower concentration of salts. The gas in the chambers is at pressures of less than one atmosphere, with the lowest pressures in the newest chambers. Pressures in the oldest chambers approximate 0.9 atmosphere. In all groups studied, the relative amounts of gas and liquid in the chambers are used to control buoyancy.

Nautilus grows very rapidly, a septum being added about every two weeks. The animal secretes watery fluid between its body and the last-formed septum; the fluid supports the body against external pressure, while the body moves forward from the last-formed septum and also while the new septum is being formed. Once the new septum is calcified, the liquid is pumped out by the siphuncle, in which porous, horny, and chalky layers surround the fleshy central stalk. The space formed by removal of the liquid is filled with gas by diffusion from the siphuncle. This sequence of events explains the progressively larger amounts of liquid and lower pressures of gas in newer chambers.

Because gas pressure in the sealed chambers is maintained at less than one atmosphere, regardless of water depth, the shell walls must bear all external water pressures encountered by the living animals. *Nautilus* descends to depths of 600 meters or more in the sea. Descent to 600 meters adds approximately 60 atmospheres of external pressure that must be resisted by the shell. Denton and Gilpin-Brown (1966) and Raup and Takahashi (1968) have subjected *Nautilus* shells (with the siphuncles sealed off) to high external water pressures in the laboratory. The shells imploded at pressures equivalent to those at ocean depths somewhat greater than the limit of *Nautilus'* range, which thus may be determined by shell strength.

Denton and Gilpin-Brown have quite reasonably concluded that the striking similarity in siphuncle and coiled-shell construction of chambered shells among fossil and living cephalopod groups indicates that buoyancy control has been achieved in the same way by all these groups (including ammonoids). Their arguments are compelling in light of the similarity of their physiologic findings for diverse Recent groups.

Let us return to the problem of the function of septal sutures. Several ideas have traditionally dominated paleontologic interpretation of the complex suture patterns of ammonoids:

One idea holds that septal fluting served to *increase shell strength* to resist differences in pressure between the internal and external surfaces of the shell. Presumably, many ammonoids lived at considerable depths and were subjected to both large external pressures and rapid pressure changes during ascent and descent through the water column.

FIGURE 8-15
Biology of *Nautilus*. A: A modern nautilus swimming backward by jet propulsion. B: A nautilus clinging to the bottom with its tentacles. C: A nautilus with part of its shell cut away and morphologic features labeled. D: Cut-away view of the shell. (A, B, and C from Fenton and Fenton, 1958; D from Denton and Gilpin-Brown, 1966.)

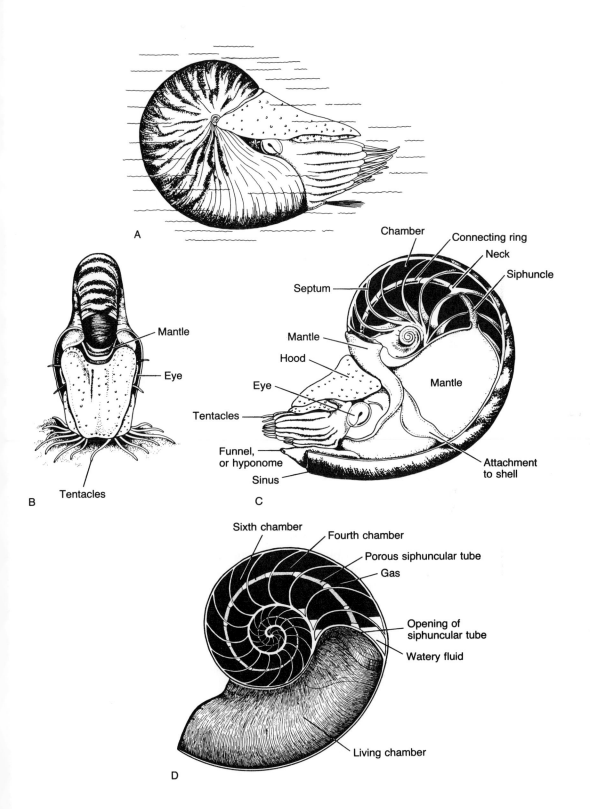

A

B

Mantle

Eye

Tentacles

C

Chamber

Connecting ring

Neck

Siphuncle

Septum

Mantle

Hood

Eye

Tentacles

Funnel,
or hyponome

Sinus

Mantle

Attachment
to shell

D

Sixth chamber

Fourth chamber

Porous siphuncular tube

Gas

Opening of
siphuncular tube

Watery fluid

Living chamber

A second hypothesis says that complex suture patterns evolved to *regulate whole-animal specific gravity*. The idea has commonly been expressed that muscles attaching the animal in its living chamber may have been arranged in such a way that the animal could regulate the volume of a gas space believed to have existed between the body and last-formed septum. The body would thus have formed a plunger that moved backward and forward to alter whole-animal specific gravity and regulate buoyancy for vertical movement through the water column. Muscles attached at *saddles* of the septum would have pulled the body forward; those attached to *lobes* would have pulled it backward (Figure 8-12).

A third hypothesis is that septal fluting was primarily an incidental consequence of mantle folding. According to this idea, mantle folding served to *increase the relative surface area for secretion of gas or liquid* in the space between the mantle and last-formed septum.

We can formulate a paradigm for each hypothesis. For the hypothesis that septal fluting was primarily for structural support of the outer shell wall, the paradigm should be a mechanical device capable of withstanding a large force with a minimum of structural material (for metabolic economy). It should give adequate support to all points on the outer shell wall. The greater the distance of a point on the outer wall from an internal supporting structure, the less support it will receive. An engineer assigned to the problem might design a system with I-beam-like braces passing inward from the shell wall at appropriate angles and converging at the center of the shell interior. Alternatively, he might employ a lattice-like support system. A mollusc, however, must lay down shell material with its fleshy mantle, which offers a two-dimensional surface for carbonate secretion. Furthermore, shelled cephalopods seem to have been required to seal off their older shell chambers from the body chamber. Even the flesh-filled siphuncle does not allow free passage of fluids between the living chamber and the older chambers.

The support-hypothesis paradigm, therefore, must be a system of solid transverse partitions secreted by the mantle at intervals during growth of the outer shell. The problem is one of supporting the walls adequately with a minimum of shell material. The effective strength of the outer shell, as a whole, will be determined by its weakest point. Engineering principles show that corrugation of the supporting partitions will offer increased support to interseptal spans of the coiled outer shell. Corrugation must be of large amplitude relative to the distance between septa to have a pronounced strengthening effect; otherwise, large unsupported spans of outer shell surface will exist. Because the postulated function of corrugation is to add support to the outer shell walls, not to strengthen the partitions themselves, it can be restricted to the peripheral portions of the partitions, where they join the walls. We can further propose that the septa might become relatively more closely spaced or the fluting patterns more complex during ontogeny, because relative shell thickness of the outer wall of ammonoids decreases during growth (Figure 8-15).

There are at least three basic factors that would be expected to have produced a considerable amount of interspecies variation in the corrugation pattern described by the paradigm: (1) different mutations among various evolving lineages should have led to different mechanical solutions to the strengthening problem; (2) taxa adapted to different depth regimes must have been subjected to different pressure effects; (3) different shapes of the coiled outer shell should have required different systems of structural support.

Let us compare the support-hypothesis paradigm with known ammonoid septal morphology. We find that the highly variable peripheral fluting pattern of ammonoid septa fits the paradigm remarkably well. Increasing complexity of suture pattern during ontogeny is a well-known phenomenon in ammonoids; it was predicted in the paradigm as a factor compensating for decreasing relative thickness of the coiled outer shell. Newell (1949) has presented data showing that suture length increases ontogenetically much more rapidly than shell diameter in a group of late Paleozoic ammonoids (Figure 8-16). The tendency for septa to become more closely spaced during ontogeny would tend to compensate in the same way, although some authors have suggested that the crowding was an incidental consequence of decreased growth rate of the coiled shell as the maximum size was approached.

One flaw in the support hypothesis should be mentioned: water pressure is transmitted to the last-formed septum by the body and an inelastic pocket of liquid. Septal fluting in ammonoids may have served to strengthen the septum *itself* in addition to strengthening the outer coiled shell. We mentioned earlier that ammonoid septa are commonly convex on the living chamber side, whereas in nautiloids they are normally concave on the living chamber side. The ammonoid construction is the stronger of the two in withstanding fluid pressure in the living chamber (recall the concave bottoms of many commercial spray cans).

We will not construct an elaborate paradigm for the second hypothesis advanced to explain septal fluting: the hypothesis suggesting that septal fluting was related to muscle attachment for movement of the animal backward and forward in the living chamber. It is sufficient to say that any paradigm constructed to represent this function would call for muscle insertion scars on the septum or inner shell wall. In view of the pressures involved, the muscles used to pull the animal toward the aperture would have to have been extremely strong and their insertion scars, very deep. The large muscles holding the living *Nautilus* in its shell are only loosely attached to the shell's interior. They must migrate anteriorly very rapidly (perhaps as much as 2–3 centimeters per month) to keep pace with the growth of the coiled shell. Neither deep nor extensive muscle scars for body attachment have been found in fossil ammonoids.

The third hypothesis was that septal fluting was an incidental consequence of increase in the surface area of the mantle, which evolved in response to the need for increased secretion of gas or liquid between the mantle and last-formed septum. In order to provide the maximum mantle surface area, however, the paradigm for this function would have to specify folding or corrugation of the *entire* septum, not just the septal margins. The paradigm, therefore, would not match the observed septal configuration of ammonoids.

FIGURE 8-16

Allometric increase of suture complexity in the ontogeny of five Paleozoic ammonoid genera. For each genus half-length of the external suture, Y, is plotted against shell diameter, X. In every genus Y increases exponentially with an increase in X. The use of logarithmic coordinates therefore yields straight-line plots. The allometric equation relating the two morphologic variables for each genus is given at the lower right. The adult suture patterns for the five genera are shown at the upper left. (From Newell, 1949.)

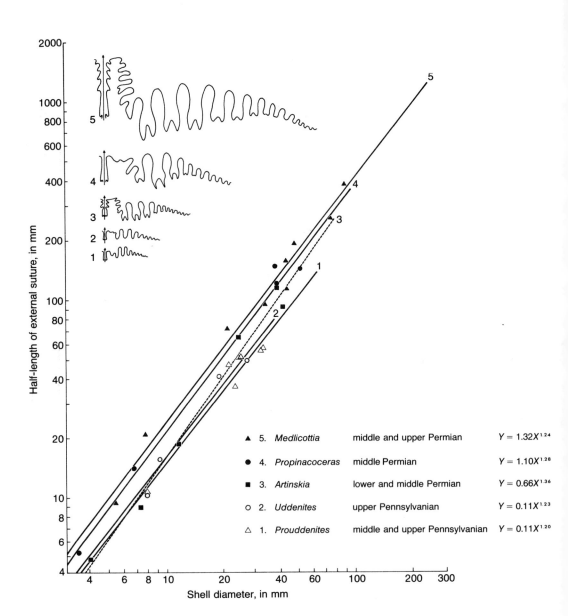

▲ 5.	*Medlicottia*	middle and upper Permian	$Y = 1.32X^{1.24}$
● 4.	*Propinacoceras*	middle Permian	$Y = 1.10X^{1.28}$
■ 3.	*Artinskia*	lower and middle Permian	$Y = 0.66X^{1.36}$
○ 2.	*Uddenites*	upper Pennsylvanian	$Y = 0.11X^{1.23}$
△ 1.	*Prouddenites*	middle and upper Pennsylvanian	$Y = 0.11X^{1.20}$

Half-length of external suture, in mm

Shell diameter, in mm

In conclusion, of the paradigms considered, the one that best fits observed septal morphology is the first one, which suggests that the function of septal fluting was to strengthen the outer wall against hydrostatic pressure. This is not to say that we have the whole answer. We have not considered all possible functions. It is quite possible that septal fluting performed more than one function in ammonoids. At the very least, the support-hypothesis paradigm should be developed in a more detailed and rigorous fashion than we have done here.

Dinosaur Jaw Mechanics

Movable skeletal elements of the vertebrate jaw commonly act as mechanical levers, with jaw muscles providing forces for movement. To interpret fossil forms, muscle systems must be reconstructed from skeletal morphology or by homology with living relatives. Recent interpretations of the jaw mechanics of the Cretaceous ceratopsian dinosaurs have shed light on their feeding habits (Ostrom, 1964). These horned quadrupeds (Figure 8-17) have traditionally been assumed to have been herbivores (plant eaters), but unlike most herbivores, which rely heavily on grinding and crushing dentition, ceratopsians employed shearing teeth (Figure 8-17).

The following account summarizes Ostrom's analysis of jaw mechanics in the genus *Triceratops,* with reference to Figure 8-18. The fulcrum about which the lower jaw of *Triceratops* rotated in life is located near the posterior end of the jaw. The principal muscle used to raise one side of the jaw is believed to have been attached to a prominent upward projection of the jaw called the coronoid process. The muscle is thought to have attached posteriorly to the large "neck shield" of the skull. Other muscles apparently played only minor roles in jaw movement and can be neglected in this analysis. The direction of force of the dominant muscle is represented by the heavy dashed arrow in Figure 8-18, D. The lower jaw of most vertebrates acts as a third-class lever because the muscular force that lifts the jaw attaches at a position *between* the fulcrum and the resistant force (the resistant force being provided by the food between the teeth). A simple third-class lever is shown in Figure 8-18, B. For each of the two forces, the leverage is determined by the length of the moment arm (the distance between the point of application of the force and the fulcrum). The **moment** of the muscle force is the product of the force and the moment arm length (b in Figure 8-18, C). If the elevating force operates at an acute angle to the lever (as in Figure 8-18, C) its moment is reduced, and is then calculated by multiplying the vertical moment by the sine of the acute angle. Thus, in Figure 8-18, C the force operates at an angle of 45° to the lever, and its moment is:

$$\text{elevating force} \times b \times \sin 45° = \text{elevating force} \times b'.$$

The smaller the angle between the muscle force and the lever, the smaller will be its moment.

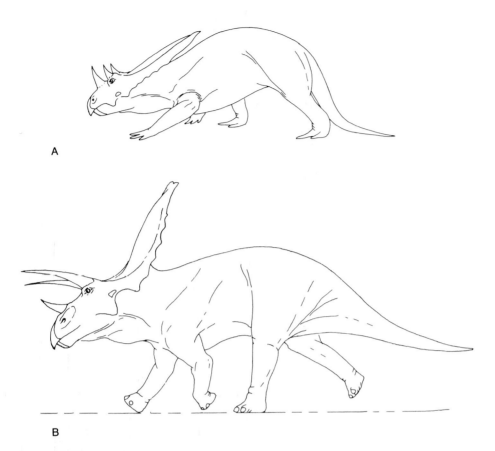

A

B

FIGURE 8-17
Reconstruction of ceratopsian dinosaurs similar to *Triceratops*. A: *Chasmosaurus*
in the sprawling posture of traditional museum and textbook ceratopsian restorations.
B: *Torosaurus* in the upright posture suggested by recent studies of dinosaur limb
mechanics. (From Bakker, 1968.)

From consideration of leverage alone, we might predict that maximum
efficiency in the vertebrate jaw could be attained by an arrangement in which
the muscle force acted at right angles to the jaw and was located as far forward as possible. Problems would be posed by such an arrangement, however: the angle of jaw opening would be inconveniently small and there
would not be room for large muscles in the facial and snout regions. In
addition, the speed of jaw closure would be decreased by having the muscles
far from the fulcrum.

There is another way that the moment arm, or leverage, of the muscle
can be increased if it operates at an acute angle to the jaw: the point of application of force may be elevated to a position above the fulcrum, thus
lengthening the moment arm on which the force operates. *Triceratops* employed such a mechanism in the form of the coronoid process. Figure 8-18, F
shows the manner in which the coronoid process lengthened the moment
arm of the principal adductor.

FIGURE 8-18
Jaw mechanics of *Triceratops*. A: Cross-section of jaw, showing shearing surface of
teeth. B: Simple third-class lever with parallel forces. C: Simple third-class lever with
nonparallel forces. D: Skull, showing location of principal adductor muscle. E: Lower
jaw. F: Lever system of lower jaw. (From Ostrom, 1964.)

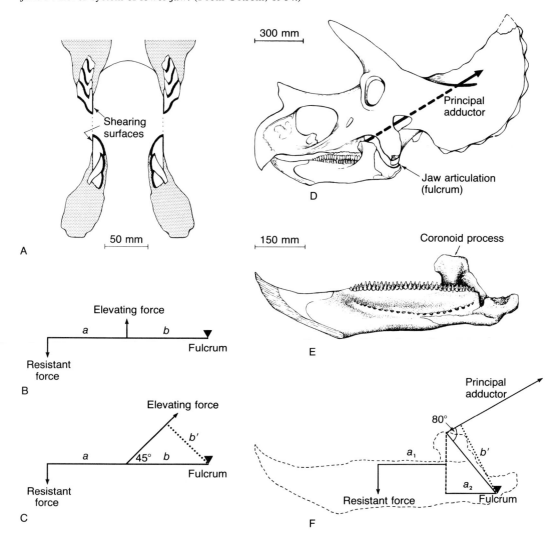

For several ceratopsian species, Ostrom calculated the percentage of the
principal adductor force that would have been available at various positions
along the jaw at which food might have provided resistance. He did so
simply by comparing the moment arm, or leverage, of the muscular force
with the moment arm of resistant food at various positions (the latter was
also the moment arm of the "equal and opposite" force applied against the

183

food). Figure 8-18, F shows that the principal adductor muscle operated at an angle of 80° to the line between its attachment site and the fulcrum. Therefore b' became its effective moment arm, and the fraction of the principal adductor force available at the position of resistant food along the jaw was:

$$\frac{\text{force applied against food}}{\text{principal adductor force}} = \frac{b'}{a_1 + a_2}.$$

Ostrom calculated that for most species about 30 percent of the principal adductor force was available at the beak, 50 percent at the front teeth, and 250–300 percent at the rear teeth, which are very near the fulcrum. Judging from the apparent size of the principal jaw muscles, these forces must have been remarkably strong.

The jaw adaptations of the ceratopsians, especially the development of a high, strong coronoid process, were for transmission of powerful shearing forces to the beak and teeth. Ostrom concluded that if the ceratopsians were herbivores, as has been traditionally assumed, they must have been specially adapted for feeding on tough, resistant plant materials, perhaps the fronds of palms and cycads, which were abundant during the Late Cretaceous.

Recent studies of dinosaur limb mechanics (Bakker, 1968) suggest that many dinosaur species that traveled on all fours were not nearly as clumsy as suggested by traditional reconstructions. In reconstructions, the front legs have often been placed in a sprawling position, with the feet positioned well to the sides of the body (Figure 8-17). It now appears that in most of these forms, including *Triceratops*, the front legs may have been positioned more directly under the body and may thus have served as efficient running limbs. Perhaps the horned *Triceratops* was a rhinoceros-like animal. Judging from Ostrom's analysis of its shearing mechanics, there is even the outside chance that it was a carnivore!

The Pelycosaur Sail

An interpretation of function is especially convincing if supported by quantitative relationships predicted theoretically, as from the Principle of Similitude (page 63). An example is the most widely accepted interpretation of the function of the dorsal "sail" of mammal-like reptiles belonging to the Permian genus *Dimetrodon*. The sail was formed of skin supported by greatly elongated neural spines (Figure 8-19). It has been proposed that the sail served as a temperature-regulating device. According to this hypothesis, *Dimetrodon* raised or lowered its body temperature efficiently by changing the orientation of the sail with respect to the direction of wind or sunlight. Rate of heat absorption or loss could also have been partly controlled internally, by regulation of blood flow to and from the sail. Romer

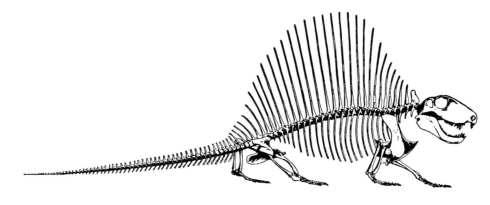

FIGURE 8-19
Skeleton of *Dimetrodon;* maximum length was about 11 feet. (After Romer, 1966.)

(1948) has shown that sail area was disproportionately large with respect to body area among larger species of *Dimetrodon*. In fact, the observed relationship can be described by a simple equation analogous to the allo-metric equation for anisometric growth (page 61):

$$\text{sail area} = (\text{body area})^{1.6} \qquad \text{(Eq. A)}$$

If the sail was a temperature-regulating device, we can predict from the Principle of Similitude that *sail area would have had to increase in propor-tion to body volume* (the volume to be heated and cooled) in order to main-tain constant efficiency during evolutionary size increase. Assuming body shape to have been constant (though it was not precisely so), body volume would have increased with the cube of linear dimensions, and body surface area, with the square of linear dimensions. In other words $(\text{body volume})^{1/3}$ would have been proportional to $(\text{body area})^{1/2}$ regardless of size. The pro-portionality becomes an equality if the same units (for example, inches or centimeters) are used to measure volume and area.

Then:

$$\text{body volume} = (\text{body area})^{3/2}.$$

Therefore, if sail area is proportional to body volume:

$$\begin{aligned}\text{sail area} &= (\text{body area})^{3/2}\\ &= (\text{body area})^{1.5}.\end{aligned} \qquad \text{(Eq. B)}$$

The close similarity between the observed relationship (Eq. A) and the theoretical prediction (Eq. B) strongly supports the idea that the sail func-tioned in temperature regulation.

Special Adaptations in Brachiopods

Rudwick has analyzed the functions of a variety of structures in the Brachiopoda. In one such study (Rudwick, 1965) he has presented an interpretation of the function of the shell morphology of two remarkably similar Mesozoic brachiopod taxa belonging to separate superfamilies (Figure 8-20). Their shells are not unusual except for the presence of four narrow projections formed by outward deflection of the shell margin. The mantle, which secreted the shell, must have underlain the projections of each valve during the organism's life. Rudwick rejects the hypothesis that the projections served to strengthen the shell or increase the surface area of the mantle for gaseous exchange. There is no evidence of any need for shell strengthening, and little mantle surface area would have been added through addition of the projections. Rudwick concludes that the projections served as a warning system, like a series of antennae. The sensory mantle would necessarily have extended out to the tips of the projections, and by its response to being touched, would have warned the animal that something was approaching the open mantle cavity. The valves could then snap shut and the animal could escape from danger. Rudwick's arguments are supported by the nearly equal length and angular separation of the projections, and by the narrowness of the projections, which minimizes the required amount of shell secretion.

Vision in Trilobites

Clarkson (1966) has taken an original approach to functional morphology in analyzing the vision of Silurian trilobites of the suborder Phacopina, in particular the species *Acaste downingiae* (Salter) (Figure 8-21). The phacopinids possess a compound eye, consisting of numerous lenses arranged in rows. The lenses are deployed in a geometric configuration that provides the closest packing possible. (There are, however, small spaces between them, as shown in Figure 8-21, E.) In life, each lens apparently lay at the exposed end of a cylinder through which light was transmitted to photoreceptive cells below.

In order to measure the orientations of the many lens axes of a single eye, Clarkson made use of a specially designed apparatus employing a stereoscopic microscope and a turntable on which the trilobite specimen was mounted. He plotted stereographic projections of the axial bearings of individual lenses relative to the animal's plane of bilateral symmetry. Together, all the lens bearings for a single eye represent its *visual field* (Figure 8-21, F, G). Because of the geometric arrangement of the lenses, *visual strips* were formed within the visual field of *Acaste,* and thus the trilobites must have seen light as a series of light and dark bands.

By comparison with compound eyes of living arthropods, Clarkson concluded that the compound eyes of phacopinid trilobites were of too low a grade of organization to have perceived form. The banding produced by the

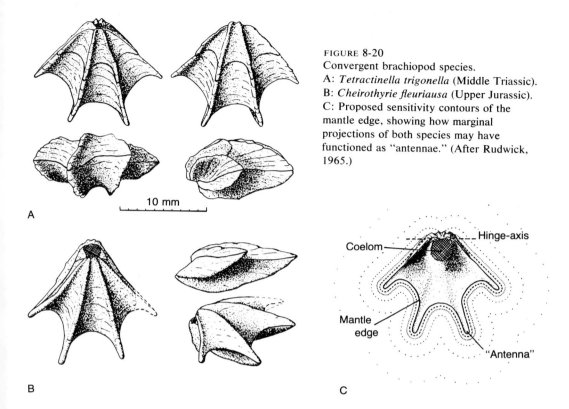

FIGURE 8-20
Convergent brachiopod species.
A: *Tetractinella trigonella* (Middle Triassic).
B: *Cheirothyrie fleuriausa* (Upper Jurassic).
C: Proposed sensitivity contours of the mantle edge, showing how marginal projections of both species may have functioned as "antennae." (After Rudwick, 1965.)

visual strips, however, could have enabled the trilobites to perceive horizontal movements of an object, and the object's size, velocity, and direction of motion. Biologic studies have shown that for movements to be perceived by eyes like those of phacopinids, there must be discontinuities in the image formed at the photoreceptors. The visual strips of *Acaste* would have produced such discontinuities. Movement of a shadow horizontally across the visual field would have been recorded successively by each of the series of visual strips. As shown in Figure 8-21, H, the animal could also have perceived movement of an approaching object by the progressive detection of the object by higher and higher lenses in the eye.

One outgrowth of Clarkson's trilobite vision study is a reasonable interpretation of the life position of the cephalon (head region) of *Acaste*. The visual fields of the two eyes of a crawling trilobite were almost certainly oriented so that the animal had 360° of lateral vision. This condition could only have existed if the anterior border of the cephalon was arched, rather than resting flat against the substratum. The cephalons of the fossil specimens of Figure 8-21, A, B, C are, in fact, preserved in this orientation although they happen to be rolled up rather than stretched out in the normal feeding or locomotion position. The animals apparently rolled up for protection, and many died in this posture.

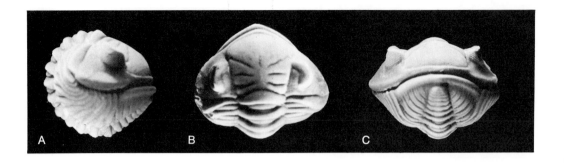

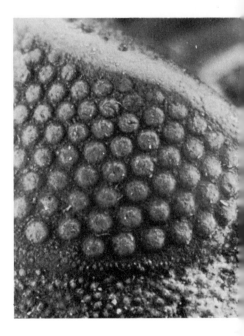

D

E

Shell Form in Bivalves

As an example of the importance of observations of living species to functional morphology, we can consider two distantly related bivalve mollusc superfamilies, the Arcacea and the Carditacea. Morphologic trends in the evolution of the two superfamilies can be related to life-habit changes. Each superfamily contains representatives of two life-habit groups (Figure 8-22). One group consists of species that attach firmly to the substratum by means of a byssus (a group of horny threads secreted by the foot). The other consists of burrowers that live unattached (or very weakly attached by a small byssus) in soft sediment. Most living arcaceans that employ a strong byssus

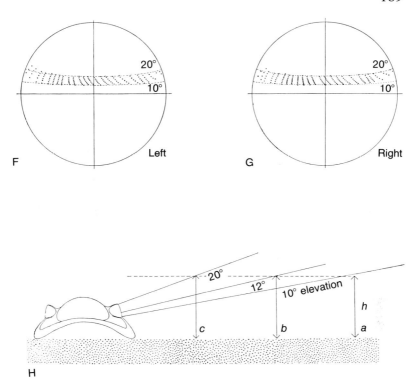

FIGURE 8-21
Vision of *Acaste downingiae*. A-C: Morphology of enrolled fossil. D and E: Right eye, showing lenses. F and G: Visual fields of left and right eyes shown by stereographic projection. H: Front view of head in the inferred life position, showing "latitudinal" limits of vision as an object of height *h* approaches from *a* to *c*. (From Clarkson, 1966.)

live attached to the substratum surface. These species tend to be much more elongate than burrowing species. Carditacean species that attach by a strong byssus are also more elongate than their unattached relatives, but some attach to objects within sediment, rather than living at the substratum surface.

The life orientations of typical burrowing species belonging to the two superfamilies are shown in Figure 8-22. The compact, equidimensional form of these animals provides them with a small ventral surface area, producing little resistance when they burrow. The ventral margin is also rounded, for easy penetration of the substratum. The elongate form of the species attached by a strong byssus would offer much greater resistance to sediment penetration. Because shallow burrowers may be exhumed by wave and current scour, it is important that they be able to reburrow efficiently.

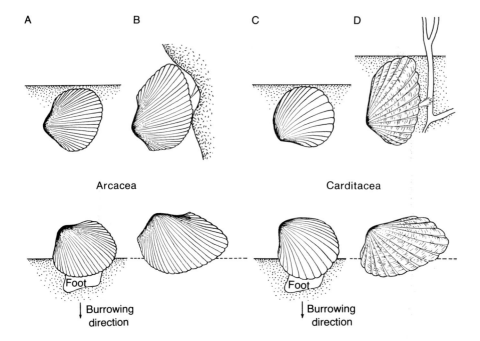

A   B   C   D

Arcacea   Carditacea

Foot   Foot

↓ Burrowing direction   ↓ Burrowing direction

FIGURE 8-22
Convergence between burrowing and byssally attached arcacean and carditacean bivalves.
A and B: The arcaceans *Anadara ovalis* and *Anadara antiquata*. C and D: The carditaceans
Venericardia borealis and *Cardita floridana*. In the upper drawings the life positions of the
four species are shown. In the lower drawings the two compact, free-living species are shown
in the act of burrowing into the sediment; the elongate shells of the other two species, which
are shown alongside the burrowing clams, would offer much greater resistance to sediment
penetration if they belonged to free-living burrowers. (Modified from Stanley, 1970.)

    Stable attachment of species having a strong byssus requires that the
ventral shell margin rest snugly against the attachment surface when the
byssal retractor muscles are contracted. The compact form and rounded
ventral margin of burrowing arcacean and carditacean species would offer
only a single point of contact if clamped against a flat surface and the shell
would be unstable.
    The life habits of fossil species belonging to the Arcacea and the Cardi-
tacea can be interpreted in light of the function of shell form in Recent representatives of the two superfamilies. Examination of the fossil record of
the Arcacea (which extends back at least to the Devonian Period) shows
that during the Paleozoic the group was represented almost exclusively by
elongate forms that must have been attached by a strong byssus in life. Compact burrowing arcaceans apparently did not arise until the Mesozoic Era.
Both life-habit groups have survived and are abundant in modern seas.

The early fossil record of the Carditacea is not well known, but the group appears to have followed a course of evolution similar to that of the Arcacea. Most early carditaceans (those found in Permian and Triassic strata) have elongate shells that were almost certainly strongly attached to the substratum in life. Only in the late Mesozoic and Cenozoic have the more compact, free-burrowing carditaceans become abundant. If the evolutionary interpretations of Heaslip (1968) are correct, some Cenozoic lineages of compact, free-burrowing arcaceans have given rise to byssally attached species, such as *Venericardia (Claibornicardia) nasuta*, which have thus reverted to the life habits of their evolutionary forebears (Figure 8-23).

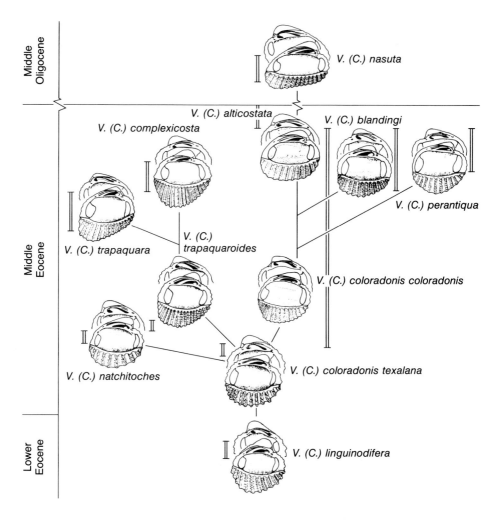

FIGURE 8-23
Cenozoic phylogeny of *Venericardia (Claibornicardia)* in the Gulf and East Coasts of North America. (From Heaslip, 1968.)

Functional morphology of the shell form of arcaceans and carditaceans is especially striking, in that both families show similar patterns of evolutionary divergence with respect to life-habit groups. Evolutionary "convergence" in both form and life habits represents one of the strongest arguments that skeletal morphology is essentially adaptive. Apparently, the marked similarity of the brachiopod taxa analyzed by Rudwick and discussed earlier (see Figure 8-20) is also a consequence of adaptive convergence.

Conclusion

Although the examples we have considered are but a small sample of many interpretations of function that have been made for fossil structures, they illustrate the great diversity of problems that invite study and the variety of analytic approaches that can be employed. Functional morphology is an extremely important area of research for two reasons:

When applied to an individual fossil species, it can greatly facilitate paleoecologic analysis — the study of the species' interaction with its environment (for example, its life habits, habitat preferences, and relationships to other species).

When applied to taxonomic groups of species, it can reveal how the morphologic trends in evolution that have been observed in the fossil record are adaptive. (Figure 8-23).

Supplementary Reading

Bonner, J. T. (1952) *Morphogenesis.* Princeton, Princeton University Press, 296 p. (An outstanding biologic treatment of growth and form.)

Dacqué, E. (1921) *Vergleichende biologische Formenkunde der fossilen niederen Tiere.* Berlin, Gebrüder Bornträger, 777 p. (The first comprehensive analysis of functional morphology of fossils.)

Gould, S. J. (1970) Evolutionary paleontology and the science of form. *Earth-Science Reviews,* **6:** 77–119.

Grant, V. (1963) *The Origin of Adaptations.* New York, Columbia University Press, 606 p.

Rudwick, M. J. S. (1964) The inference of function from structure in fossils. *Brit. Jour. Philos. Sci.,* **15:**27–40. (A provocative discussion of the logical problems encountered in deducing function from morphology.)

Thompson, D'A. W. (1942) *On Growth and Form.* Cambridge University Press, 1116 p. (The classic reference on problems of form and function.)

CHAPTER 9

Paleoecology

Ecology is the study of interrelationships between organisms and their environment. Paleoecology is the same kind of study for fossil species.

The reasons for studying paleoecology are many. A major goal of geology is to unravel the history of the earth and its inhabitants. Although most detailed paleoecologic analyses are limited to short geologic time intervals and geographic areas whose boundaries are measured in meters, kilometers, or tens of kilometers, the accumulation of information from many such analyses gives us a general history of life and environments on earth. Because there are gaps in the stratigraphic record and because fossils do not provide complete biologic information, we can never reconstruct this history completely. Nevertheless, paleoecology can provide us with considerable knowledge about the history of life on earth, and we are only in the beginning stages of exploring its possibilities.

A dangerous pitfall to paleoecologic research involves circular reasoning. In any paleoecologic study it must be made clear whether (1) an environment is being reconstructed from nonpaleontologic data and is being used as a framework for interpretation of fossil species or (2) the organisms themselves are being used to reconstruct the past environment. Both approaches may be used in a single study, but must not be intermingled to the point at which circular reasoning destroys the validity of final interpretations.

The first step in paleoecologic interpretation is to determine whether the fossils being studied have been preserved where they lived, and, if not, how far they have been transported before burial. We will consider this problem later in the chapter. At the outset, a few generalizations are in order.

As discussed in Chapter 1, marine animals, most of which are invertebrates, inhabit a vast depositional basin in which an organism's remains are generally more likely to be preserved than on land. The majority of terrestrial organisms are insects, which are not easily preserved. In order to be preserved, the remains of most preservable terrestrial species (chiefly plants and vertebrates) must be transported from their native habitats to depositional environments such as rivers or lakes. Most fossil assemblages include remains of species transported from more than one habitat. Because of mixing and transport, few terrestrial fossil assemblages have been subjects of paleoecologic study. They have been used extensively for regional faunal and floral studies, however, and have contributed much to our understanding of past climates and biogeography. Marine organisms are not only less likely to undergo transport after death, they are more likely to be preserved. Shelled marine invertebrates have also existed in greater abundance and diversity in the geologic past than preservable terrestrial plants and animals. Lake-dwelling biotas are intermediate between terrestrial and marine biotas in their paleoecologic utility. They are not likely to undergo considerable post-mortem transport, but lakes cover only small areas of the earth's surface and lake deposits run a high risk of being destroyed by erosion. Because shelled marine invertebrates have been the subjects of most paleoecologic studies, they will be emphasized in our discussion.

Fundamental Ecologic Principles

The largest unit of study considered in ecology is the *ecosystem,* which consists of a chosen portion of physical environment plus all of the organisms contained in it. The ecosystem includes all the interactions—chemical, physical, and biologic—that occur within the chosen physical boundaries. Thus we might consider as an ecosystem the biosphere or just as reasonably, a tiny puddle of rainwater containing two or three protozoan species.

The *habitat* is the environment in which an organism lives. It might be a rocky seashore, a grassland, or, for a parasite such as a flea, a host species, such as a dog. There may be more than one habitat in an ecosystem. The *ecologic niche* is often defined as the organism's position in the habitat, including its way of life and the role it plays in the ecosystem. Some workers prefer to define ecologic niche in terms of the features of the environment that permit a species to exist. What we are really concerned with is the interaction between a species and its environment. The two types of definition are not fundamentally different, but simply differ in their emphasis. The first stresses the physiologic and behavioral adaptations of the species to the environment, and the second stresses the environmental limits of adaptation of the species. The first commonly makes description of a particular niche easier, but the second commonly facilitates comparison of two or more similar niches. Most habitats are occupied by several species, each with its own ecologic niche. Usually each species is represented by two or more individuals, which constitute a *population* (Chapter 4). Populations of two or more species occupying a habitat are commonly referred to as a *community.* Commonly a community is named for one or two of its most conspicuous and abundant species. Familiar examples might be a spruce-fir forest community of northern latitudes and a barnacle-mussel community along a rocky ocean shore. Just as there may be more than one habitat in an ecosystem, there may be more than one community. The problem of rigorously defining the term "community" in ecology remains unsettled: we will discuss it in more detail later in the chapter.

Within any ecosystem many constituents, both living and nonliving, interact. Among the most fundamental interactions are those in which energy and materials are transferred. The idealized pattern for the flow of materials through a simple but typical ecosystem is shown in Figure 9-1. In most communities the organic compounds are synthesized from the environment by *producers* in the form of photosynthetic plants. These are

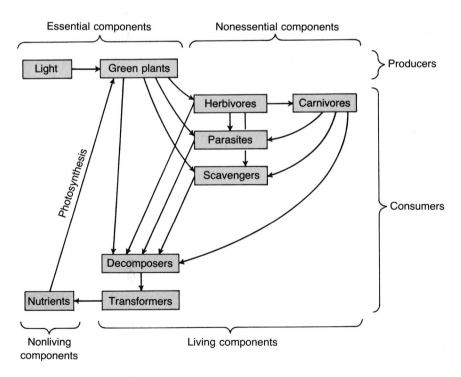

FIGURE 9-1
Diagram of the flow of materials through a typical ecosystem.
(Modified from Clarke, 1965.)

consumed by *herbivores,* some of which are devoured by *carnivores. Parasites,* which feed on living organisms without necessarily killing them, and *scavengers,* which feed on dead organisms, may enter into the system in a variety of ways. The sequence of species from producers through unpreyed-upon carnivores in any ecosystem is termed a *food chain* or *food web.* Organic materials of producers, herbivores, and carnivores not assimilated by species higher in the food web are broken down by organisms called *decomposers,* which are chiefly bacteria. Finally other organisms called *transformers,* also mostly bacteria, chemically alter certain decomposition compounds to render them utilizable once again by producers. Bacteria are thus extremely important in effecting completion of the cycle; without them, ecosystems as we know them could not exist.

It should be understood that the cycle shown in Figure 9-1 uses materials. Energy, though also moving clockwise through the cycle, is not recycled. The total amount of living matter, including stored food, in the ecosystem at any time is termed the *biomass.*

Flow of materials or energy through a food chain or web may be represented by a simple diagram. Figure 9-2 shows what is called an energy pyramid and represents rate of energy flow. The ultimate source of energy for most ecosystems is the sun because most producers use solar energy to synthesize compounds necessary for life. Producers are fed upon by herbivores, which are fed upon by carnivores, which in turn are fed upon by "top carnivores." The slope of the pyramid represents the energy loss at each successive step, resulting primarily from inefficiency of metabolic systems and decay of unpreyed-upon individuals. A similar pyramid can be constructed for biomass (at a single time) within an ecosystem. Most, but not all, carnivores are larger than their prey. Therefore, the biomass at the top of a food pyramid is not only smaller, but tends to be divided among fewer animals. Most food pyramids are formed of only three or four steps. The fewer the steps in a food pyramid, the less the energy loss tends to be.

FIGURE 9-2
Energy pyramid for an ecosystem at Silver Springs, Florida. Trophic levels:
P = producers, *H* = herbivores, *C* = carnivores, *TC* = top carnivores,
D = decomposers. Each bar represents the total energy flow through a given trophic level. The darker portion of each bar represents energy locked up in biomass in the area studied; the lighter portion represents energy lost through respiration or movement downstream and out of the study area. (From Odum, 1959.)

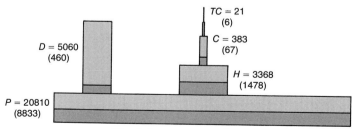

Kilocalories per square meter per year

Determining the flow of energy and materials through ecosystems is virtually beyond the reach of paleoecology. The paleoecologist can, however, reconstruct at least a partial picture of food chains for certain fossil assemblages. He can, for example, commonly recognize predatory species and their prey.

The Marine Ecosystem

The present-day marine ecosystem is used by most paleoecologists as a general model for interpretation of ancient marine sediments and faunas. One drawback in this approach is that climates have been considerably warmer than at present for about two-thirds of the time period since the beginning of the Cambrian (Dorf, 1960). Also, continents are relatively emergent today; during most of the Paleozoic, Mesozoic, and Cenozoic, larger areas of continental crust were submerged beneath shallow epicontinental seas. The Baltic Sea and Hudson's Bay are examples of modern epicontinental seas, but both are situated in cold northern latitudes.

Figure 9-3 is an idealized block diagram of the edge of a modern-day continent. The submerged border of the continent is called the **continental shelf**. The continental margin lies at a depth of about 200 meters. From it the **continental slope** extends to a depth of about 5000 meters, where it reaches the ocean basin floor.

FIGURE 9-3
Idealized block diagram of the edge of a modern continent.
The vertical scale is exaggerated and distorted.

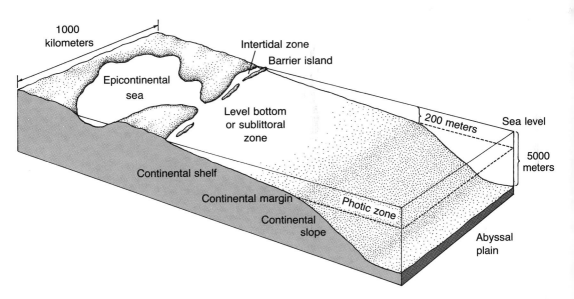

FIGURE 9-4

Life habits of marine invertebrates. A: A snail (gastropod mollusc). B: A sea urchin (echinoid echinoderm). C: A jellyfish (scyphozoan coelenterate). D: Squids (cephalopod molluscs). E: A starfish (asteroid echinoderm). F: A crab (crustacean arthropod). G: A sponge (poriferan). H: Mussels (bivalve molluscs). I: A cockle (bivalve mollusc). J: A chaetopterid worm (polychaete annelid). K: A macomid clam (bivalve mollusc). L: A trumpet worm (polychaete annelid). M: A sea cucumber (holothurian echinoderm). Animals C and D are pelagic (C is planktonic and D is nektonic).

The marginal marine region alternately covered and uncovered by the tides is called the *intertidal,* or *littoral, zone;* the continental shelf surface is called the *sublittoral zone* or *level bottom* and the surface of the continental slope forms the *bathyal zone.* The ocean floor forms the *abyssal plain,* which is in places interrupted by deep trenches, towering submarine mountain chains, and other less conspicuous topographic features. The *photic zone* is the portion of water that is penetrated by light. The lower limit of the photic zone varies from place to place, depending primarily on water clarity, but usually more-or-less coincides with the margin of the continental shelf. Only the upper half, that is, the upper 100 meters, has sufficient illumination for substantial photosynthesis. In the diagram an arm of the ocean extends inland to form a shallow epicontinental sea. Like the continental shelf farther offshore, it is floored by what is called a sublittoral or level bottom substratum. Most marine deposits we now encounter on continents were deposited in epicontinental seas of this type; few were deposited at abyssal depths.

Marine organisms are commonly classified according to a simple scheme, depending on where they live and whether they are capable of self-propulsion (Figure 9-4). Bottom-dwellers are called *benthos* (from which is derived the adjective *benthonic,* or *benthic*). They may be *epifaunal,* which means they live *on* the substratum, or *infaunal,* which means they live *in* the substratum. Benthonic forms capable of locomotion are called *vagile* and immobile forms are called *sessile.* Organisms living in the water above the bottom are described as *pelagic,* those whose major movements are accomplished by swimming being *nektonic* and those that are transported primarily by waves and currents, *planktonic. Phytoplankton* are plants and *zooplankton,* animals. Figure 9-5 shows a simplified version of the food cycle in the oceans. Food webs in the marine ecosystem tend to be relatively uncomplicated. The principal producers in the modern-day seas are single-celled phytoplankton, mainly diatoms and dinoflagellates. In offshore waters, phytoplankton are the sole photosynthetic producers, but on the continental shelf, where the photic zone reaches the sea bottom, benthonic plants, including marine grasses, augment food production.

The primary consumers in offshore waters are small zooplankton. Most of the adult organisms among the zooplankton of modern seas belong to two crustacean groups. In addition, a large percentage of the zooplankton in

The rest of the animals are benthonic, inhabiting a rocky substratum (right) and soft sediment. Of the benthonic animals, I, J, K and L are infaunal; the rest are epifaunal. Only the sponge, G, is entirely sessile; the chaetopterid, J, and the mussels, H, are capable of very little locomotion and can be considered sessile; the rest of the benthonic animals are vagile, although some are more active than others. A and B are grazers, feeding on benthonic algae; D, E, and F are carnivores that may also scavenge on dead animals; G, H, I and J are suspension feeders; K, L and M are deposit feeders; arrows show feeding currents.

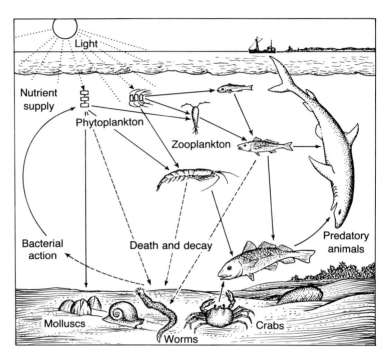

FIGURE 9-5
The food cycle of the oceans. (From Clarke, 1965.)

many areas consists of tiny larvae of invertebrates that are benthonic in their adult stages. On the ocean floor benthonic invertebrates are the primary consumers. Metabolic and decay products of plants apparently form much of the food of both zooplankton and primary benthonic consumers. The absence of plants on the abyssal plain suggests that the animals there feed for the most part on inanimate organic matter.

The biomass pyramid in many marine settings tends to be inverted (Figure 9-6). The reason frequently given to explain why zooplankton typically outweigh phytoplankton is that there is a rapid turnover of phytoplankton (a high *rate* of production). Some workers, however, believe that the metabolic products of phytoplankton constantly being released into the water form a large proportion of the food used by zooplankton and benthonic consumers.

Carnivores feeding on zooplankton are largely nektonic fishes. In shallow water, zooplankton are joined by large numbers of benthonic invertebrates, which also feed on phytoplankton. Benthonic invertebrates also live at bathyal and abyssal depths, but in less abundance because of the reduced food supply.

Marine paleoecologists concern themselves primarily with benthonic organisms because species living on or in the substratum are much more likely to be preserved with little or no post-mortem transport than are pelagic species. Pelagic species must be transported to the bottom from their native habitat to be preserved at all, and the substratum on which they fall may bear no relation to the environment in which they lived.

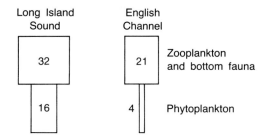

Long Island
Sound

English
Channel

32

21

Zooplankton
and bottom fauna

16

4

Phytoplankton

FIGURE 9-6
Inverse biomass pyramids for two shallow marine water bodies. Biomass figures (in gr/m²) are for the water column and sea floor beneath one square meter at the sea surface. (From Odum, 1959.)

Hunt (1925) proposed a classification of benthonic consumers that has been widely adopted (see Figure 9-4 for examples). He recognized three feeding groups of soft-substratum dwellers. *Carnivores* are "animals which feed mainly upon other animals, whether living or as carrion." (Because carrion-eating or "scavenging" is engaged in by most predators and because most scavengers eat some live animals, strict separation of scavengers and predators is unwarranted.) *Suspension feeders* are "animals which feed by selecting from the surrounding water the suspended micro-organisms and detritus." *Deposit feeders* are "animals which feed upon the detritus deposited on the bottom, together with its associated micro-organisms." To Hunt's three categories we might add a fourth, *grazers,* which feed by selectively removing organic surface films (chiefly algal coatings) from the substratum. Most grazers inhabit hard substrata; none were present in the soft-bottom fauna that gave rise to Hunt's classification.

Limiting Factors and Local Habitats

Our discussion thus far has been largely restricted to Recent ecology. In this and the following sections we will provide examples to show how ecologic data and principles can be applied to the fossil record. A principal goal in paleoecology is to reconstruct accurately the environments in which preserved animals lived. The physical, chemical, and biological properties of an environment that limit the distribution of any given species are commonly referred to as *limiting factors.*

Most benthonic invertebrates reproduce by releasing gametes into the water, where fertilization takes place. The fertilized egg develops into a planktonic larva, which aids in dispersal of the species over large areas. Thorson (1950) has stressed that the planktonic larval stage is commonly the stage in a life cycle most sensitive to limiting factors. The larva is commonly less hardy than the adult.

Food supply must ultimately be a limiting factor for all consumers. At present, however, we know little about the appetites of most suspension feeders, deposit feeders, and grazers. Thus, we lack sufficient information to understand food supply as a limiting factor for most noncarnivorous marine species. Substratum type and salinity are the primary factors that have been demonstrated to control the distribution of particular benthonic populations.

SUBSTRATUM TYPE

Hard, rocky substrata are only rarely a part of the fossil record. They tend to be swept free of sediment and normally represent sites of erosion rather than of deposition. Soft substrata are the preservation medium for most fossils. They are especially important in paleoecology because, excepting the fossils themselves, the character and distribution of sediments are the sole source of information for reconstructing ancient habitats.

The fabrics of sedimentary rocks commonly reflect depositional processes, and the geometric configurations of sedimentary rock units commonly reflect depositional settings. Sedimentary particles range in grain diameter from clay size (< 0.004 mm) through silt size (0.004–0.0625 mm) and sand size (0.0625–2 mm) to pebbles, cobbles, and boulders. Mud includes clay-size and silt-size particles. Grain size commonly reflects the degree of wave or current agitation in the depositional environment. Fine-grained sediments are usually deposited in quiet-water areas; well-sorted, coarse-grained deposits are most common in areas of strong wave and current activity. Most marine sediments are one or the other of two types: (1) terrigenous sediments are land-derived and consist primarily of silicate fragments, especially of quartz, feldspar and clay particles, and (2) calcium carbonate particles are primarily skeletal debris and aragonite needles precipitated by algae or inorganic processes. As yet terrigenous and carbonate sediments have not been shown to have differing effects on the distribution of benthonic organisms. The main effects of substratum are associated with *grain size*.

One of the most important relationships between substratum and the distribution of benthonic organisms concerns feeding mechanisms. Sanders (1956) has shown that the relative abundance of deposit feeders is greatest in muddy sediments, whereas the relative abundance of suspension feeders is greatest in sandy sediments (Figure 9-7). Sanders has explained this relationship primarily on the basis of food supply. Fine particulate organic matter for deposit feeding settles to the bottom in quiet-water areas where mud also accumulates, but is winnowed out of coarse-grained sands that are subjected to greater agitation. Sanders also postulates that suspended food supplies are not replenished rapidly enough in quiet-water areas for extensive feeding. D. C. Rhoads and D. Young, more recently, have argued that adequate food supplies actually are suspended in quiet-water areas, and have suggested that the soft mud clogs feeding apparatuses of many suspension feeders to limit their occurrence to sandy substrata.

Relationship between sediment type and feeding mechanism can be seen well back into the Paleozoic record, in which, for example, nuculoid clams (Figure 9-8) are found in many mudstones and shales. Modern nuculoid clams depend largely on deposit feeding for food gathering and live predominantly in muddy sediments. Because basic morphologic features of the superfamily Nuculacea are associated with the group's deposit-feeding mechanism, homology as well as position in the stratigraphic record suggests that most nuculoid clams of the Paleozoic were deposit feeders.

Substratum is also important relative to method of attachment and locomotion of benthonic animals. We will consider four common types of sub-

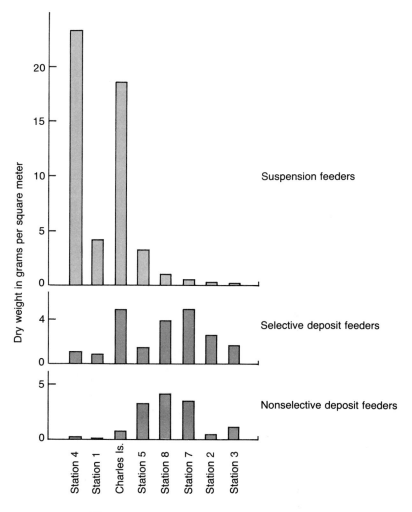

FIGURE 9-7
Abundance (measured by total dry weight of tissue) of feeding-group representatives at bottom stations in Long Island Sound. The stations are arranged from left to right in order of increasing percentage of silt plus clay. (From Sanders, 1956.)

stratum in this light. Rocky bottoms are primarily sites for the attachment of epifaunal species. Only a few infaunal species bore into rock or nestle in pre-existing crevices. Because rocky surfaces are seldom horizontal and are commonly scoured by currents and waves, attachment is necessary for most species that colonize them. Of the types of soft substrata, shifting sands are the least easily colonized by benthonic organisms. They tend to be inhabited by a few, highly mobile infaunal species that can reburrow rapidly when exhumed by currents or waves. That fossil remains are absent from or very scarce in many beach deposits of the geologic record reflects not only un-favorable conditions for preservation, but also the unfavorable conditions

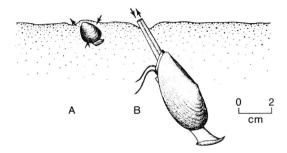

FIGURE 9-8
Nuculoid clams: A, *Nucula,* and B, *Yoldia,* both feeding by appendages of the lip region called palp proboscides; C, *Clinopisthia* (dorsal and right lateral views), a Carboniferous genus whose life habits were probably similar to those of *Nucula;* D, *Polidevcia* (dorsal and right lateral views), a Carboniferous genus whose life habits were probably similar to those of *Yoldia.* (A and B from Stanley, 1968; photographs of C and D courtesy of A. L. McAlester and E. G. Driscoll.)

C   D

for *life.* Muddy bottoms in quiet-water areas tend to be soft and soupy. Their shelled infauna consists largely of small animals, many of which have thin shells. Large, thick-shelled species would sink into soft mud; the feeding and respiratory mechanisms of many would be unable to function properly. Figure 9-9 is a plot of animal weight versus percent mud for clams from many different environments. Large, heavy species are restricted to sandy sediments. Because of problems of attachment and flotation on soupy bottoms, few epifaunal species inhabit muddy substrata.

Substrata composed of mud-sand mixtures usually differ from shifting sand and quiet mud bottoms in being both stable and firm. Commonly such

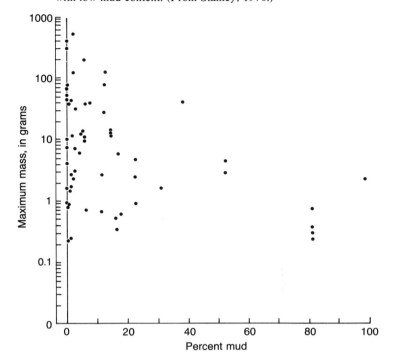

FIGURE 9-9
Maximum individual mass of bivalve species from a variety of substratum types. Large individuals are restricted to sediments with low mud content. (From Stanley, 1970.)

substrata are inhabited by a wide variety of species. Animals that form permanent dwelling burrows (tubes or channels) are largely restricted to muddy sand because of its cohesiveness.

SALINITY

Salinity, which is a measure of the dissolved salt content of natural waters, is usually measured in parts per thousand. The value for undiluted sea water is about 35 parts per thousand. Table 9-1 shows the names generally applied to natural waters of various salinities.

TABLE 9-1
Classification of Salinities in Natural Waters

Descriptive Term	Salinity in parts per thousand
Fresh water	0–0.5
Brackish water	0.5–30
Sea water	30–40
Hypersaline water	40–80
Brine	> 80

Source: Kinne, 1964

The effects of salinity on organisms vary from species to species, with respect to both the mean and the range of salinities tolerated. A species with a narrow tolerance range is said to be *stenohaline,* and one with a broad range, *euryhaline.*

The most striking ecologic effect of salinity is on local species diversity. The largest number of aquatic species are marine forms. Many of these are stenohaline, often living in offshore areas where they are never subjected to appreciable salinity changes. A smaller number of aquatic species are stenohaline freshwater forms. Relatively few species can live in brackish or hypersaline water, but many species that tolerate either tolerate both. Brackish and hypersaline water is found in nearshore bodies of water (such as bays and lagoons), which do not undergo thorough or rapid enough mixing with the open ocean to reach a stable salinity of about 35 parts per thousand. (The Baltic Sea, Figure 9-10, is a good example.) The salinity fluctuates with tidal movements, floods, storms, and seasonal climatic changes. Inhabitants of such unstable and stressful nearshore environments must be euryhaline. Apparently there have been relatively few euryhaline species at any time in geologic history. The few species that live in most brackish and hypersaline environments usually are present in great abundance because of reduced interspecific competition for ecologic niches. Most brackish-water species have evolved from marine, rather than from freshwater, forms. Few are capable of living in freshwater though many can tolerate normal marine salinity.

Lowered species diversities owing to brackish and hypersaline environments can commonly be recognized in the fossil record on the basis of stratigraphic evidence documenting nearshore depositional environments. Especially important are contiguous terrestrial, lake, or river deposits, which can often be recognized by fossil content, sedimentology, and regional stratigraphy.

Most hypersaline water bodies exist in warm, arid climates. Their sediments commonly include minerals, such as halite, gypsum, and anhydrite, that were formed by chemical precipitation due to evaporation. Such sediments are sometimes associated with dune deposits. Nearby river delta deposits, plant remains, or lake beds may be diagnostic evidence for brackish rather than hypersaline conditions.

Sometimes the salinity of a shelled marine species' habitat is recognizably correlated with shell form. With decreasing salinity, the European cockle *Cardium edule* grows a thinner shell with fewer ribs. This relationship may derive from increasing difficulty of calcium carbonate secretion, but is not well understood, and we should hesitate to apply it to the interpretation of fossil species. For an extinct species the reverse relationship might hold! Furthermore salinity tolerance is a much less fundamental trait in evolution than basic morphologic adaptations related to feeding or locomotion, for example.

Extrapolating backwards in time from a living species to a related fossil species with respect to salinity tolerance is also dangerous. One genus that appears to have been euryhaline for many millions of years is the inarticulate

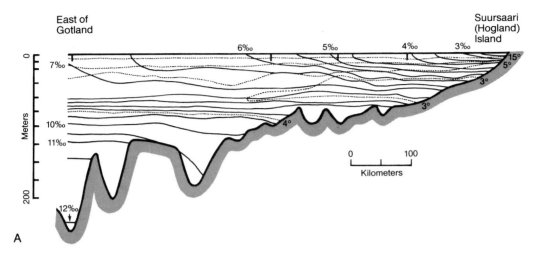

East of
Gotland

Suursaari
(Hogland)
Island

A

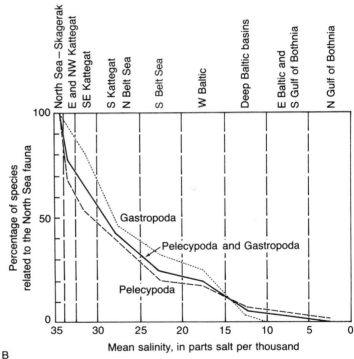

B

FIGURE 9-10
Relation of marine species diversity to salinity in the Baltic Sea.
A: Schematic cross section of the Baltic Sea from inner reaches
of the Gulf of Finland to Gotland Deep, July, 1933. Solid
contours show salinity in parts per thousand (‰) and dotted
contours show temperature in degrees Centigrade. (From Jurva,
1952.) B: Molluscan percentages and salinity gradient along
a slightly different transect. (From Sorgenfrei, 1958.)

brachiopod *Lingula* (Figure 9-11), which today lives in nearshore habitats, often in brackish water. The fossil record of this "living fossil" extends back at least to the Ordovician Period. A wealth of evidence suggests that *Lingula* occurs throughout its stratigraphic range primarily in nearshore faunas, which are characterized by low species diversity.

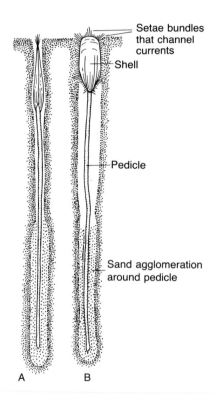

FIGURE 9-11
Lingula in its burrow. A: Lateral view. B: Dorsal view. (From Hyman, 1959; after P. Francois.)

Hudson (1963) has related faunal distribution and diversity to salinity for the Jurassic Great Estuarine Series of England. He has delineated several faunal assemblages, each representing a group of species that tend to occur together in certain rock types. Hudson's conclusions are summarized in Figure 9-12. The assemblages do not form a simple stratigraphic sequence and they have not been traced laterally to determine how they may inter-grade, but two independent lines of evidence have pointed to Hudson's interpretation. One is the salinity tolerances of certain of the fossil genera that survive today. *Unio*, a freshwater clam, and *Viviparus*, a freshwater snail, are both abundant in many regions today. *Mytilus* and *Modiolus* are common mussels in modern seas, inhabiting brackish and normal marine waters, but most *Modiolus* species cannot tolerate salinities as low as those tolerated by *Mytilus* species. *Liostrea* is similar to Recent oyster species that occupy restricted, low-salinity bays. Hudson assumes there has been no

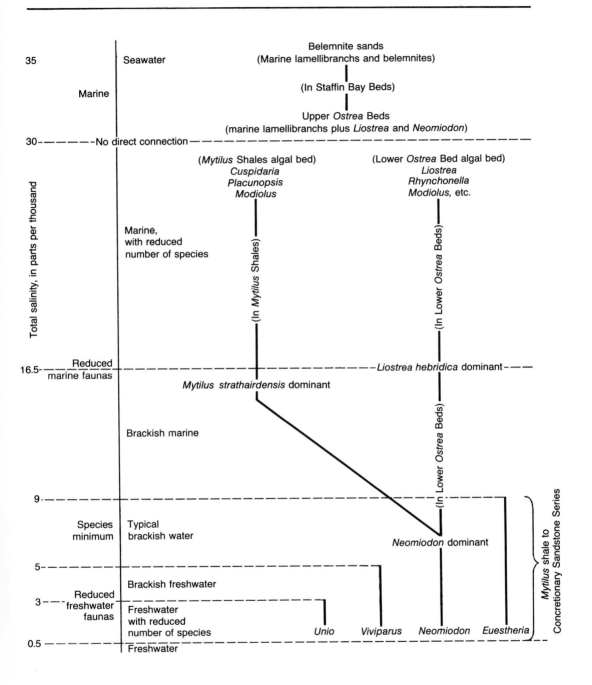

FIGURE 9-12

Salinity-determined faunal assemblages in the Great Estuarine Series of the British Jurassic. (From Hudson, 1963.)

major change in the absolute salinity of the ocean since the Jurassic. Arrangement of still-living genera in a sequence according to their present-day salinity tolerances (and inclusion of associated fossil species) produces the pattern shown in Figure 9-12. The scale of approximate salinities, which is determined by comparison with living relatives, yields important supporting evidence. The minimum species diversity falls in the lower brackish range (5–9 parts per thousand on the chart), and the maximum diversity falls in the higher brackish and marine range. Taxonomic diversity data suggest that the genera with living representatives have not tended to alter significantly their salinity preferences in the course of about 150 million years.

Certainly some aquatic groups, throughout their geologic history, have been largely stenohaline and have lived in normal marine water, while others have tended to be euryhaline. It is generally thought, for example, that most corals, cephalopods, articulate brachiopods, and echinoderms have been stenohaline marine. Ostracods, bivalves, and gastropods have tended to inhabit both freshwater and marine environments and many species of these groups have been euryhaline. Arenaceous foraminifera today are especially common in nearshore, low-salinity environments and are thus widely used as paleoecologic indicators.

From our knowledge of *Lingula* and the genera discussed by Hudson, we might argue that all species of a genus have approximately the same salinity tolerance. Still we must be wary of oversimplification. Numerous living genera contain species with a variety of salinity tolerances. It is certainly not safe to make generalizations about salinity tolerances within a family. Many mollusc families, for example, show considerable variation in salinity tolerance. Some, such as the Ostreidae (oysters), Mytilidae (mussels), and Neritidae (intertidal snails), have demonstrated a general tendency to produce brackish-water taxa, but all have also produced stenohaline marine forms. Strict taxonomic arguments should be combined with species diversity data and sedimentologic and stratigraphic arguments in any attempt to determine approximate salinities of ancient depositional environments. In most cases only crude conclusions can be reached even when all approaches are combined. We are doing well simply to distinguish between fossil assemblages from freshwater, brackish water, marine, and hypersaline environments. The farther back in the fossil record we venture, the poorer is our resolution.

WATER DEPTH

Although it is one of the most valuable parameters for reconstruction of environments of the past, water depth is very difficult to determine, even in an approximate way, from marine fossil assemblages. Ancient water depth is important to know because it defines the configuration of marine basins and contributes to our picture of paleogeography for particular intervals of geologic time. Water depth affects organisms most directly through hydrostatic pressure, but exerts relatively little direct control over the distribution of marine life, except over larger depth intervals than are commonly represented in paleoecologic studies. Furthermore, water depth has little effect on solid skeletons, so that adaptation to pressure is not directly correlated with skeletal strength. Rapid pressure change does affect soft tissue. Most living animals brought rapidly upward from the abyssal plain in bottom-sampling devices suffer from the rapid decrease in pressure. Although many taxonomic groups have diversified to inhabit a variety of water depths, taxonomic and functional morphologic relationships to water depth are rare.

Indirectly, water depth may operate to affect species distribution through such depth-dependent parameters as light intensity, salinity, temperature, dissolved oxygen, and food supply. For example, many living benthonic species inhabit deeper water toward the equatorward part of their geographic range than toward the poleward part because the shallow waters toward the equator are too warm for them.

Because no food may be produced by photosynthesis below the photic zone, food supply decreases markedly with depth below about 200 meters. It has long been recognized that the abundance of organisms decreases downward across the continental slope to the abyssal plain. Recently, Sanders, Hessler, and co-workers (1965; 1967) at the Woods Hole Oceanographic Institution have confirmed this trend in a large-scale study with modern sampling devices, but have demonstrated greater abundance of benthonic life in the deep sea than was previously recognized. In addition they have found taxonomic diversity in the deep sea to be much higher than suggested by earlier workers. Implications of this high diversity will be discussed in a subsequent section (pages 241–245). A sidelight of the Woods Hole study is the demonstration of a marked dominance of deposit-feeders in deep-sea deposits. The low concentration of suspended food makes suspension feeding relatively unprofitable.

Light intensity is obviously important in controlling plant distribution in the marine realm, as suggested by the very presence of the photic zone, in which light intensity decreases with depth to zero. Light toward the red end of the wavelength spectrum is more rapidly absorbed than light toward the blue end. While most plants can apparently utilize some light of all wavelengths, different species have different wavelength optima. Among the marine algae, the taxonomic group called the "green algae" tends to absorb radiation preferentially at the red end of the spectrum and is therefore largely restricted to intertidal and shallow subtidal depths. "Red algae" extend much deeper into the subtidal realm. Both groups contain species that secrete calcium carbonate skeletons and that are represented in the fossil record. The presence of untransported fossil green algae generally indicates ancient water depths of less than 60 or 70 meters.

For marine paleoecology one of the most important ecologic effects of light intensity is on the depth distribution of reef-building corals. There are two ecologic groups of modern corals (the Scleractinia, or hexacorals). Ecologic differences between the two groups are related to the presence of certain algae, called *zooxanthellae,* in the soft tissues of one group. The precise physiologic relationship remains uncertain, but there is evidence that the algae may aid the corals by assimilating metabolic wastes, by helping in secretion of the calcareous skeleton, or both. At any rate, large, massive colonial skeletons are restricted among hexacorals to those containing zooxanthellae. This group, called **hermatypic** ("mound-variety"), are the frame-builders of tropical coral reefs. Species lacking zooxanthellae are called **ahermatypic** corals. The light requirements of zooxanthellae restrict hermatypic corals to shallow water. Figure 9-13 shows that, although hermatypic corals occasionally inhabit water depths greater than 200 feet, they seldom build reefs at depths exceeding 150 feet (their reefs seldom flourish at depths greater than about 90 feet). True fossil coral reefs are therefore among the best indicators of shallow-water depositional settings. Modern reefs are formed by large numbers of coral species, in association with calcareous algae and a rich invertebrate fauna. The deep-water coral "banks" of Figure 9-13 are built by a few ahermatypic species in association with a low-diversity invertebrate fauna and no calcareous algae. On this basis, Teichert (1958) has concluded that what were formerly thought to have been coral reefs in the Late Triassic of southern Alaska, comprising skeletal material from only eight coral species, must actually have been comparable to the modern-day deep and cold-water coral banks off the coast of Norway.

We must be careful in applying the limiting factors of modern-day coral reefs to fossil reefs. Enough Late Triassic and Early Jurassic taxa survive today to warrant the assumption that the earliest (mid-Triassic) reef-building hexacorals contained zooxanthellae and had approximately the same depth limits as modern forms. Major Paleozoic reef-building groups, like the tabulate corals and stromatoporoids, were taxonomically distinct from modern hexacorals. We may never know whether they lived in association with zooxanthellae, but the bulk of independent evidence suggests that Paleozoic reef and reef-like structures, which exhibit high species diversity, grew in shallow water. Many such structures restricted water circulation in back-reef areas, suggesting that they grew to, or nearly to, sea level.

Fossil sponges have been shown to be useful indicators of water depth, though they are abundant in relatively few sedimentary rocks. Sponges are sessile, epifaunal suspension feeders (Figure 9-4, G). Living sponges of the class Calcarea, in which the supporting skeletal elements (called spicules) are calcareous, are most common on the sea bottom at depths of less than 100 meters, although a few species live in deeper water. Living sponges of the class Hexactinellida, in which the spicules are siliceous, are most common in the depth range of 200–300 meters; few species therefore occur on continental shelves and almost none live at depths of less than 10 meters (Reid, 1968). Finks (1960) has shown that the same ecologic relationships existed as long ago as the late Paleozoic. In the Permian rocks of Texas he has recognized shallow-water "shelf" sponge faunas, in which species of the Calcarea abound, and deep-water "basin" sponge faunas, in which the Calcarea are lacking but hexactinellids are numerous.

One of the sharpest ecologic breaks in the bathymetry of the oceans lies between the intertidal and subtidal zones. Species in the intertidal zone are

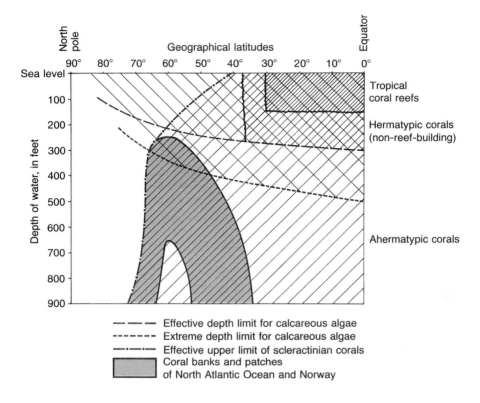

FIGURE 9-13
Depth and latitude ranges of most modern coral reefs, hermatypic corals, and deepwater coral "banks," (From Teichert, 1958.)

subjected to rapid and extensive fluctuations in the physico-chemical environment, especially fluctuations in temperature and salinity. On soft bottoms many intertidal species (especially species of annelid worms, crustaceans, and bivalve molluscs) have become deep burrowers for protection. Evidence that such species lived in a certain place may be provided in the fossil record by preserved burrows. Figure 9-14 shows the deep burrow of the crustacean *Callianassa,* which lives today along intertidal sandy shores and has left a record of burrows extending back at least into the Cretaceous. Throughout its geologic range the genus is an excellent indicator of intertidal deposits. Seilacher (1964) and Rhoads (1967) have contrasted deep intertidal burrows with the shallow horizontal burrows of deep-water species, which have less to gain from deep burial. Seilacher suggests that most deep-sea burrows belong to deposit feeders, as we might expect (page 211), and follow patterns that permit systematic coverage of sediment layers containing food (Figure 9-24).

FIGURE 9-14
Callianassid decapod crustaceans. A: A living *Callianassa* exposed in its burrow.
B: External surface of a Recent burrow (×1). C and D: Similar borrows found in the
Cretaceous Fox Hills Sandstone of the Denver Basin. (Photos provided by R. J. Weimer
and J. H. Hoyt. A, B, and D from Weimer and Hoyt, 1964.)

A

B

C

D

Stromatolites, which are typically layered cabbage-like structures formed by a succession of algal mats (Figure 9-15), are usually taken to indicate intertidal deposits. A single mat is commonly produced by one or more species of blue-green algae. The algae are thread-like structures covered with sticky mucilage, which traps detrital calcium carbonate (they do not secrete a true skeleton). After the surface of the mat is coated, filaments grow through the sediment layer, to be covered again. Stromatolites are very hardy plant assemblages. Their fossil record extends well back into the Precambrian, which makes them especially useful to paleoecology. Their external form commonly reflects such environmental factors as local current or wave activity.

Determination of water depth of deposition for ancient sedimentary deposits and fossil assemblages must be approached through many lines of evidence. Frequently the final designation reflects general distance from shore (such as "nearshore" or "offshore") rather than actual water depth. Intertidal deposits can often be recognized from physical and chemical evidence, such as mudcracks and fine-grained dolomite, as well as from biologic information. Regional stratigraphy and associated evidence commonly permit recognition of shallow-water lagoonal deposits. Sandy-beach deposits may be recognized by a variety of characteristics, including sediment grain size, bedding features, and low species diversity. Deep-sea deposits may also be identified by certain sedimentologic and biologic features. Fine resolution, however, is seldom possible. It is especially difficult to estimate depth of deposition for level-bottom sediments and faunas of epicontinental-sea and continental-shelf deposits, which unfortunately, form the bulk of our sedimentary rock record.

Spatial Distribution of Populations

We have already noted that populations tend to be distributed in characteristic spatial patterns.

Many benthonic marine species tend to be aggregated in clusters of individuals. There are two main causes for this type of distribution pattern: inhomogeneities in the physical environment and the reproductive and social behavior of the species. The second of these causes appears to be the more important. A striking illustration is provided by species of marine grass that carpet shallow-water areas in many parts of the world (Figure 9-16), propagating partly by means of runners that spread laterally beneath the sediment surface. Once an individual plant grows from a seed in a previously barren area its runners may produce a discrete grass patch. Many submarine grass patches are separated by areas of barren sediment. Grass alters the bottom environment by stabilizing the sediment and commonly acts as a baffle-like trap for fine-grained sediments and also offers new substrata (plant stems and roots) for colonization by a variety of organisms. Certain species of clams, snails, sea urchins, and worms are largely restricted to the "grass-flat" habitat.

A

B

C

D   E

FIGURE 9-15
Living and fossil stromatolites. A: Recent stromatolites, Shark Bay, Western Australia.
The stromatolites grow in the littoral zone of a hypersaline lagoon with their tops at
the high-water mark. Small, actively growing immature forms are in the right foreground.
The stromatolites are well laminated internally and are composed of calcareous oolites
and foraminifera, seen as rippled sand between the stromatolites and derived from the
offshore sublittoral shelf. The sand is driven onshore by waves and then trapped and bound
by the sticky mucilagenous sheaths of the filamentous blue-green (nonskeletal) algae
covering the stromatolites. The stromatolites are rapidly cemented by inorganically
precipitated aragonite. These stromatolites occur around an exposed headland, where they
are subjected to strong wave action. The scale in the center-left middle-ground is five
feet long. B: Recent stromatolites from the same area as those in A, but occurring along
a somewhat more protected coastline and oriented parallel to the direction of wave
surge (perpendicular to the shoreline). The shovel in the right center gives the scale.
C: Bedding plane view of columnar stromatolites. Elongation is parallel to the ancient
current direction and perpendicular to the ancient shoreline. Proterozoic, Great Slave Lake
area, Canada. Compare with B. D: Cross section of branching columnar stromatolites
normal to bedding; part of a sixty-foot thick bed of columnar stromatolites that now consist
of dolomite. The vertical part of the scale is one foot. Proterozoic, Great Slave Lake area,
Canada. E: Cross section of connected, dome-shaped stromatolites now composed of
dolomite and chert. Divisions on the scale are tenths of feet. Proterozoic, Coppermine River
area, Canada. (Provided by Paul Hoffman.)

Mode of formation of local populations is another factor causing patchy distribution of benthonic invertebrates. Most species produce large numbers of larvae that spend variable periods of time (averaging about three weeks) as members of the plankton before settling to the bottom to metamorphose. There is much debate about the ability of larvae to choose suitable sub-strata for settling, and also about the effects of unfavorable benthonic con-ditions in delaying settling. At any rate, larvae of a given species often tend to be transported together, as a population, by currents and to settle in one area. Often, too, they are gregarious when settling and are attracted to adult animals of their species, which have obviously enjoyed some degree of success where they settled. Thus, successful settlement is often in clumps whose distribution appears to be unrelated to basic inhomogeneities in the botton environment.

Benthonic species that lack a planktonic larval stage are even more likely to be found in patchy distributions. The ontogeny of some such species includes "brooding" of eggs and larvae and release of juveniles near the parent.

Finally, some vagile species tend to aggregate as adults, apparently in areas in which food supply or other factors make the environment especially suitable for habitation.

Some aggregation of benthonic species is caused by inhomogeneities of the physical benthonic environment. Substratum, rather than salinity and temperature, is especially important here. The common mussel *Mytilus* attaches to hard objects by a byssus. Even in sandy areas a few "pioneer" mussels can settle on scattered shell fragments or pebbles. Their shells commonly form attachment sites for additional mussels. The result may be an aggregated distribution of mussels over large areas (Figure 9-17). Clumps of Paleozoic brachiopods seem to have formed in the same way, as docu-mented by Ziegler and co-workers for certain Silurian groups. Accumula-tions of shells of dead brachiopods nucleated colonies of several species, some of which attached to the shells of other living individuals. The Pale-ozoic brachiopod genus *Pentamerus* left numerous colonies of closely packed individuals preserved in life position (Figure 9-18). It is uncertain whether adults were attached by **pedicles** (fleshy stalks possessed by many brachiopod genera) or rested free on the sediment. At any rate their orienta-tion with the pedicle region directed downward suggests that they did not attach to each other, but simply aggregated during larval settlement. (The adult brachiopod had no means of locomotion.)

In environments in which patchy marine populations have been preserved in place, their fossil assemblages may retain the original distribution pat-tern. A common modifying factor is time. In areas in which the location of plankton, and hence that of the larval stage of the organisms, shifts from year to year and sedimentation rates are low, the record of preservation over many years may tend to even out inhomogeneities in the distribution of single year classes.

It is important to understand that while there must always be environ-mental factors explaining the success of a certain species in a particular

FIGURE 9-16
Marine turtle grass in the Mediterranean Sea off the coast of Spain. Note the roots exposed at the margin of the area where currents have scoured out the bottom. Such scoured areas usually support fewer benthonic animal species than are found in adjacent grassy areas. (Photograph supplied by E. A. Shinn.)

setting, the absence of the species in other similar settings may be largely the result of chance, principally because of the vagaries of larval transport and the transient, short-lived nature of many local populations. A uniform depositional environment does not necessarily support a uniformly distributed population or produce a uniformly distributed fossil assemblage.

Life Habits

We have been concerned with limiting factors and habitats of living and fossil benthonic species. We will now consider the life habits or modes of life of species within their respective habitats. For fossils only certain types of information may be within our grasp: What was the organism's orientation in life? Did it move, and if so, how? Was it attached to the substratum or free-living? How did it obtain food? How did it reproduce?

FIGURE 9-17
Clumps of the mussel *Mytilus edulis* on a broad intertidal sand flat
at Barnstable Harbor, Massachusetts.

DIRECT EVIDENCE THROUGH PRESERVATION

The most direct way to learn about the life habits of extinct fossil species is to observe them preserved in the midst of some life activity. An example is shown in Figure 9-19. Many such examples represent unique preservational circumstances, such as sudden, catastrophic burial.

HOMOLOGY

A less direct approach to understanding life habits is to infer them by homology (page 166). It has often been said, for example, that were it not for the existence of living turtles, we would have no concept of the life habits of fossil turtles. We might suspect that the turtle shell served a protective function for some bizarre creature, but we certainly would never be able to deduce from their fossil remains the differences in life habits of a largely

FIGURE 9-18
Undersurface of a cluster of the brachiopod species *Pentamerus oblongus* preserved in life position, Silurian Red Mountain Formation, Alabama (×0.45)
(From Ziegler, Boucot, and Sheldon, 1966.)

FIGURE 9-19
The Devonian cystoid *Adocetocystis williamsi* preserved in life position,
attached to an ancient erosion surface within the Shell Rock
Formation, Iowa (×1.3). (From the first published report of fossil cystoids
unequivocally attached by stems—Koch and Strimple, 1968.)

terrestrial box turtle and a fresh-water snapping turtle. Certainly the deductions described in Chapter 8 for ammonites could never have been made without information gained from study of the Recent *Nautilus*. Many years of endeavor would probably have been required simply to ascertain that ammonoids were cephalopods. Only recently have preserved cephalopod jaws or "beaks" been found associated with fossil ammonites.

FUNCTIONAL MORPHOLOGY

Functional morphology, a subject discussed more fully in Chapter 8, represents a major source of evidence for the reconstruction of life habits, when coupled with inferences about habitat derived from sedimentologic evidence. An example is the study by Grant (1966) of the brachiopod genus *Waagenoconcha,* found in the Permian of Pakistan. Juvenile animals bear convergent attachment spines near the beak that were apparently used for clinging to stalk-like benthonic plants, sponges, or bryozoans on which the larvae apparently settled (Figure 9-20). Adults lived free on fine-grained sediment. From direct preservational evidence and from the great weight of the structure, Grant concluded that the spiney ventral valve lay undermost. The spines apparently served primarily to support the suspension-feeding animal on a soupy mud bottom, although Grant suggested that they also served an anchoring function. Like snowshoes, they prevented a dense object from sinking into a soft underlying medium. The plight of the juveniles, had they not been attached to plants, is predictable. Their small size would scarcely have elevated their commissure line above the mud. Any major disturbance of the bottom would probably have clogged their feeding mechanisms, interrupted normal activities, and led to the early demise of most individuals.

INTIMATE SPECIES ASSOCIATIONS

Many life habits of fossil organisms can be deduced from preservation showing physical associations between species. An example is a specimen of the ammonite *Buchiceras* described by Seilacher (1960) from the Cretaceous of Peru (Figure 9-21). The specimen is encrusted on both sides and on the outer whorl with oysters. The location of the oysters shows that they did not attach to the dead shell on the ocean bottom, for they are evenly spread over both sides. They show a preferred orientation with respect to

FIGURE 9-20
Reconstruction of the life habits of the productoid brachiopod *Waagenoconcha abichi* from the Permian of Pakistan. Spat are attached to idealized algae. (From Grant, 1966.)

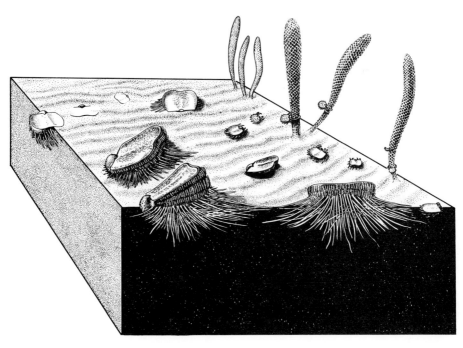

FIGURE 9-21
Fossil specimen of the ammonite *Buchiceras bilobatum* with attached oysters.
A: Left lateral view. B: Right lateral view. C: Rear view. D: Life orientation
suggested by growth direction of oysters attached to left side (solid arrows)
and to right side (broken arrows). (From Seilacher, 1960.)

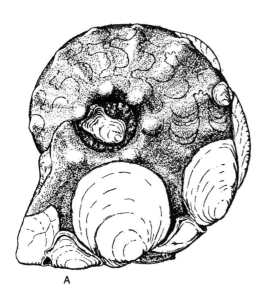

A

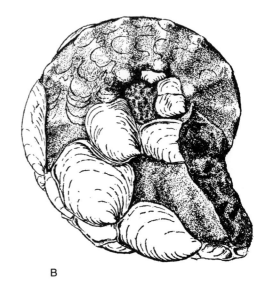

B

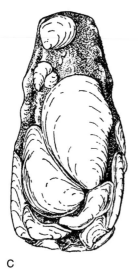

C

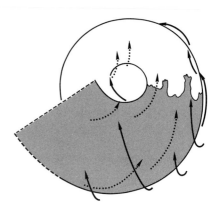

D

slope of the attachment surface, indicating a life orientation for the ammonite that agrees with the one suggested for similar forms on a theoretical basis by Trueman (1941). Furthermore, the location of oysters on the undersurface of the shell rules out the crawling, benthonic mode of life suggested for the genus by an earlier worker. The oysters would have been frequently disturbed, and probably damaged, had their host bumped along at the sediment surface in life. Seilacher's conclusion is that the genus was a swimmer whose life position was similar to the living *Nautilus*.

Organisms attaching to other organisms are called **epibionts, epiphytes** being plants with this habit and **epizoans,** animals. The chief problem in using epibionts as indicators for the life habits of their host species is one of establishing whether attachment took place before or after death of the host. As in the example above, both the location and orientation of epibionts may be useful in this regard. Commonly suspension-feeding epizoans, such as worms secreting calcareous tubes, occupy positions near the region of the host where water is drawn in; here they take advantage of the increased current flow (Figure 9-22). Regions of current flow are the only parts of the shells of some aquatic animals that are exposed above the sediment surface in life; sometimes epibionts therefore indicate not only current-flow region but life position of the host species.

FIGURE 9-22
Spirorbid worm tubes preferentially attached to the current-flow regions of nonmarine Carboniferous bivalve shells. Rows of tubes tend to follow growth lines, suggesting that juvenile worms attached at the shell margin. (From Trueman, 1942.)

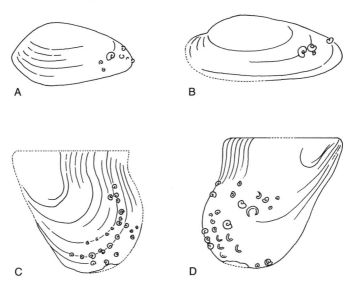

A

B

C

D

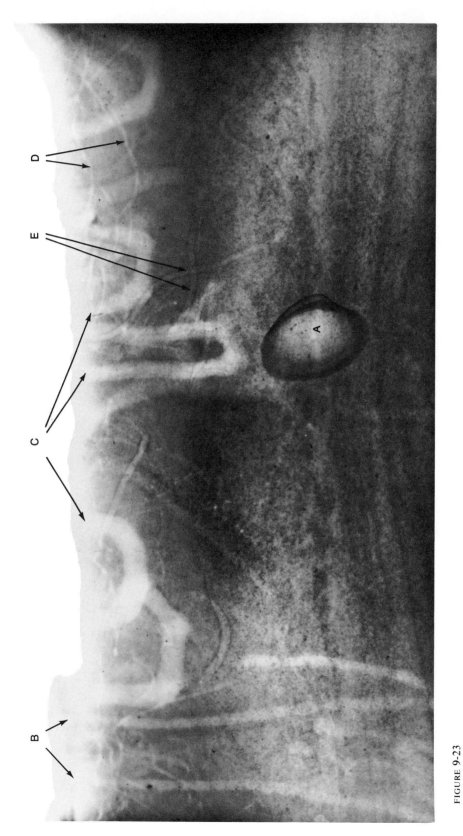

FIGURE 9-23

X-radiograph of a benthonic fauna in native sediment (muddy sand) collected from 40 feet of water, Quissett Harbor, Massachusetts, illustrating the paleontologic importance of burrow preservation (×1). A: The bivalve *Pitar* (shell preservable). B: Arms of a brittle star feeding at the sediment surface (skeleton not likely to be preserved intact). C and D: Burrows of amphipods (noncalcareous skeletons not likely to be preserved). E: Burrows of a polychaete (soft-bodied). (Courtesy of D. C. Rhoads.)

EVIDENCE OF BIOLOGIC ACTIVITY

Even when not represented by skeletal parts, many organisms have left fossil evidence of their presence and way of life in the form of tracks or trails. The Germans, who have contributed most to the study of these features, call them *Lebensspüren* (traces of life). Trace fossils were produced by movements and can therefore reveal much about the organism's behavior. Formation of many types of trace fossils can be studied in the Recent, both in nature and in the laboratory. One fruitful method makes use of X-ray photographs of animals burrowing in thin aquaria. Progressive movements of animals can be recorded by time lapse (or motion picture) photography (Figure 9-23).

One pioneer student of trace fossils, Rudolf Richter, analyzed the form shown in Figure 9-24. It was produced by a worm-like deposit feeder that moved in a special pattern through fine-grained deposits to systematically ingest sediment, without passing the same material through its system twice. Four "rules" were necessary to produce the observed pattern. The animal had to: (1) tunnel horizontally through a single sediment layer, (2) make a U-turn after moving a certain fixed distance, (3) move close to one or more previous segments of its tunnel system, and (4) maintain a certain minimum distance from previous tunnel segments. The two final "rules" tended to produce a pattern in which segments of the tunnel system were separated by uniform distances. The "rules" must have taken the form of genetically coded instructions. The required instructions can be programmed for a computer to simulate the observed pattern. Presumably, animals in nature carry out their genetic instructions by responding to chemical and tactile stimuli in the surrounding medium.

Some Lebensspüren mark resting places of animals, others represent more-or-less permanent shelters (Figure 9-14), and some are merely crawling tracks.

Interesting stories are also told by tracks of terrestrial vertebrates. Much of the thick Triassic redbed sequence of the Connecticut Valley region was deposited in swampy lowland areas. Tracks on bedding surfaces reveal the presence of several dinosaur species (Figure 9-25).

FIGURE 9-24
The trace fossil *Helminthoidea labyrinthica*, found in Cretaceous and Eocene sediments in the Alps and Alaska. (From Seilacher, 1967, after R. Richter.)

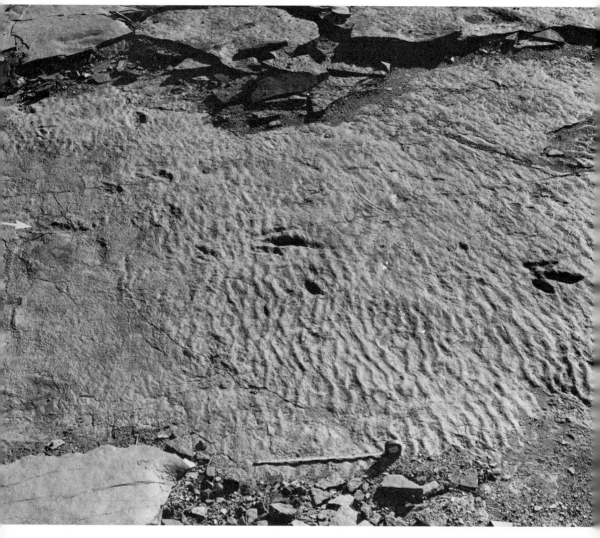

FIGURE 9-25
Triassic dinosaur tracks from Rocky Hill, Connecticut. The animal walked from a firm
mud bank, where rain prints are visible (arrow points to shallow track), into a pond
or stream whose soft bottom-mud was covered by ripple marks (note the two deeper
tracks). (Courtesy of J. H. Ostrom.)

   Trace fossils are especially important in providing evidence of past life
where no skeletal remains are found. Connecticut Valley Triassic rocks
yield virtually no dinosaur bones, yet abundant tracks attest to the existence
of a diverse dinosaur fauna. The warm humid climate of the region in which
the redbeds formed was apparently unfavorable for skeletal preservation,
despite the fact that the region was a depositional basin receiving thousands

of feet of sediment over a period of several million years. Otherwise barren marine sedimentary rocks also commonly show evidence of abundant benthonic life through the presence of trace fossils.

Fossil Communities

We have considered life habits and local distributions of populations belonging to single species. Commonly these topics are united under the heading *autecology,* the study of the interactions between single species and their environments. The study of interrelations between two or more species and their environment is called *synecology.* Actually, part of the environment of any species is biologic, consisting of other species. Autecology and synecology are, therefore, not clearly separable, but represent different emphases in the study of ecology.

Consideration of species living in association with each other has led to the concept of the biologic *community.* Most modern workers speak of a community as any natural assemblage of species living in a given area. This definition will be adopted here.

Since it has long been recognized that certain groups of plants and animals tend to occur together in nature, some workers have specified that to be a community, an assemblage must recur in many areas with little variation in species composition. We will instead consider such an association to be a special type of community called a *recurrent community.*

Finally, some ecologists have suggested that the label "community" should imply interaction and interdependence among component species. In some instances, strong interaction among neighboring species is obvious. In others, evidence for interaction of certain species pairs is lacking. It is generally agreed that interaction should be excluded from the definition of "community."

In terrestrial settings and along boundaries between land and water, *ecologic succession* sometimes takes place. Succession is a continuous sequential change in the communities inhabiting a given area, ideally from an initial "pioneer" community to a final "climax" community. With the appearance of the climax community, an approximate equilibrium is reached and there is then little further biotic change until some major external environmental change interferes. Usually succession is seen in a region following a major, rapid environmental change (such as that brought about by a forest fire or withdrawal of a glacier), which leaves it barren for colonization by a "pioneer" community. The major reason for ecologic succession appears to be that certain communities tend to alter their environment so that it becomes better suited for other communities. The phenomenon is especially common among rooted plants, which alter the soil, and commonly interfere with the path of sunlight.

For several reasons, ecologic succession is not common in the marine realm. In the first place succession must occur in a fixed area. Pelagic species swim and drift from place to place in a fluid medium. On the sea floor the environment is stationary, but most benthonic species have short life spans, and in many areas populations appear and disappear from year to year, depending on many factors, including the vagaries of larval transport. The important producers in most areas are not rooted plants, but single-celled phytoplankton that do little to alter the benthonic environment physically. Ecologic succession of benthonic communities apparently occurs only where organisms like submarine grass (page 215) physically alter the bottom environment in a major way. Perhaps the most dramatic examples of marine ecologic succession are provided by organic reefs, which we will consider in some detail.

ORGANIC REEFS

Marine paleoecology emerged as a distinct branch of paleontology during the 1950's. Many of the early marine paleoecologic studies dealt with organic reefs, partly because of commercial interest in ancient reefs as traps for petroleum and partly because of scientific curiosity about reefs themselves.

The term "reef" has been defined and redefined by geologists. The most important characteristic included in most definitions is the presence of a rigid organic framework that stands, or once stood, above the adjacent ocean bottom. Organic reefs today are produced largely by frame-building hermatypic corals in association with encrusting calcareous algae and many minor constituents, and are among the most spectacular organic communities in the beauty and diversity of their component species (Figure 9-26).

Ancient reefs offer three major advantages for paleoecologic analysis: (1) most of the reef-forming species have rigid skeletons that are readily preserved; and (2) the skeletons, in forming a rigid framework, binding the framework together, or being trapped in it, are commonly preserved without post-mortem transport, many being preserved in life position; and (3) reefs form a continuous record of community composition through time. Reefs are also important geologically in that they alter the marine environment, often in neighboring areas as well as where they grow. Most reefs grow upward to, or very near to, sea level (page 212) and shelter back-reef areas from waves and currents, permitting fine-grained sediments to accumulate. Commonly zonation in species composition is evident across a reef from the windward to the leeward side, because the reefs' obstruction of waves and currents produces an "energy gradient" (Figure 9-27).

Many reefs grow for thousands of years, forming their own fossil record. Often the reef community changes as upward growth approaches sea level. Lowenstam (1950) studied Silurian reefs in the Great Lakes region of the United States that were formed by groups of organisms which are now extinct. The reefs were mound-like structures in which ecologic succession occurred during growth. As shown in Figure 9-28, the quiet-water stage, built by the pioneer community, consisted of relatively few groups of organisms. As the reef grew upward into rough water, the faunal composition changed and diversified; more robust species took over. A climax community formed the final wave-resistant stage near sea level. Ecologic succession is virtually never seen in the fossil record for non-reef-forming communities.

FIGURE 9-26

A coral reef knoll in a few meters of water near North Rock, Bermuda. The scale is 50 centimeters long. The large head of brain coral at the lower right and the one at the left center belong to different species of the genus *Diploria*. The sheet-like encrusting coral colony below the scale belongs to the genus *Montastrea*. The branching, tree-like forms are alcyonarian sea whips and sea fans. (Photo supplied by Peter Garrett.)

FIGURE 9-27
Cross section of the windward reef of Eniwetok Atoll, showing current velocity gradient and species zonation. (From Odum, 1959, and Odum and Odum, 1955.)

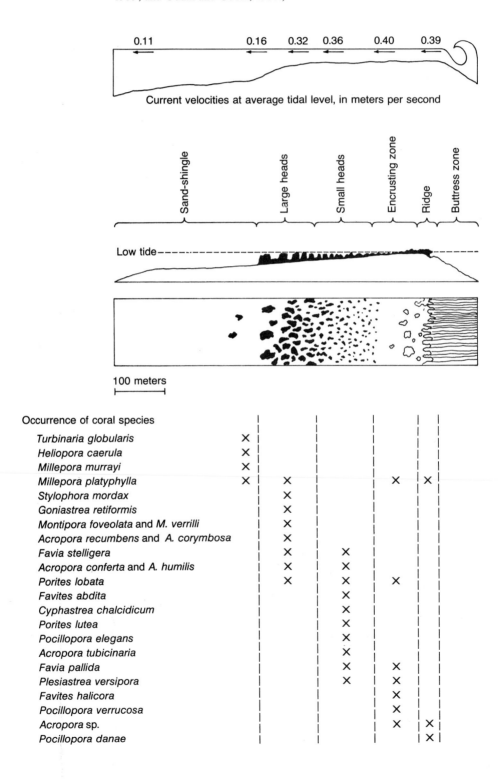

Current velocities at average tidal level, in meters per second

0.11 0.16 0.32 0.36 0.40 0.39

Low tide

100 meters

Occurrence of coral species

	Sand-shingle	Large heads	Small heads	Encrusting zone	Ridge	Buttress zone
Turbinaria globularis	X					
Heliopora caerula	X					
Millepora murrayi	X					
Millepora platyphylla	X	X		X	X	
Stylophora mordax		X				
Goniastrea retiformis		X				
Montipora foveolata and M. verrilli		X				
Acropora recumbens and A. corymbosa		X				
Favia stelligera		X	X			
Acropora conferta and A. humilis		X	X			
Porites lobata		X	X	X		
Favites abdita			X			
Cyphastrea chalcidicum			X			
Porites lutea			X			
Pocillopora elegans			X			
Acropora tubicinaria			X			
Favia pallida			X	X		
Plesiastrea versipora			X	X		
Favites halicora				X		
Pocillopora verrucosa				X		
Acropora sp.				X	X	
Pocillopora danae					X	

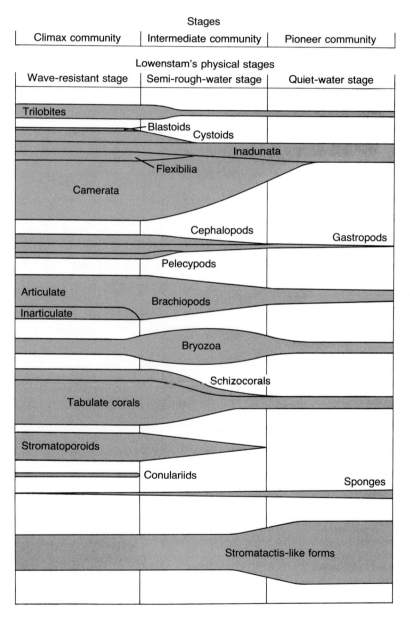

FIGURE 9-28
Three stages in the ecologic succession recognized by Lowenstam
for Silurian reefs of the Great Lakes region of the United
States. (From Nicol, 1962.)

SOFT-BOTTOM COMMUNITIES

The idea of recurrent benthonic communities was established by Peterson (1913), based on extensive studies in the North Sea. Peterson observed that species tend to cluster into groups, each of which is *recurrent*—that is, occurring in many separate areas on the sea bottom. Peterson emphasized that the recurrent communities he recognized were merely statistical entities. He avoided any implication that each of the recurrent communities existed because of ecologic interdependence among the species forming it. His view was that the species forming a recurrent community are simply species that have similar environmental requirements. Some workers have suggested that there is inadequate statistical evidence for the presence of recurrent communities, pointing out that just as no two benthonic environments are identical, no two communities have identical species compositions. Most workers now believe that the concept of recurrent communities is valid for at least some soft-bottom environments, but more data than have been gathered will be needed for the concept to be evaluated definitively.

We would expect the recurrent-community concept to be most valid for communities with considerable species interdependence, such as reef communities, in which many species depend on others as substrata. Many ecologists believe, as Peterson did, that relatively little interdependence exists among consumers in soft-bottom communities, and that most benthonic consumers exist together only because they happen to prefer similar environmental conditions. A lack of attached producers, the shortness of food chains, and year-to-year fluctuations of species abundance all suggest that, except for predator-prey and symbiotic relationships, relatively little interdependence exists among species of soft-bottom communities. The Russian worker Turpaeva (1957) studied interaction through competition for food. She found that of all species that feed by a particular method in a single habitat, one species tends to outweigh the others by a large amount, in terms of biomass. This situation illustrates the ecologic postulate often called the *competitive exclusion principle,* which states that if two coexisting species require an environmental resource that is available in limited supply, the species that is the superior competitor for the resource will greatly overshadow the other. Given enough time, the inferior competitor may be eliminated from the habitat. In a particular community at any given time there may exist small numbers of inferior competitors that have not yet been excluded. For the communities described by Turpaeva, food appears to be an environmental resource that is in limited supply. Many of the species listed in Table 9-2 that constitute a small portion of the biomass for each feeding group may be inferior competitors for food. In many soft-bottom benthonic communities a very few species comprise 90 percent or more of the total biomass. The reason appears to be that most habitats offer favorable opportunities for feeding by only a few methods. With one species tending to predominate among those employing each feeding method, only a few

TABLE 9-2

Dominant Group of Species in the *Astarte crenata* Biocoenosis
of the Barents Sea (mean of 49 stations)

Species	Feeding Type[a]	Biomass, in grams per square meter	Number of Individuals per square meter
Astarte crenata	Filterer-A	60.80	23
Spiochaetopterus typicus	Collector	5.08	106
Phascolosoma margaritaceum	Swallower	5.00	3
Cardium ciliatum	Filterer-A	3.94	0
Ctenodiscus crispatus	Swallower	2.49	3
Macoma calcarea	Collector	1.84	13
Bryozoa	Awaiter	1.54	—
Saxicava arctica	Filterer-A	1.47	1
Nephthys sp.	Swallower	1.45	6
Ophiocantha bidentata	Collector	0.66	2

[a] *Filterers-A* are suspension feeders that filter the extreme bottom layer of water. *Collectors* are selective deposit feeders that collect detritus from the sediment surface. *Swallowers* are non-selective deposit feeders that ingest sediment. *Awaiters* are suspension feeders that trap food from water currents; they lack a pumping system of their own.

species will flourish. Lacking rooted plants, complex food chains, and a stable environment, many soft-bottom communities never reach the state in which inferior competitors are totally excluded.

A pioneering study in the field of community paleoecology is that of Ziegler et al. (1968) for Silurian deposits of Great Britain. This study recognized five communities that were apparently recurrent in space and time. They are recognized both on the basis of species composition and geographic distribution (Figure 9-29). The five communities, which intergrade, represent an approximate nearshore-offshore sequence, numbered 1 through 5. While communities 1 and 2 or 4 and 5 may have several species in common, 1 and 4 or 2 and 5 are mutually exclusive in species composition.

The community nearest shore is named the *Lingula* community (Figure 9-30). *Lingula* was described on page 208 as having been a predominantly nearshore genus since the early Paleozoic.

If we analyze this Silurian *Lingula* community in terms of Turpaeva's feeding-group model, we observe that the most abundant species, the articulate brachiopod *"Camarotoechia" decemplicata,* was an epifaunal suspension feeder. *Lingula,* the second-most-abundant species, was an infaunal suspension feeder that fed at the sediment surface. *Palaeoneilo,* a nuculoid bivalve, was, by homology with similar living forms (Figure 9-8), a deposit feeder. The three species obviously belonged to distinct feeding groups.

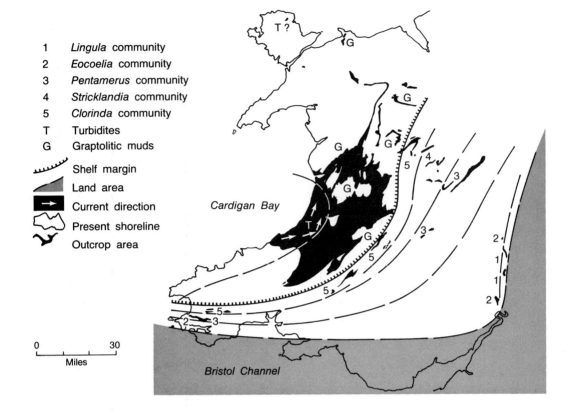

1   *Lingula* community
2   *Eocoelia* community
3   *Pentamerus* community
4   *Stricklandia* community
5   *Clorinda* community
T   Turbidites
G   Graptolitic muds

Shelf margin
Land area
Current direction
Present shoreline
Outcrop area

Cardigan Bay

0   30
Miles

Bristol Channel

FIGURE 9-29
Early Upper Llandovery (Silurian)
paleogeography of Wales, showing
distribution of five recognized benthonic
communities in zones parallel to the
ancient shoreline. (From Ziegler, 1965.)

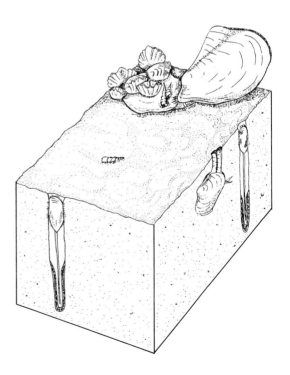

FIGURE 9-30, A
Artist's reconstruction of the *Lingula*
community of Figure 9-29.

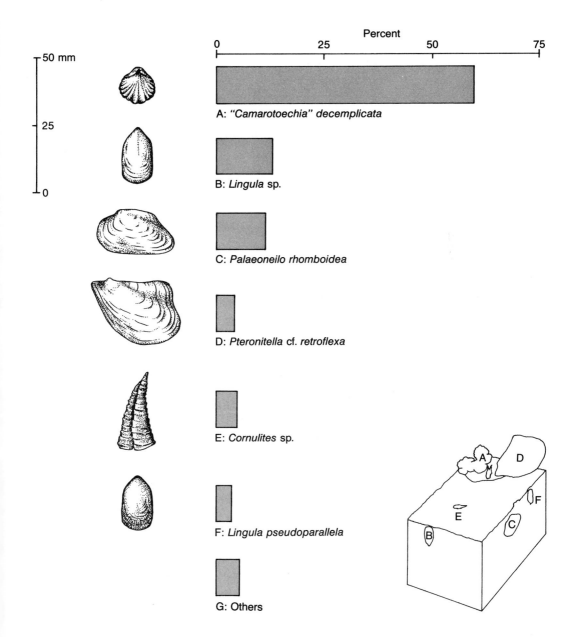

Percent

FIGURE 9-30, B
Relative species abundance in the *Lingula* community of Figure 9-29
(Both A and B from Ziegler, Cocks, and Bambach, 1968.)

Even if we are reasonably certain that a fossil assemblage represents a true fossil community, two important sources of error may invalidate the results of our analysis: feeding-group information may be inadequate (we may not know the water zone in which certain brachiopods fed, for example); or the preserved "community" may not include any evidence of soft-bodied species that actually were among the dominant forms. The example of the *Lingula* community certainly fits the Turpaeva model well. The feeding-group assignments for the three species are almost certainly accurate, and their abundances are great enough to suggest that they were actually among the dominant species. Still, we are treading on thin ice because of preservation problems. Furthermore, an analysis based on Turpaeva's work would be more difficult for the offshore communities of the Silurian study because of their greater diversity and our limited knowledge of the life habits of their component species.

The Silurian *Lingula* community has additional significance. By homology, we can suggest that the Silurian *Lingula* species lived in a nearshore setting that had a fluctuating salinity (page 208). The lower species diversity of the *Lingula* community relative to the four offshore communities also suggests stressful salinity conditions (page 206). These conclusions are supported by stratigraphic evidence. The *Lingula* community occurs on paleogeographic maps (Figure 9-29) adjacent to a land area, where fluctuating salinity might be expected.

Temperature and Biogeography

What controls the distribution of species over large areas of the earth's surface? What restricts some species to equatorial regions, and others to polar regions? What limits certain species to a single continent or ocean basin? We are concerned here with *biogeography* — the study of distribution patterns of entire species.

The primary limiting factor in geographic distribution is temperature. Temperature has only minor effects on *local* distribution, except where rapid and pronounced temperature fluctuations occur (as in the intertidal zone). Its main effect is on *latitudinal* species distribution. One of the few environments in which temperature has little effect on biogeography is the deep sea, which everywhere remains near freezing.

Important biogeographic factors other than temperature are discussed elsewhere in this text because they are especially important to our understanding of organic evolution and biostratigraphy. For the present, we shall largely confine our discussion to temperature-related problems.

Most paleoclimatic interpretations are based on nonbiologic evidence. Glacial deposits, dune deposits, tropical limestones, and many other rock types are useful climatic indicators.

Mammals and birds are capable of maintaining nearly constant body temperatures, but nearly all other taxonomic groups assume their environmental temperature. Both "warm-blooded" and "cold-blooded" organisms tend to be restricted to certain temperature ranges, which are usually determined by limits for survival or successful reproduction. Species that are tolerant of a broad range of temperature are called *eurythermal,* and those that are restricted to a narrow range, *stenothermal.*

Several potential sources of error hinder climatic and biogeographic reconstruction for the geologic past. Recent studies of paleomagnetism have persuaded many geologists that continents have shifted on the earth's mantle in the course of the earth's history. At present, patterns of past continental movements are so poorly understood that it is virtually impossible to piece together worldwide climatic schemes for pre-Cretaceous, and perhaps even for pre-Cenozoic time intervals. Still, local climatic conditions for any age can be interpreted from appropriate rocks and fossils.

Such features as major ocean currents commonly disrupt simple climatic patterns. Ocean temperatures along the southeastern coast of the United States are much warmer than temperatures at corresponding latitudes along the West Coast. Because of the Coriolis Effect, the warm Gulf Stream sweeps north along the Atlantic Coast, while the cold California Current sweeps south along the Pacific Coast.

TEMPERATURE AND MORPHOLOGY

Temperature is an environmental parameter similar to salinity in having little effect, for the most part, on skeletal morphology. Still, the average size of individuals belonging to a single species tends to be larger in the cold part of the species' geographic range than in the warm part. Appendages, such as ears and tails of mammals, also tend to be shorter in the colder part of a species' range. These relationships, known respectively as Bergmann's Rule and Allen's Rule, have not been extensively applied to fossils because of preservation biases. Both relationships are apparently associated with increase in the volume to surface ratio of species inhabiting cold climates.

Nearly all spiney molluscan groups and many groups with thick shells inhabit tropical and subtropical regions. This relationship may well reflect the relative ease of calcium carbonate secretion in warm water.

TEMPERATURE AND HOMOLOGY

Most paleoclimatic studies that have used fossil data and homological interpretations have dealt with faunas and floras of the Cretaceous and Cenozoic. Most have made use of fossil plants, which are excellent climatic indicators for as far back as the abundant fossil record of *flowering* plants extends (mid-Cretaceous).

Fossil floras indicate that the Pacific Coast of the conterminous United States, at least as far north as Oregon, had a subtropical climate in the early Cenozoic. Alaska supported a temperate or perhaps even a subtropical flora. Younger floras attest to a general cooling up to the Pleistocene and Recent (Figure 9-31). The fact that many species shifted geographically together as temperatures changed supports the conclusions based on homology for single species. One potential pitfall, however, stems from the fact that floras of high elevations at middle latitudes are similar to those in low elevations at high latitudes. Thus, a local fossil flora might erroneously suggest that a region formerly had a temperate climate when, in fact, the given flora was from a high-altitude environment surrounded at lower altitudes by tropical floras that would actually be the proper indexes to the former climate of the region. Fortunately, low-altitude floras, which tend to be the best indicators, are the more likely to be preserved, by the very topography of their environment.

Coral reefs grow only in areas where the temperature seldom falls below 18°C. Nearly all living reefs lie within about 30° of the equator (Figure 9-32). Their fossil counterparts have been used extensively as paleoclimatic indicators of former tropical, or near-tropical, climates. Even Paleozoic reefs, formed by diverse communities in regions of limestone deposition, are commonly taken to indicate warm seas. Durham (1950) amassed a huge volume of data on the geographic distributions of Cenozoic fossil reef corals and molluscs of the Pacific Coast of North America. His conclusions (Figure 9-33), based on homology (mostly at the genus level), corroborate the paleobotanical evidence cited in the previous paragraph.

FIGURE 9-31
Northern limits of the subtropical climatic zone in North America during various time intervals of the Cenozoic Era. The positions of the limits, which reveal a general trend of climatic cooling, have been estimated from study of fossil floras. (Altered from Dorf, 1960.)

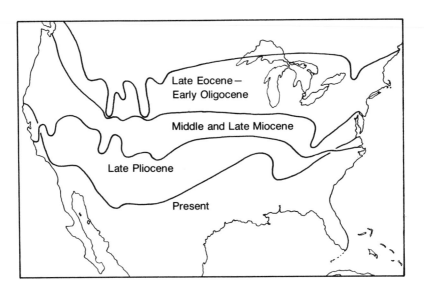

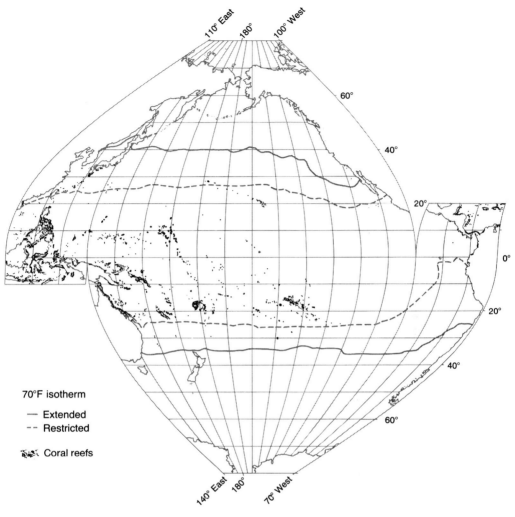

FIGURE 9-32
Latitudinal limits of living coral reefs (dashed lines) and hermatypic corals (solid lines) in the Pacific Ocean. (After Wells, 1957.)

TAXONOMIC DIVERSITY GRADIENTS

Taxonomic diversity gradients are perhaps the most striking worldwide biogeographic patterns related to temperature. Species diversity for most higher taxonomic categories increases markedly from the poles toward the equator (Figure 9-34).

Fischer (1960b) has pointed out that the nonbiologic environment offers no more niches in tropical regions than in temperate regions, although diversification may be somewhat more rapid in the tropics. He and others have stressed the idea that diversity is the product of evolution, and that interruptions in evolutionary diversification prevent environments at all latitudes

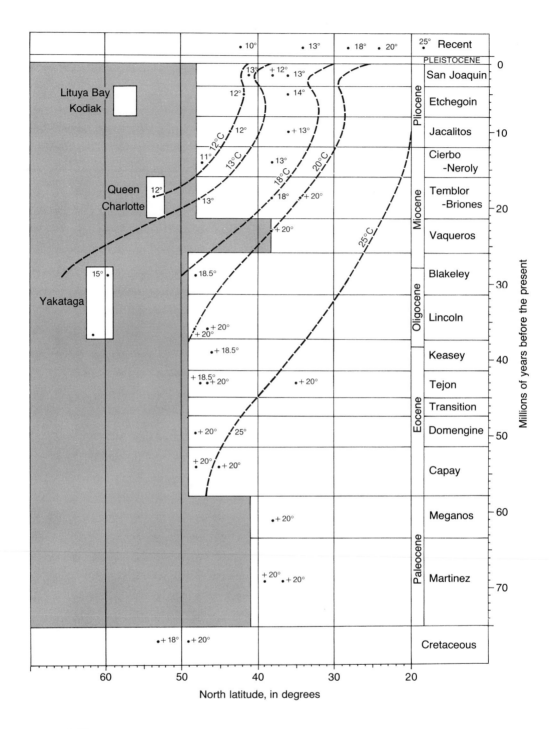

FIGURE 9-33
Past positions of marine winter isotherms (lines of equal temperature) along the
Pacific Coast, estimated from fossil occurrences of molluscs and reef corals that have
close relatives alive today. The general shift of the isotherms toward the equator
during the Cenozoic reflects a general cooling of the climate. Compare Figure 9-31.
(From Durham, 1950.)

from reaching saturation in terms of species diversity. Interruptions, chiefly due to climatic changes, are far more frequent and destructive in cold climates than in the tropics. Major extinctions have also perhaps been more frequent at higher latitudes. Through more continuous, uninterrupted evolution, tropical communities more closely approach maturity and tend to maintain higher species diversities. The idea that environmental stability, rather than temperature *per se,* is the major factor controlling species diversity has received support from recent studies of deep-sea diversity (page 211). Diversity of benthonic species at abyssal depths is not low, as previously believed, but comparable to that of shallow-water tropical regions! Although deep-sea temperatures are only slightly above that of freezing, they are extremely uniform and have apparently permitted relatively uninterrupted diversification for many millions of years.

Still, the origin of latitudinal diversity gradients has not been explained to the satisfaction of all workers. Pianka (1966) has reviewed the environmental stability hypothesis and others, and stressed that a combination of factors may be responsible.

FIGURE 9-34
Diversity gradient of shallow-water gastropods species along the East Coast of North America. Each line represents ten species. The ranges of twelve species continue through the Artic Ocean and into the North Pacific. (From Fischer, 1960b.)

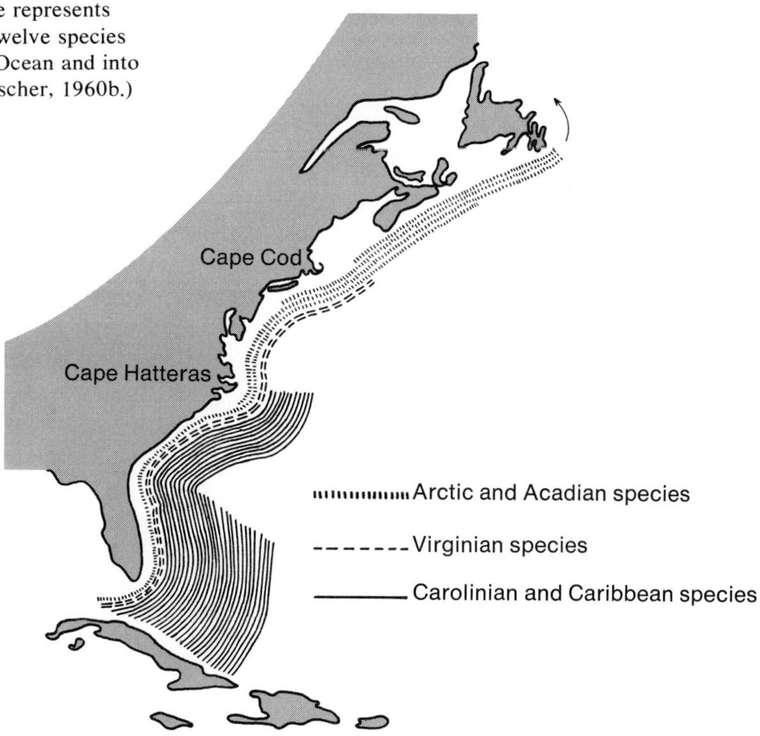

Cape Cod

Cape Hatteras

⸿⸿⸿⸿⸿⸿ Arctic and Acadian species

– – – – – – Virginian species

———————— Carolinian and Caribbean species

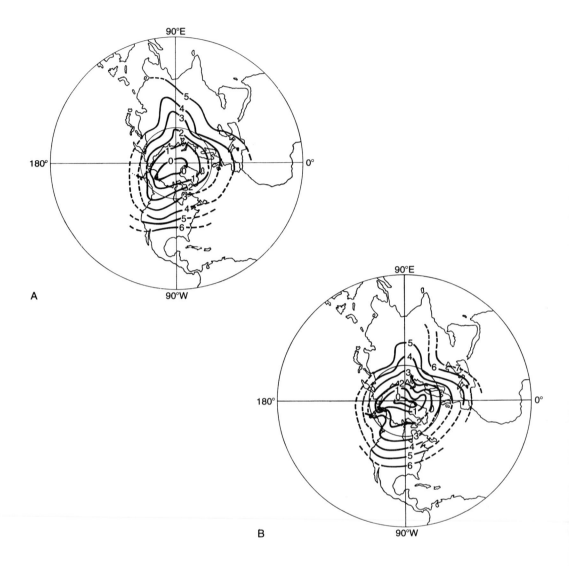

FIGURE 9-35
Estimated positions of the Permian North Pole using diversity gradients
of brachiopod genera belonging to two families. Diagram A is for terebratuloids
and B is for orthotetaceids. The concentric lines in each diagram pass
through regions of equal generic diversity, based on data from the known
fossil record. The number for each line indicates number of fossil genera.
The lines are dashed in regions now covered by the ocean and therefore not
providing data. The pole positions suggested by the diversity gradients for the
two brachiopod groups are nearly identical and nearly coincide with the present
pole position. (From Stehli, 1964.)

Valentine (1968) has introduced the idea that latitudinal diversity gradients, whatever their cause, may have produced higher *worldwide* species
diversity when strong temperature gradients have prevailed than when
worldwide equitable climates have existed. At the latter times, particular
faunas may have ranged over large geographic areas; with steeper gradients,
narrower belts of latitudinally restricted faunas may have existed. A surprising corollary of Valentine's model is that worldwide fluctuations in
climate should have produced increased diversity, and times of climatic
warming should have been times of large-scale extinction (of cold-adapted
faunas).

Stehli has attempted to use taxonomic diversity gradients as indicators
of ancient positions of the poles. By compiling diversity information for
particular taxa at specified times in the geologic past, mathematical analysis
can be used to predict ancient pole positions, assuming that latitudinal gradients existed then as now (Figure 9-35). Diversity data must be judiciously
gathered. Numbers of species recorded for rock units may be a function of
former local environment rather than former latitude. They may also reflect
thoroughness of collection and degree of taxonomic "lumping" or "splitting." In addition, recent geophysical evidence favoring the idea of extensive
continental drift in the geologic past may complicate our attempts to make
use of diversity gradients for fossil taxa.

Post-mortem Information Loss

It is convenient for us to divide the history of an organism from its birth to
our discovery of its fossilized remains into time intervals, and to give different names to the study of the various intervals (Figure 9-36). In its strict
sense, paleoecology concerns only the interval between the organism's birth
and death. Although the names given here for the study of the later intervals
have not been widely adopted, they do present a useful classification. Instead of "Post-mortem Information Loss," this section of the chapter might
have been named "Taphonomy."

FIGURE 9-36
Classification of paleontologic disciplines
concerned with environments of fossil
organisms between their time of birth or
hatching and discovery as fossils.
(From Lawrence, 1968.)

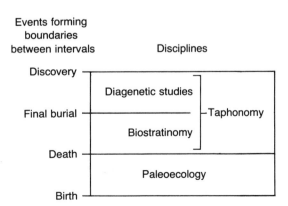

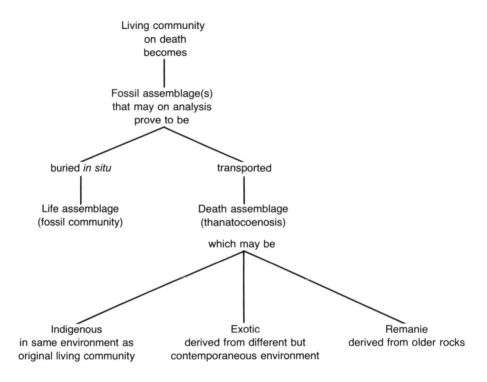

FIGURE 9-37
Classification of fossil assemblages according to mode of origin.
(From Craig and Hallam, 1963.)

We have used "fossil assemblage" to mean a group of fossils (of one or more species) found together in the stratigraphic record. A further qualification should be added. In order to be evaluated as a unit, the assemblage should be nearly homogeneous or uniformly heterogeneous in species composition. In the vertical dimension assemblages are usually restricted to a single stratigraphic bed or bedding plane, but may extend through several beds. Horizontally, they may extend for centimeters, meters, or (rarely) kilometers. Vertically and horizontally, an assemblage may give way, abruptly or gradually, either to barren strata or to other assemblages. The homogeneous, or uniformly heterogeneous, composition of a fossil assemblage results from its unique ecologic and preservational history.

Many systems of classification have been used to distinguish between various types of fossil assemblages. One such system is shown in Figure 9-37. We will refer to a fossil assemblage composed entirely of species belonging to a single community and preserved in the environment where they lived as a *life assemblage*. If a fossil assemblage is composed entirely of species transported from the environment where they lived, even if they lived together, we will refer to the fossil assemblage as a *death assemblage*. Assemblages that contain species that lived in two or more habitats will be termed *mixed death assemblages*.

Two items of information are generally sought in studying the taxonomic composition of a fossil assemblage. The first is what species are present. The second is the relative abundance of these species at the time when they lived. Studies on invertebrates, vertebrates, and plants have indicated that the second item can seldom be determined even approximately.

As discussed in Chapter 4, many criteria have been used to determine whether fossils are preserved where they lived. Preservation in life position is a sure clue for ruling out post-mortem transport. An attribute that makes both coral reefs and trace fossils especially valuable in paleoecologic interpretation is that they represent in-place preservation. Occasionally, other kinds of communities or populations may suddenly be buried in a blanket of sediment with the result that many individuals are preserved in life position.

That fossils in marine benthonic settings were subjected to post-mortem transport can often be recognized from the fossils themselves. Among the most important effects of transport are disarticulation, breakage, wear, and size sorting. Johnson (1960) analyzed these effects for three preservation models, which are presented in Box 9-A. The table lists the preservation features postulated for each model. The models, of course, represent positions in a spectrum. Most real fossil assemblages differ from all three, but many resemble one more closely than the other two.

A fruitful attempt to circumvent the difficulties imposed by the fact that many terrestrial vertebrate assemblages are mixed death assemblages has been made by Shotwell (1958) for Pliocene faunas of Oregon. Shotwell analyzed assemblages from several collecting sites in what are apparently nearly contemporaneous portions of a single stratigraphic unit. He assumed that all species found at a single collecting site lived in the region, but that some were preserved near where they lived, while others were transported considerable distances. For each species completeness of skeletal preservation was used as an index of transport distance. In Shotwell's method completeness of preservation for a species at one site is calculated as:

$$\frac{\text{Total number of skeletal elements}}{\text{Minimum number of individuals preserved}}$$

where the minimum number of individuals is the smallest number that could account for all the skeletal elements collected. For example, if nine elements were found for a species and all were different except two, the minimum number of individuals would be two. Ratios for the various species are standardized by adjusting them for the number of preserved elements that are potentially recognizable for that species. A fossil assemblage can then be represented by a pie diagram (Figure 9-38, A). The number of degrees of the sector representing a species indicates the species' relative abundance. The percentage of the radius in the blackened portion of the sector indicates the corrected number of specimens per individual (completeness of preservation). A somewhat arbitrary division of species into those that lived in the vicinity of the collecting site and those living some

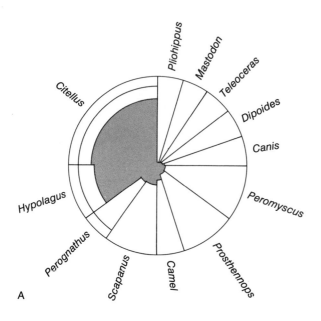

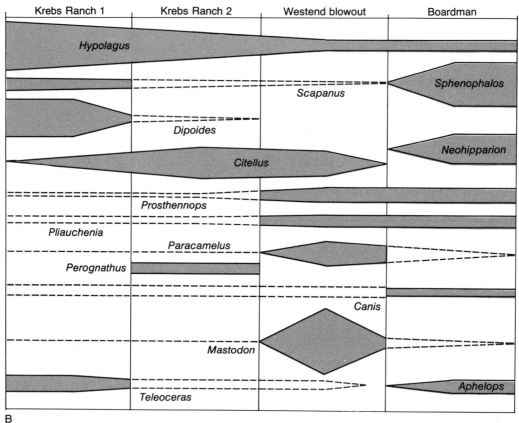

FIGURE 9-38
Pliocene terrestrial vertebrate communities of Oregon. A: Pie diagram in which the concentric line inside the margin distinguishes local inhabitants of the collecting site (Krebs Ranch 2) from species whose remains were transported to the site from a considerable distance. B: Communities at four neighboring sites. Solid bars represent common local inhabitants, and dotted lines represent occasional visitors. (From Shotwell, 1958.)

FIGURE 9-39
Interpretations of major habitats represented by the faunas of Figure
9-38. (From Shotwell, 1958.)

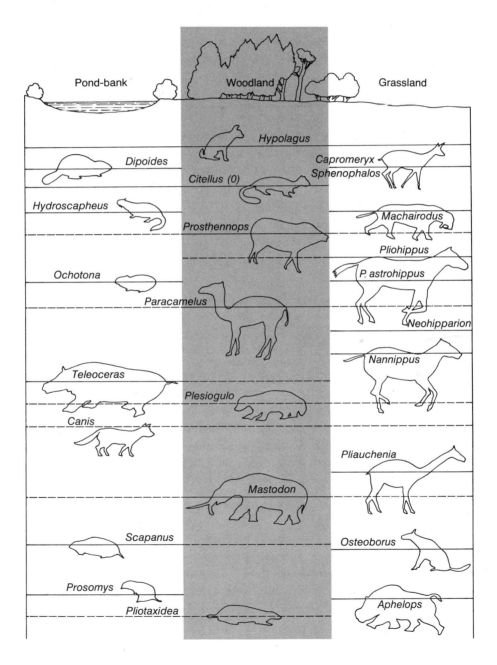

distance away is made according to whether the number of skeletal elements per individual of a species is smaller or greater than the mean for the entire group. The concentric line inside the circle margin indicates the species considered on this basis to have been local inhabitants.

Figure 9-38, B shows the general pattern of local species distribution among four major collecting sites. Nearly all of these fossil taxa have close living relatives. By homology, the life habits and habitats of these Pliocene genera can thus be established with a high degree of certainty. Such reconstructions support the analysis summarized in Figure 9-38, A by suggesting distinct habitats for the local communities, with a geographic transition from a pond and pond-bank habitat to a woodland habitat, and finally to a grassland habitat (Figure 9-39).

It is difficult to witness the formation of terrestrial vertebrate fossil assemblages in the Recent. A look at the famous Pleistocene La Brea tar pits of California is instructive, however. The tar pits formed a local death trap for animals. Certain species came to prey on dead or dying animals, and many of these then became mired in tar. Fossil remains of certain carrion-eaters thus form an inordinately large percentage of the fossil assemblage. Among them were huge "dire wolves" and saber-tooth cats, which were certainly active carnivores in other habitats. Skeletons of many of the saber-tooth cats show signs of injury, which suggests that they may have been incapacitated and thus forced to feed on carrion at the tar pits. The La Brea deposit has greatly biased our fossil sample in terms of relative abundance. Taken at face value the fossil assemblage spuriously suggests an inverted food pyramid (Figure 9-40).

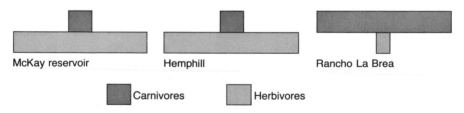

FIGURE 9-40
Relative proportions of carnivores and herbivores at two Pliocene collecting sites (McKay Reservoir, Oregon, and Hemphill, Texas) and the Pleistocene Rancho La Brea tar deposit of California. The latter spuriously suggests an inverted food pyramid. (From Shotwell, 1955.)

It is true that the local Pliocene faunas of Oregon analyzed by Shotwell exhibit normal food pyramids, with carnivores being far less abundant than herbivores. While this gross picture may approximately represent relative feeding-group abundance in life, the relative abundance of individual fossil species may well represent a strong preservational bias. As with invertebrates, preservability of species has varied according to life habits and skeletal durabilities.

BOX 9-A   Preservation Models

MODEL I: A community lives in a restricted area of the sea bottom below the low-tide mark but above the maximum wave base. The water mass moves over the bottom at a low velocity which is occasionally raised to erosional competency. The substrate is composed of clastic sediments. Intermittently, small amounts of similar sediments are introduced into the area together with minor amounts of the durable remains of organisms. The abundant and diverse life consists of soft-bodied organisms and organisms bearing hard parts. Continually and under the conditions of normal mortality, elements of this community die and become potential fossils. These decompose; some are buried after varying periods of exposure. Then, suddenly, the entire community is buried and killed by the rapid introduction of a large amount of sediment not unlike the material of the former substrate. The sedimentary body containing the fossil assemblage is gradually compacted as a result of further accumulation of sediment at the site.

MODEL II: A community lives in an environment similar to that described for Model I. Continually and under the conditions of normal mortality, elements of the community die and become potential fossils. These decompose; some of the remains are carried away, whereas others of durable composition are buried after varying periods of exposure. In time, the local environment changes and the

community is eventually replaced by another of different composition. The fossil assemblage of interest becomes more deeply buried, and the sedimentary body becomes gradually compacted as a result of further accumulation of sediment at the site.

MODEL III: A community lives in a restricted area of the sea bottom below the low-tide mark but above the maximum wave base. The water mass moves at a moderate velocity over a bottom of clastic sediments consisting in large part of the durable remains of organisms. Frequently the velocity of the water mass is high enough to move sediment and organic detritus through the area. The hydrodynamic circumstances favor the accumulation of organic remains at the site. The sparse fauna consists of a few epifaunal species of scavengers, boring and encrusting organisms. Continually and under the conditions of normal mortality, elements of this community die and become potential fossils. These decompose; some of the remains are carried away, whereas others of durable composition are buried after varying periods of exposure. Eventually, the rate of accumulation of organic remains at the site decreases as changes occur in the source areas of the sediment and debris. In time, the zone containing the high concentration of durable remains is buried and gradually compacted as a result of further accumulation of sediment at the site.

(continued)

BOX 9-A *(continued)*

Relative Expressions of Features Developed under the Conditions of the Models

Only the most distinctive or most probable development is shown.

Feature	Expression		
	Model I	*Model II*	*Model III*
Faunal composition	Ecologically coherent assemblage of species	As in Model I	Not necessarily as in Models I and II
Morphologic composition	Delicate structures and heterogeneous shapes and sizes may be preserved. Suites of shapes represented are ecologically consistent.	As in Model I	May consist only of the durable parts of the species present; may be homogeneous in shape and sizes
Density of fossils	Wide range of densities possible	As in Model I	High
Size-frequency distribution	Many species exhibit a size-frequency distribution conforming to an ideal distribution for an indigenous population	Some species as in Model I	Not as in Models I and II
Disassociation	High proportion of articulated remains; disassociated parts represented in appropriate relative numbers for a species	Moderate proportion of articulated remains; disassociated parts as in Model I	Not as in Models I and II
Fragmentation	Low proportion of remains are fragments.	Moderate proportion of remains are fragments.	High proportion of remains are fragments.
Surface condition of fossils	Surfaces of preserved structures as in life.	Various states of wear represented	As in Model II
Chemical and mineralogical composition	No general expectations are warranted by present knowledge in this area.	As in Model I	As in Model I
Orientation	Some species may retain orientation as at time of death.	Majority of fossils oriented with long axis parallel to bedding plane (some exceptions)	As in Model II (some exceptions)

Feature	Expression		
	Model I	Model II	Model III
Dispersion	Articulated remains of some species may retain pattern of dispersion as in life.	Not as in life	As in Model II
Sediment structure and texture	Consistent with inferred tolerances of the fauna and with relatively quiet waters	As in Model I	Not necessarily consistent with inferred tolerances of the fauna; consistent with relatively turbulent waters

Source: Modified from Johnson, 1960.

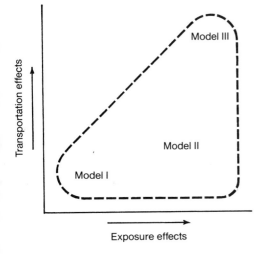

Hypothetical diagram representing the relative alteration of the fossil assemblages resulting from three modes of formation. Model I represents the sudden burial of a community; Model II, the gradual accumulation and burial of the remains of organisms living at the site of deposition; and Model III, an assemblage comprised almost entirely of transported remains. Exposure effects include degree of abrasion, solution, fragmentation, decomposition, encrustation and boring. Transportation effects include spurious association of species and size sorting. Most fossil assemblages probably occur within the area of the dotted line. Assemblages formed in quiet waters should cluster near Models I and II while those of turbulent water should resemble Model III.

Supplementary Reading

Ager, D. V. (1963) *Principles of Paleoecology.* New York, McGraw-Hill, 371 p. (A comprehensive documentary treatment with a valuable bibliography.)

Hecker, R. F. (1965) *Introduction to Paleoecology.* New York, Elsevier, 166 p. (A translation from Russian dealing primarily with practical methodology.)

Hedgpeth, J. W., ed. (1957) *Treatise on Marine Ecology and Paleoecology, 1, Ecology.* Geol. Soc. Amer. Mem. 67, 1296 p. (A comprehensive reference volume with chapters and bibliographies on many aspects of biologic oceanography.)

Imbrie, J., and Newell, N. D. (1964) *Approaches to Paleoecology.* New York, Wiley, 432 p. (A collection of papers by authorities on various aspects of paleoecology.)

Ladd, H. S., ed. (1957) *Treatise on Marine Ecology and Paleoecology, 2, Paleoecology.* Geol. Soc. Amer. Mem. 67, 1077 p. (Chapters emphasizing the regional biostratigraphic approach and bibliographies on paleoecology of fossil groups.)

Laporte, L. F. (1968) *Ancient Environments.* Englewood Cliffs, Prentice-Hall, 116 p. (A compact summary of biologic and physical approaches for reconstructing environments of the past.)

Moore, H. B. (1968) *Marine Ecology.* New York, Wiley, 493 p. (A summary of habitats, limiting factors, and biologic adaptations in the sea.)

Schafer, W. (1965) *Aktuo-Paläontologie, nach Studien in der Nordsee.* Frankfurt, Waldemar Kramer, 666 p. (A summary of many years' research on animal-sediment relationships, Lebensspüren formation, and fossilization; soon to be translated into English.)

CHAPTER 10

Evolution and
the Fossil Record

Since the publication of Darwin's *On the Origin of Species,* in 1859, the concept of organic evolution has become one of the most important ideas bearing on modern man's intellectual view of himself in the universe.

The mechanism of evolutionary change has been discussed earlier, especially in Chapters 4–6. Our concept of this mechanism, in which genetic mutation and natural selection are fundamental, has been derived from many neontologic sources of evidence, all of which deal with organisms during a very short time—a brief moment of geologic history. The fossil record, on the other hand, contributes little to our understanding of the basic mechanism of evolution, but is our sole source of factual *documentation* for the large-scale evolution of life. Without the fossil record, large-scale organic evolution would remain a hypothesis instead of being an accepted fact.

Study of the fossil record has contributed five principal types of information to our understanding of evolution: (1) phylogenetic relationships among major taxonomic groups, (2) times of appearance of major adaptations, and (3) rates, (4) trends, and (5) patterns of evolution. One man, George Gaylord Simpson, has contributed far more than any other to our understanding of the last three. Simpson has shown paleontology and neontology to be firmly welded together by the bond of evolution.

Both morphologic and taxonomic criteria are used to measure evolutionary change. ("Morphologic" here is used in a broad sense to include genetic and biochemical, as well as structural, features.) Genetic composition is the most basic parameter that can potentially be used to measure morphologic change, but even the biologist lacks extensive information on the coded DNA messages carried within chromosomes. In the fossil record we are limited by having to work with an incomplete complement of structural features and with a tiny, often altered, sample of the original chemical composition of the organism. In studying taxonomic change in fossil organisms, we are forced to use as parameters certain categories from an artificial system of classification that we have imposed on an evolutionary continuum — a classification in which, commonly, there are large phylogenetic gaps between recognized categories. Despite its imperfection, however, the fossil record permits us to reach many significant conclusions concerning evolution, extinction, and phylogeny.

Geographic Speciation

Because of stratigraphic gaps and inadequate fossil preservation, very few examples of geographic speciation have been well documented in the fossil record. Most opportunities for such documentation arise through study of Cenozoic fossils. Perhaps the best example is the analysis by Waller (1969) of the phylogeny of the scallop genus *Argopecten* in North America (Figure 10-1).

During the Late Miocene *Argopecten comparilis* is thought to have lived in both open-marine and enclosed-bay environments, ranging from Virginia southward to the Caribbean and through seaways into the Pacific. With the emergence of the Isthmus of Panama during the Pliocene, the gene pool of *A. comparilis* became geographically divided into an Atlantic and a Pacific gene pool. Geographic isolation resulted in evolutionary divergence. The two Pacific descendents, *A. circularis* and *A. purpuratus,* remained ecologically generalized, living in a variety of environments. The Atlantic descendents became ecologically specialized, with some species becoming adapted to open-marine environments, and others to enclosed bays. It seems reasonable to speculate that the greater amount of speciation among Atlantic descendents has been a result of the continental shelf areas of the Atlantic and Gulf Coasts being much broader than those of the Pacific Coast and being bordered by more lagoons, which lie behind barrier islands along coastal plains. In contrast, the Pacific Coasts of North and South America are adjacent to mountain chains, and in most areas their narrow continental shelves give way to steep continental slopes.

Detailed morphologic and stratigraphic studies are necessary for reconstructing detailed phylogenies like the one of Figure 10-1. Not only can such reconstructions demonstrate the importance of geographic speciation, but they also justify the use of species in paleontology, however arbitrary must be the placement of species boundaries within evolutionary lineages.

FIGURE 10-1
Cenozoic phylogeny of scallops of the genus *Argopecten* in North America. Numbers 1 through 8 identify chronologic planes spaced according to absolute time; each plane represents a stratigraphic unit that has contributed fossils and other data used to construct the phylogeny. The straight solid line on planes 5, 6, and 8 symbolizes the geographic barrier separating the Atlantic Ocean, Gulf of Mexico, and Caribbean from the Pacific Ocean after the Miocene. The curved dotted line on planes 3, 5, 6, and 8 symbolizes the ecologic barrier separating enclosed bay environments (upper right) from open marine environments (lower left) on the Atlantic side. (From Waller, 1969.)

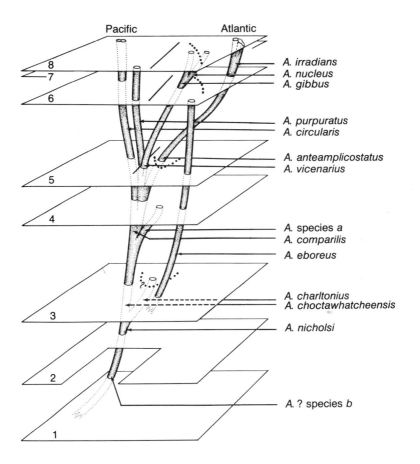

	Age	Years ago
1	Early Middle Miocene	19,000,000
2	Middle Middle Miocene	16,000,000
3	Late Middle Miocene	13,000,000
4	Late Miocene	8,000,000
5	Early Pliocene	6,000,000
6	Early Pleistocene	1,800,000
7	Late Pleistocene	250,000
8	Present	0

Rates of Evolution

A rate of evolution is simply a measure of biologic change with respect to time. Ideally, evolutionary rates are measured in terms of absolute time (usually in millions of years). When precise lengths of time are not known, amount of change is commonly related to sedimentary thickness or relative geologic time. We might, for example, consider the amount of evolutionary change per thousand feet of sediment in a continous sedimentary sequence, or the amount of evolutionary change per epoch in a given geologic period or era.

In our discussion of evolutionary rates a distinction will be made between two similar adjectives. *Phylogenetic* means pertaining to a phylogeny, which may contain a single evolutionary lineage or two or more related lineages. *Phyletic* has a more restricted definition. It means pertaining to a *single* lineage.

We will consider three basic types of evolutionary rates commonly measured from fossil data:

> Phylogenetic rates based on morphologic criteria
> Phylogenetic rates based on taxonomic criteria
> Taxonomic frequency rates

Simpson (1953) has stressed the importance of distinguishing between phylogenetic rates and taxonomic frequency rates. Phylogenetic rates measure change within a single lineage (phyletic change) or within multiple, closely related lineages. For example, in the Mesozoic coral family Rhipidogyridae (Figure 10-2), we could measure the rate of *morphologic change* in the generic lineage from *Codonosmilia* to *Placogyra,* or from *Codonosmilia* to *Acanthogyra* using various structures. Another approach would be to consider *taxonomic change,* as expressed by the average duration of genera (in millions of years per genus) in each lineage. Taxonomic frequency rates refer to the rate of origination of new taxonomic groups within a higher taxon. We might choose the suborder Caryophylliina (to which the Rhipidogyridae belong) or the entire coral order Scleractinia (Figure 10-3). We could then, for example, plot a graph showing numbers of new genera or families per million years for each Mesozoic and Cenozoic period. In Figure 10-4 such a graph is shown for the entire order. Later in the chapter we will analyze this graph.

PHYLOGENETIC RATES BASED ON MORPHOLOGIC CRITERIA

It is extremely difficult to recognize a true lineage of three or more taxa in the fossil record. Most published diagrams purporting to present such phyletic lineages are based on incomplete evidence and represent oversimplifications of complex phylogenies. In addition, most morphologic trends that are shown in simple published lineages represent time intervals

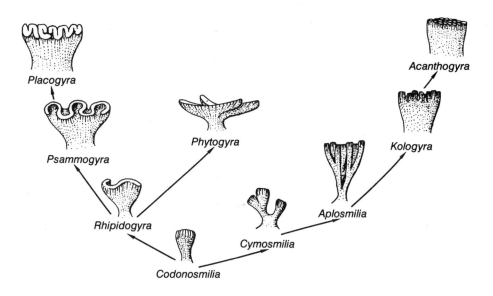

FIGURE 10-2
Evolution of the Mesozoic coral family Rhipidogyridae. (From Wells, 1956.)

so short that absolute ages are not well enough known to permit measurement of rates in terms of absolute time. These problems have hindered attempts to determine morphologic rates of evolution.

One of the simplest types of morphologic evolutionary change is change in size. Figure 10-5, B presents size data for a generic lineage of ceratopsian dinosaurs. There is a marked increase in animal length in this lineage, as in most other lineages shown in the diagram (Figure 10-5,A). The change in slope of the line on the graph is a measure of the rate of size increase. In this instance, the rate appears to decrease, but the simple curve drawn through the three points may well be an oversimplification.

The three ceratopsian genera of Figure 10-5, B exhibit another important type of morphologic change. While overall length increased in the lineage, body proportions changed. In Figure 10-6, length of the back is standardized and the relative proportions of other structures are compared. A relative reduction in limb and tail length is apparent, while the relative size of the skull is nearly constant.

Although many similar examples of differing morphologic rates of evolution have been recognized, few have been plotted graphically against time. One example is shown in Figure 10-7, which is a plot of two tooth dimensions for horses, revealing different rates of evolutionary change for two lineages.

Brinkmann (1929) studied change in two characters of the ammonite genus *Kosmoceras* (Jurassic of England). He collected specimens from a sedimentary unit 13 meters thick, keeping track of the distance above the base for each specimen. He recognized four principal phyletic sequences,

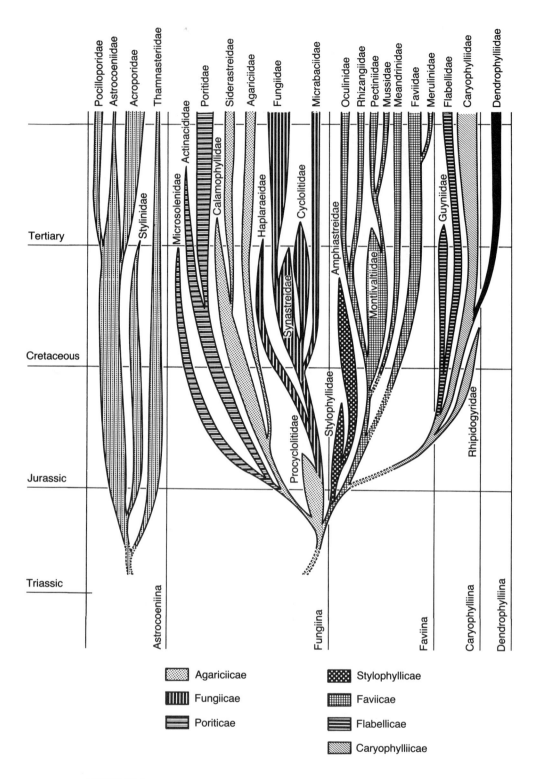

FIGURE 10-3
Evolutionary pattern of the scleractinian corals (hexacorals). Suborders (five in number), superfamilies (represented by patterns), and families are shown. (From Wells, 1956.)

or lineages, in which species divisions are somewhat arbitrary (Figure 10-8). It is now believed that the lappets (protuberances at the shell apertures in lineages 3 and 4) are sexual characters, probably of male animals. This suggests that there are two, rather than four, lineages, the animals of lineages 3 and 4 being the male counterparts of forms in 1 and 2.

Brinkmann detected phyletic increase in the ratio of outer ribs to marginal spines. A plot of this ratio for specimens of one lineage against their position above the base of the sedimentary sequence (Figure 10-9, A) revealed a morphologic discontinuity at the 1093.5 centimeter level, where a thin layer of concentrated ammonite fragments occurred. Brinkmann interpreted the morphologic break as representing a time of nondeposition, during which fine sediments were perhaps winnowed out and shells were broken by current agitation. By separating the two segments of the plot until they could be connected by a straight line (Figure 10-9, B), Brinkmann estimated that the period of nondeposition was equivalent to that required for deposition of about 80 centimeters of sediment (assuming a constant rate of deposition during times of sediment accumulation).

Subsequent studies have supported Brinkmann's observations, showing that the phyletic changes he recognized took place over a wide geographic area. The nektonic habit of most ammonite species gave them extensive distributions and made their entombment in sediments (they sank to the bottom after death) independent of local benthonic conditions, except those affecting preservation. The importance of Brinkmann's study is that it documents morphologic and taxonomic phyletic change at the species level.

Lerman (1965) has estimated that the 13-meter sequence studied by Brinkmann represents a time interval of approximately 1.5 million years. His estimate is based on the sedimentary thicknesses of Jurassic ammonite zones and the time intervals they represent and is subject to major errors, however.

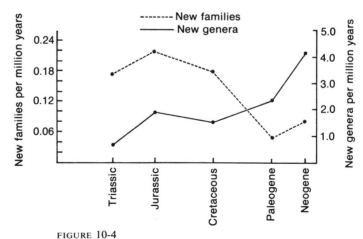

FIGURE 10-4
Rates of evolution of new families and new genera for scleractinian corals (hexacorals). (Data from Wells, 1956.)

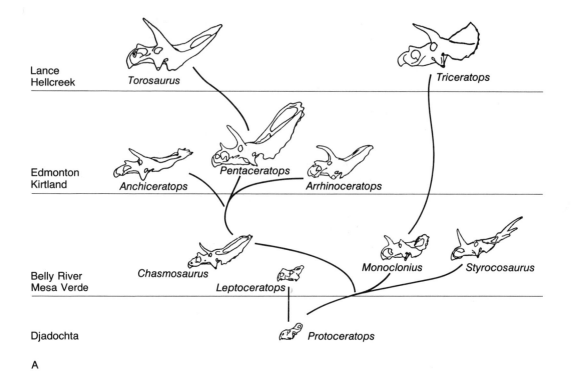

Lance
Hellcreek — *Torosaurus* — *Triceratops*

Edmonton
Kirtland — *Anchiceratops* *Pentaceratops* *Arrhinoceratops*

Belly River
Mesa Verde — *Chasmosaurus* *Monoclonius* *Styrocosaurus*
Leptoceratops

Djadochta — *Protoceratops*

A

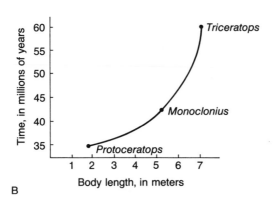

B

FIGURE 10-5
Evolutionary increase in body length of a generic lineage of Cretaceous ceratopsian
dinosaurs. A: Phylogeny of North American ceratopsians. B: Increase in body length
in the generic lineage between *Protoceratops* and *Triceratops*. (From Colbert, 1948.)

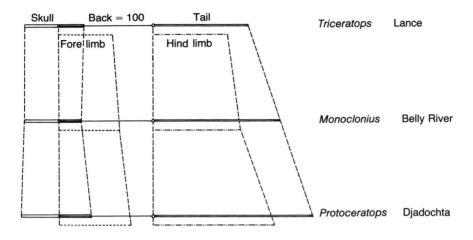

FIGURE 10-6
Skeletal proportions in the ceratopsian genera of Figure 10-5B. (From Colbert, 1948.)

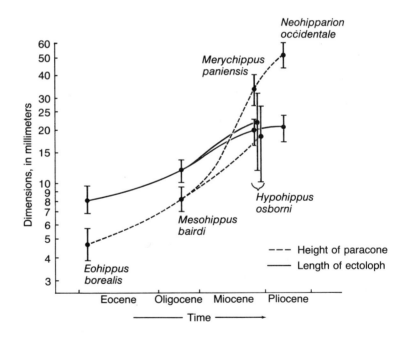

FIGURE 10-7
Plot of two tooth dimensions against time for a branching horse phylogeny.
Paracone height increased more rapidly relative to ectoloph length in
the lineage leading to the grazing genus *Neohipparion* than in the lineage
leading to the browsing genus *Hypohippus*. (From Simpson, 1953.)

It is often desirable to compare rates-of-change for *several* unit characters in a single lineage. This procedure can demonstrate differences among the rates for the chosen characters and can also provide an estimate of the mean rate-of-change in whole-organism morphology. A subjective but useful technique has been applied to fossil lungfish by Westoll (1949), who scored 26 skeletal features for fossil genera based on comparison with the oldest, most primitive lungfish and the youngest, most specialized forms. The total score for a genus is the sum of its scores for the 26 features. Simpson (1953) has modified the scoring system and plotted the graphs shown in Figure 10-10. Lungfish evolved rapidly early in their evolutionary history, but have undergone little change since the Late Paleozoic.

FIGURE 10-8
Nearly continuous lineages of the genus *Kosmoceras* in the Jurassic Oxford Clay, Peterborough, England. (From Woodford, 1965; after Brinkmann, 1929.)

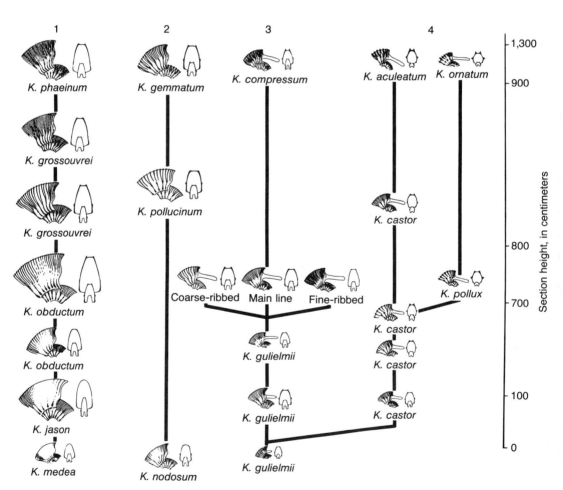

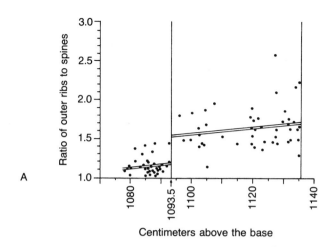

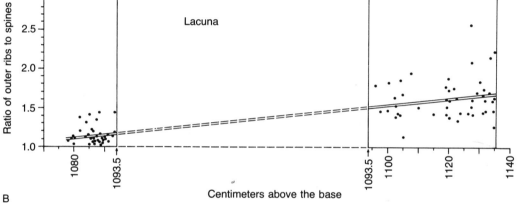

FIGURE 10-9
Graphs for one of the *Kosmoceras* lineages diagrammed in Figure 10-8. A: Ratio of outer surface ribs to marginal spines plotted against stratigraphic position. B: The same graph separated at the stratigraphic discontinuity to show the relative time of nondeposition. (From Woodford, 1965; after Brinkmann, 1929.)

PHYLOGENETIC RATES BASED ON TAXONOMIC CRITERIA

In using taxonomic criteria to measure phylogenetic rates of evolution, we face some of the same problems encountered in the use of morphologic criteria: true evolutionary lineages are often difficult to recognize and, when recognized, are difficult to relate to absolute time.

Phylogenetic change in the *Kosmoceras* ammonite lineages studied by Brinkmann can be measured by taxonomic as well as by morphologic criteria. Working at higher taxonomic levels with multiple lineages of closely related taxa, we can commonly measure rates in terms of absolute time with moderate accuracy. Figure 10-11 shows Simpson's plots of the duration of bivalve and carnivorous land mammal genera. Although Simpson believes

FIGURE 10-10
A: Degrees of modernization of fossil lungfish genera based on
mean scores for degree of modernization of 26 characters. B: Rate
of modernization of fossil lungfishes, derived from graph A by
plotting the first derivative, or slope, at each point on the graph
against time. (From Simpson, 1953; data from Westoll, 1949.)

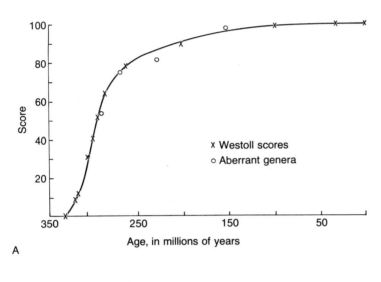

A

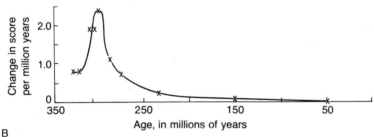

B

that the curves may exaggerate the difference, mean duration of carnivore
genera (about 8 million years) has been considerably less than that of bivalve genera (about 78 million years). The difference demonstrates a more
rapid average rate of evolution for the carnivores.

One of our chief interests in studying evolution is to learn about rates of
speciation. Species duration of Recent and near-Recent species can be
crudely examined by compiling data on the geologic antiquity of Recent
species or on the percentage of fossil species of a given geologic time that
survive today. In either case, we are looking at only a part of the duration
of many species (those that are not yet extinct). Many fossil plant and invertebrate species of the Miocene Epoch belong to species that are living
today, but few mammal species of the Miocene or Pliocene are still extant.

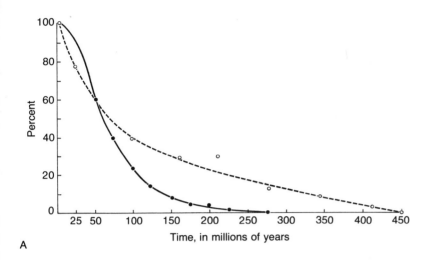

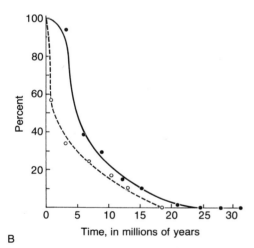

A

B

FIGURE 10-11
Survival times of living and extinct genera. The plots are cumulative (the points indicate percentages of genera whose survival times are equal to or greater than the times shown on the horizontal axis). Solid lines are for extinct genera and dashed lines, living genera. A: Bivalves (mean survival time for living plus extinct genera is 78.0 million years). B: Carnivorous land mammals (mean survival time for living plus extinct genera is 8.1 million years). (From Simpson, 1953.)

The data gathered by Simpson for land carnivores and bivalves (Figure 10-11) suggest that mammals have evolved more rapidly than invertebrates during the Cenozoic. As discussed in Chapter 5, rates of geographic and phylogenetic speciation depend both on environmental factors and on inherent reproductive and genetic characteristics of evolving populations. Since large-scale extinction of any group would obviously make available the ecologic niches that it occupied, increased rates of speciation would often be expected to follow major extinctions. The glacial advances of the Pleistocene Epoch, by causing rapid environmental changes, may have affected average rates of speciation for many Cenozoic taxonomic groups, so we must be careful in our interpretation of evolutionary rates for taxa that existed during the Pleistocene Epoch.

As discussed in the first chapter, Simpson (1952) estimated that mean species duration for all members of the animal kingdom has been about 2.75 million years. He has suggested 5 million years and 0.5 million years as reasonable upper and lower limits for this mean. Studies by Marks (1952) and Weisbord (1962, 1964) show that many Recent molluscan species of Venezuela have lived for 10 million years or more, and a small percentage have lived for 20–30 million years (Figure 10–12). Many fossil species must have existed considerably longer than Simpson's estimate for the mean.

TAXONOMIC FREQUENCY RATES

Taxonomic frequency rates are the most commonly measured rates of evolution. We have already plotted numbers of new genera and families per million years for scleractinian corals (Figure 10-4). Figure 10-13, A is a plot of the number of scleractinian genera existing in each epoch; in it genera inherited from the previous epoch are distinguished from new genera. This graph provides the raw material from which evolutionary rates can be calculated. An alternative method of presenting the information of Figure 10-13, A is shown in Figure 10-13, B, in which rate of appearance and extinction are plotted separately; the two curves are remarkably similar, the three peaks representing times of rapid turnover in scleractinian evolution. Many other plots and plot combinations are possible. The best choice depends on such considerations as whether emphasis is to be placed on rates of diversification or on rates of extinction.

FIGURE 10-12
Plot of percentage of molluscan species alive today that are found in the Tertiary of Venezuela against time. (Data from Marks, 1952, and Weisbord, 1962, 1964.)

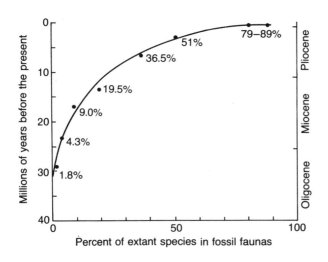

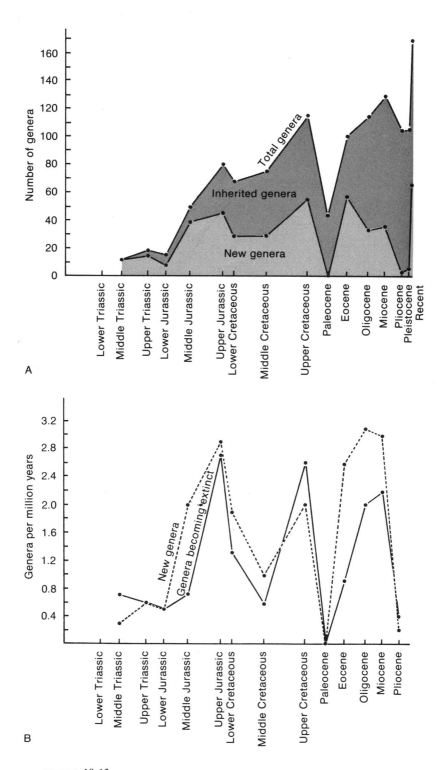

FIGURE 10-13
Evolution (A) and extinction (B) of genera of scleractinian corals
(hexacorals). (Data from Wells, 1956.)

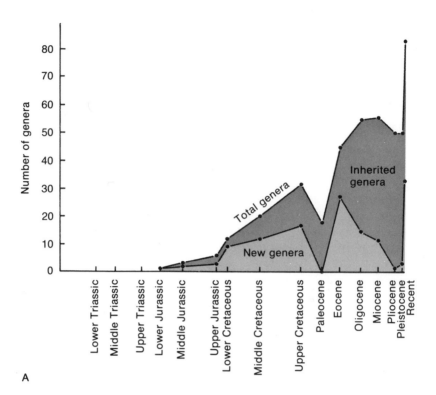

A

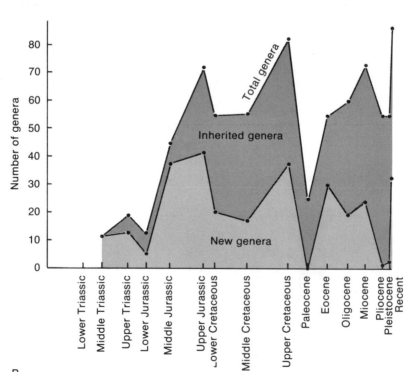

B

The plots of Figure 10-4 and 10-13 are oversimplifications in light of the ecologic division of the Scleractinia into hermatypic and ahermatypic taxa (page 212). Most families contain exclusively hermatypic or ahermatypic genera, but some contain both. The evolution of hermatypic genera might be expected to reflect climatic change, or other environmental changes that affected reef growth, whereas ahermatypic corals, which are less restricted ecologically, are less likely, as a group, to have been sensitive to environmental change.

In Figure 10-14, the rates at which genera of hermatypic and ahermatypic corals have evolved are plotted separately. Hermatypic corals arose before ahermatypic ones and the number of hermatypic genera increased greatly long before ahermatypic groups diversified extensively. In fact, hermatypic genera account almost entirely for the Upper Cretaceous peak in Figure 10-13.

Except for an apparent Paleocene reduction in rate of appearance, the number of ahermatypic genera has increased from their time of origin to the present (Figure 10-14, A). The record of hermatypic genera (Figure 10-14, B) is much less consistent, showing three diversity peaks, which coincide with those in Figure 10-13. The stratigraphic record shows that the first two peaks coincide with times of extensive reef building, which not only left an extensive record of hermatypic corals but favored their rapid diversification.

In Figure 10-13, the dominant trends of the evolution of hermatypic Scleractinia have largely masked the trends of ahermatypic genera, even though there are nearly half as many recognized ahermatypic fossil genera as hermatypic. Clearly, choice of taxonomic subgroups to be included is extremely important in the interpretation of taxonomic frequency rates of major animal groups.

Biases in the fossil record must also be taken into account in interpreting graphs. The virtual lack of known first appearances and extinctions of genera in the Paleocene is probably a consequence of the poor marine sedimentary record of this very short epoch: it is possible that some extinctions listed as Upper Cretaceous and first appearances listed as Eocene actually occurred in the Paleocene. Another striking feature is the apparent marked increase in the number of coral genera since the Pleistocene (Figure 10-14), which is largely a consequence of our knowing more about Recent faunas than about those of the Pleistocene. Many of the hermatypic coral fossils of the Pleistocene that have been found in exposed reefs cannot be identified because of poor preservation. Many ahermatypic genera are restricted to deep water, and a large percentage of Pleistocene genera may thus be absent from sediments that are accessible for study. In addition, many genera of both ecologic groups may be accessible and well preserved, but as yet undiscovered.

FIGURE 10-14
Evolution of genera within the two major ecologic groups of scleractinian corals (hexacorals). A: Ahermatypic Scleractinia. B: Hermatypic Scleractinia. (Hermatypic and ahermatypic assignments cannot be made with certainty for a small percentage of the genera.) (Data from Wells, 1956.)

Phyletic Trends

We will define a **trend** as a direction of adaptive change within a lineage. The problem of recognizing true fossil lineages is not as great a problem in studying trends as in studying rates of evolution because many trends can be recognized from quite fragmentary fossil documentation of lineages.

ORTHOGENESIS

In the past certain workers propounded the concept of orthogenesis, or phyletic change that proceeds in a constant direction. Orthogenesis is sometimes described as straight-line evolution. An example would be evolution toward increased size. Some workers claimed that orthogenetic evolution, once begun, would proceed in the same direction even if the resulting changes ceased to be adaptive and might finally bring about the extinction of a lineage. Such an idea is inconsistent with the concept of adaptive change through natural selection.

A classic example often cited in arguments over orthogenesis is the evolution of *Megaloceras*, the genus that included the giant Irish "elk," which lived during the Pliocene and Pleistocene. This animal had antlers that were gigantic even for its large body size (Figure 10-15). The antlers were shed and regrown annually, requiring a huge expenditure of energy. During the evolution of the genus, body size increased, but antler size increased even more rapidly. Various early workers suggested that the progressive trend toward huge antler size was inadaptive and led to the extinction of the genus. Julian Huxley (1932) showed that there is a relative increase in antler size of the present-day red deer during its ontogeny. Since this increase is approximately allometric (Figure 10-16), Huxley (1931) proposed that relative increase in antler size during the evolution of the Irish "elk" may have followed a similar pattern. If body size and antler size were genetically interrelated during its evolution, selection pressure toward increased body size would have incidentally produced gigantic antlers, which may initially have been advantageous or disadvantageous. If at any time the increase in antler size became disadvantageous, evolutionary size increase would end only when selection pressure *against* antler size increase balanced selection pressure *for* body size increase. There is no evidence to suggest that some mysterious orthogenetic mechanism was driving the body-antler complex beyond this condition to an inadaptive state. Huxley pointed out, however, that if the size of a form like *Megaloceras* should increase to a point that "is verging on the deleterious" it might so specialize the taxon that an environmental change could tip the balance in such a way as to bring about its extinction. In the rapidly shifting climatic zones and biotic realms of the Pleistocene, the genus *Megaloceras* may thus have met its end.

Another important idea regarding antler size in *Megaloceras* is that allometric growth of antlers during ontogeny would have produced gigantic antlers only in old animals. Antlers could, in fact, have become cumbersome

FIGURE 10-15
Megaloceras, the giant Irish "elk" of the
Pleistocene. Antler spread is about 2.5
meters. (From Romer, 1966; after Reynolds.)

and detrimental in old males that were excluded from the breeding population (through physiologic aging or defeat in rutting combat with younger
males) without affecting evolution.

Some workers have claimed that the shell coiling of gryphaeid oysters of
the Mesozoic increased to a point at which the shell could not open, producing extinction. The huge canine teeth of Cenozoic saber-tooth cats have
likewise been attributed to inadaptive evolution that led to extinction. The
arguments advanced to counter the "inadaptive" explanations for the Irish
"elk" antlers can also be applied to these examples.

In its milder form, orthogenesis simply recognizes "straight-line" evolution as the common pattern of evolutionary change, without reference to
external driving forces or inadaptive trends. Many workers, most notably
G. G. Simpson, have refuted even this form of the concept. Early in the
twentieth century, many workers believed that the descent of the modern
genus *Equus* from the "dawn horse," *Hyracotherium* (formerly *Eohippus*),
of the Eocene followed a linear path. Simpson has shown, however, that
trends such as those toward increase in overall size and tooth height (see

Figure 10-7) and decrease in number of toes occurred in the overall evolution of horses but not as a single phyletic trend. In fact, the changes occurred sporadically and at differing rates in various lineages of a complex, branching pattern. Even the phylogeny shown in Figure 6-5 is an oversimplification, although the multiple lineages are divisible into two major phylogenetic groups that differ in tooth evolution. The early horses were browsers, and had teeth suitable for feeding on soft, leafy vegetation, but in the mid-Tertiary, apparently in response to the spread of abrasive siliceous grasses that began in the early Tertiary, some lineages developed more complex molars, adapted to grazing on abrasive, siliceous grasses. The primary changes in the molars were increase in the complexity of crown patterns, increase in the height of the crowns, and development of cement. The grazing horse was one of many new grazing lineages that evolved in the middle and late Cenozoic; a conservative, less successful phylogenetic group of browsing horses continued on into the Pliocene.

As Simpson has emphasized, virtually no valid examples of straight-line phylogenetic change have been documented in the fossil record. In fact, any new adaptation, by definition, marks a directional change in the course of evolution.

RECAPITULATION AND HETEROCHRONY

Early in the nineteenth century, von Baer noted that some stages in the embryologic development of higher animals resemble those of lower animals. A human embryo, for example, passes through successive stages in which it bears superficial resemblance to the embryos of fish, amphibians,

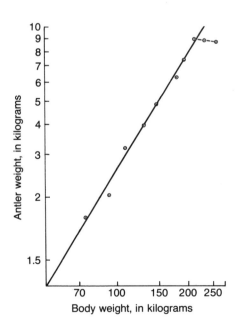

FIGURE 10-16
Logarithmic plot showing allometric relationship between antler weight and body weight for the red deer *Cervus elephus*. (From Huxley, 1931.)

reptiles, and lower mammals. This pattern became known as the Biogenetic Law. Haeckel and other early evolutionists later expanded the idea into what became known as the Principle of Recapitulation. This "principle" asserted that higher animals pass through stages resembling successive *adult* stages in their evolutionary ancestry. This idea, usually expressed by the statement "ontogeny recapitulates phylogeny," is largely in error. Von Baer's original generalization is closer to the truth, in that the development of an advanced group frequently resembles the *development* of ancestral forms. Still, there is usually considerable alteration of developmental rates and sequences during evolution. For understanding such changes it is useful to think of evolution as a *sequence of ontogenies* (or partial ontogenies, depending on when reproduction takes place).

At any stage of ontogeny, evolution may operate in ways that have little direct effect on later stages. For example, marine invertebrate larvae commonly exhibit elaborate features that have evolved semi-independently of any changes in the adult form.

Evolution has also commonly produced profound changes in adults through alteration of the ontogenetic sequence through which their ancestors had developed. For example, the sequence in which two organs develop may be reversed, or development of one may be retarded to the point at which it becomes a mere vestige or disappears altogether. Evolutionary changes in the sequence of development of organs are referred to as **heterochrony**. The great variety of recognized patterns of heterochrony has given rise to a large and cumbersome vocabulary that will not be reviewed here; the interested reader is referred to an excellent summary by deBeer (1958). We will, however, consider several examples.

During the eighteenth and early nineteenth centuries, many students of Mesozoic ammonites attempted to apply Haeckel's recapitulation theory to all ammonite species, believing that the course of ammonite evolution could thus be read from ontogenetic changes in shell ornamentation and suture patterns. In 1901, Pavlow invalidated the strict recapitulation concept by showing that in certain Jurassic lineages of ammonites new evolutionary features arose in the early stages of ontogeny; not until later in the evolutionary history of their respective lineages were these changes retained in the adult stages. In other words, ontogenetic development of the new features was retarded, relative to time of reproductive maturation.

Certain graptolite lineages exhibit heterochrony in colony formation (which is not a true ontogenetic sequence, because the colony is composed of a series of many individuals, each with its own ontogeny, formed by successive budding events.) Bulman (1933) noted that many structural changes that occurred during graptolite evolution made their appearance in the early stages of colony formation. Through subsequent evolution they were retained in the later stages of colony development (Figure 10-17).

Heterochrony may have played an important role in the origin of certain higher categories. Schindewolf and Cloud have suggested such a mechanism for the origin of modern hexacorals (Scleractinia) from the Paleozoic tetracorals (Rugosa) that apparently became extinct near the end of the Permian.

The graptolite *Monograptus argentus,* in which only the first-formed cups, or thecae (lower portion of the drawing), exhibited a new (hooked) shape. Arrows indicate the directions in which the apertures of thecae point. Through evolutionary loss of the late stages of colony growth, the hooked shape was passed on to all stages in some species that descended from this one. (From Bulman, 1933.)

The oldest known hexacorals are of Middle Triassic age. As shown in Figure 10-18, the septa (blade-like vertical partitions) in hexacorals are arranged radially in cycles of six or multiples of six. In tetracorals, however, septa are arranged in four quadrants in the adult. Nevertheless, the adult pattern in tetracorals developed through modifications after the emplacement of six primary septa. It is possible that by heterochrony the six primary tetra-coral septa came to remain dominant to the adult stage, and tetracorals thus gave rise to hexacorals. Morphologies of an intermediate type are, in fact, shown by certain late Paleozoic tetracorals and early Mesozoic hexacorals. The origin of the hexacorals from the tetracorals is still uncertain, however; their skeletons differ from those of tetracorals not only in symmetry but in mineralogy, being aragonite rather than calcite. Hexacorals may have arisen separately from naked anemone-like ancestors.

COPE'S RULE

In the nineteenth century the famous vertebrate paleontologist E. D. Cope observed the common tendency of taxonomic groups to evolve toward greater size in the course of phylogeny. The tendency is neither universal nor orthogenetic but is, nevertheless, widespread.

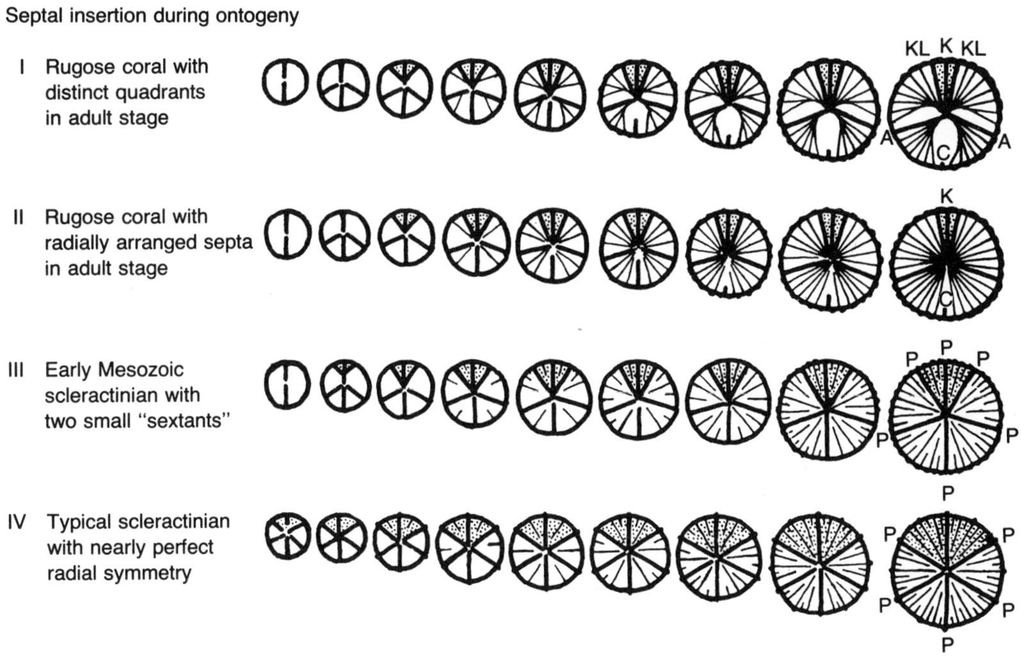

Septal insertion during ontogeny

I Rugose coral with distinct quadrants in adult stage

II Rugose coral with radially arranged septa in adult stage

III Early Mesozoic scleractinian with two small "sextants"

IV Typical scleractinian with nearly perfect radial symmetry

FIGURE 10-18

Diagram showing possible derivation of the Scleractinia (hexacorals) from the Rugosa (tetracorals). The drawings are of coral cups viewed from above. Ontogenetic patterns of septal insertion are sketched for a typical rugose coral (I), a typical scleractinian coral (IV), and corals that are somewhat intermediate in form (II and III). The six primary septa in all forms are represented by heavy lines. C: Cardinal septum. A: Alar septa. K: Counter septum, KL: Counter lateral septa (Rugosa), P: Primary septa (Scleractinia). The minor septa, which are emplaced between major septa, are shown as finer lines. The evolutionary transition from tetracorals to hexacorals is suggested to have occurred by insertion of minor septa in the stippled areas. (From Moore, Lalicker, and Fischer, 1952.)

The ceratopsian dinosaurs discussed earlier show phyletic trends toward increased size (Figure 10-5). Similarly, the overall picture of horse evolution displays size increase, as can be seen by comparing modern *Equus* and its Quaternary relatives with the Eocene dawn horse, *Hyracotherium*, which was about the size of a fox terrier. Horses did not increase in size until after the Eocene, and several lineages that decreased in size after the Eocene are known. Newell (1949) has discussed several examples of apparent phyletic size increase among invertebrates.

The reasons for an increase in size being advantageous to animals undoubtedly vary among lineages. Some animals may gain strength that permits them to attack other animals, or to ward off attacks, through size increase. Heat production in warm-blooded animals is proportional to volume of flesh (that is, the *cube* of their linear dimensions), but heat loss is proportional to exposed surface area (that is, the *square* of their linear dimensions). The result (an example of the Principle of Similitude described on page 63) is that large animals keep warm in cool climates more efficiently than small animals. Another advantage of size increase is that, although large animals cannot run appreciably faster than small animals, they have greater endurance and can maintain a given speed for a longer period of time. Because the greater time required for muscle contraction in large animals is offset by the greater length of each movement (for example, the length of stride in running), the speed of locomotion of small and large animals tends to be approximately the same, but a large animal will make fewer movements in covering the same distance and will therefore have proportionally greater endurance. This factor may have been important in the phylogenetic size increase of grazing horses.

It is important to realize that size increase also poses certain *problems*, as would be predicted by the Principle of Similitude. For example, the increase in the size of grazing horses must have posed feeding problems since tooth surface area for chewing would have increased with the *square* of linear dimensions, but the volume of living tissues to be sustained would have increased with the *cube* of linear dimensions. Simpson has interpreted the complex evolutionary modifications of horse molars (Figure 10-6) as adaptive trends helping to solve this problem. In general, size increase is self-limiting. A man who grew to a height of 100 feet could not walk and would have difficulty supporting his own weight because his weight would have increased with the cube of his linear dimensions, while his strength, being proportional to the cross-sectional area of muscles, would have increased only with the square of his linear dimensions. Many surface-dependent activities also operate less effectively as size increases. For example, food requirements increase with tissue volume, which is proportional to the cube of linear dimensions, but food collection by ciliated surfaces in such groups as the brachiopods, bryozoans, and bivalve molluscs, increases only with the square of linear dimensions. Similarly, oxygen requirements increase with the cube of linear dimensions, but animals absorb oxygen along two-dimensional surfaces such as those of gills or lungs.

Instead of thinking of Cope's Rule as describing a tendency toward evolutionary size increase, we can reverse our viewpoint and think of it as describing evolution from small size. It is well known that many higher taxa have arisen from ancestral groups that are of relatively small size. In large part this is because such ancestral groups tend to be relatively unspecialized, and small individuals are much more likely to be unspecialized than large ones. Most large species within higher taxa have evolved so many special mechanisms for coping with their size that it has become very unlikely that they could give rise to a large new group whose morphologic features would be different enough for the group to constitute a major new taxon. For example, it is highly unlikely that an elephant genus could give rise to an entirely new quadruped order. The modifications of evolving elephants, including greatly thickened limbs for supporting their massive bodies, a shortened neck to support the head, and a trunk to take the place of a long neck (Rensch, 1960), greatly limit their future evolutionary potential. New groups thus tend to arise from small unspecialized ancestors, rather than from large specialized ancestors, and this may be why Cope's Rule seems like a good description of nature.

ADAPTIVE SIGNIFICANCE OF MORPHOLOGIC TRENDS

The adaptive significance of many morphologic trends recognized in the fossil record remains a mystery. An example is the change of ornamentation documented by Brinkmann for the *Kosmoceras* ammonite lineage (page 261). Because of their abundant occurrence in depositional marine environments, invertebrates leave fossil records of nearly continuous phyletic change more commonly than do vertebrates. One of the most thoroughly interpreted fossil lineages is found within the echinoid genus *Micraster* in the Cretaceous "Chalk" of England. Phyletic trends in *Micraster* were first recognized by the amateur paleontologist Rowe (1899). Later, Kermack (1954) confirmed Rowe's basic observations biometrically, though rejecting some details of his account. More recently, Nichols (1959a, 1959b) has analyzed the adaptive significance of the recognized morphologic trends in *Micraster*. The combined work of these men represents a classic evolutionary contribution.

Echinocardium is an irregular sea urchin belonging to the group of echinoids that have altered their ancestral radial symmetry to bilateral symmetry. This change is associated with unidirectional movement of the animals on or within sediment. A typical burrowing species of the Recent Epoch is shown in Figure 10-19. The animal occupies an open burrow, and moves forward by excavating sediment with movable spines, chiefly on the ventral surface. Connection with the overlying water is maintained through a vertical "funnel," formed by special respiratory tube feet. The animal is a deposit feeder that excavates organic-rich sediment with special tube feet that surround its mouth. Some food is also passed down to the mouth from the dorsal surface via the anterior groove (Figure 10-20). Spines and tube feet surrounding the anus excavate a posterior sanitary drain, where feces are left in the animal's wake.

The morphologies of three *Micraster* species and three Recent heart urchin species are shown in Figure 10-20. Spines are almost never preserved in attached position, but much information about their morphology and location can be inferred from their knob-like attachment sites, or tubercles, on the test surface. In nearly all heart urchin species, certain regions of the test bear bands of tubercles to which heavily ciliated spines are attached in life. These regions are termed "fascioles." Cilia of the inner fasciole are largely responsible for drawing water down the respiratory funnel. Cilia of the anal fasciole provide currents to pass feces away from the body and downward, where cilia of the subanal fasciole aid other body cilia in producing currents that carry waste into the sanitary drain.

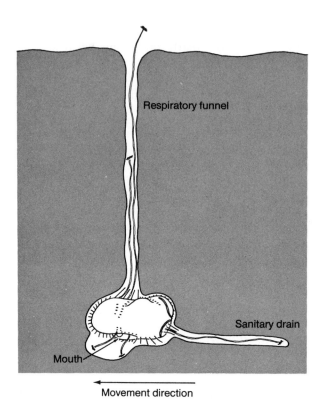

Respiratory funnel

Sanitary drain

Mouth

← Movement direction

FIGURE 10-19
Echinocardium, a typical burrowing heart urchin
of the Recent. (From Nichols, 1959a.)

Figure 10-21 shows the distribution of several recognized *Micraster* species of the Cretaceous Chalk in relation to time and inferred depth of burrowing in life. The species intergrade and are separated arbitrarily. The original species, *M. leski,* is thought to have been a shallow burrower. It gave way to a range of forms, including both shallow and deep burrowers, that are arbitrarily divided into two species, *M. corbovis,* and *M. cortes-tudinarium.* Only representatives of the latter species survived into the Senonian Age, when they gave rise to the lineage given the species name *M. coranguinum.* The shallow-burrowing species, *M. (Isomicraster) seno-nensis,* is thought to have entered the region of study in the early Senonian from a geographically separate region, probably having been isolated from the main stock for a substantial period of time. It had apparently not attained sufficient genetic isolation to prevent interbreeding with *M. coranguinum* (although paleontologists have designated the two as separate species); intermediate coexisting forms suggest that there was interbreeding between the two groups.

Nichols inferred the life habits of the *Micraster* representatives from their adaptive morphology by homology with Recent forms. The supposed burrowing depths of the three *Micraster* species and the known burrowing depths of three Recent species are shown in Figure 10-20. Some of the adaptive features that were progressively developed in the *M. cortestudinarium — coranguinum* lineage, leading Nichols to recognize the trend toward deeper burrowing, are as follows (see Figure 10-20 for illustrations):

1. A broader test, with the broadest and highest region shifting posteriorly. This change apparently reflected the need for streamlining and elimination of the need for the dorsal surface to shed sediment, as it must for organisms that live only partly buried in sediment.
2. Increased ornamentation in the region of the respiratory tube feet, apparently to increase the area of ciliated surface for drawing strong water currents down through the long respiratory funnel.
3. Increase in the granularity of the ventral surface, apparently to increase the area of ciliated surface for movement of fine-grained particles posteriorly in locomotion.
4. Widening of the subanal fasciole, apparently for improved sanitation in the confines of a deep burrow. The test of *M. (Isomicraster) senonensis,* the alleged shallow burrower or surface "plower," lacks a subanal fasciole.

Most evolutionary trends recognized in the fossil record and interpreted in terms of functional morphology are more poorly documented and less striking than those in the *Micraster* lineage. Major trends in horse evolution, for example, are recognized from phylogenetic sequences based on interpretation of a discontinuous fossil record. Recognition and interpretation of nearly continuous fossil lineages is a major goal of paleontologic research, but a goal that is seldom attained.

Patterns of Evolution

In our discussion of phyletic trends, we have been concerned with individual lineages, or phylogenies that approximate lineages. The existence of an evolutionary *pattern* can be determined by a comparison of rates or trends in two or more lineages.

EVOLUTIONARY DIVERGENCE

If there are any characteristic features of evolutionary change, they are *diversification* and *divergence* rather than orthogenesis. Diversification is produced primarily by geographic speciation, or splitting of a single lineage into two or more. Divergence is simply change in a manner that increases the differences between two or more lineages; use of the term implies that the lineages shared a common ancestry. We may choose to analyze a branching pattern on any taxonomic scale, using whatever taxonomic unit we wish.

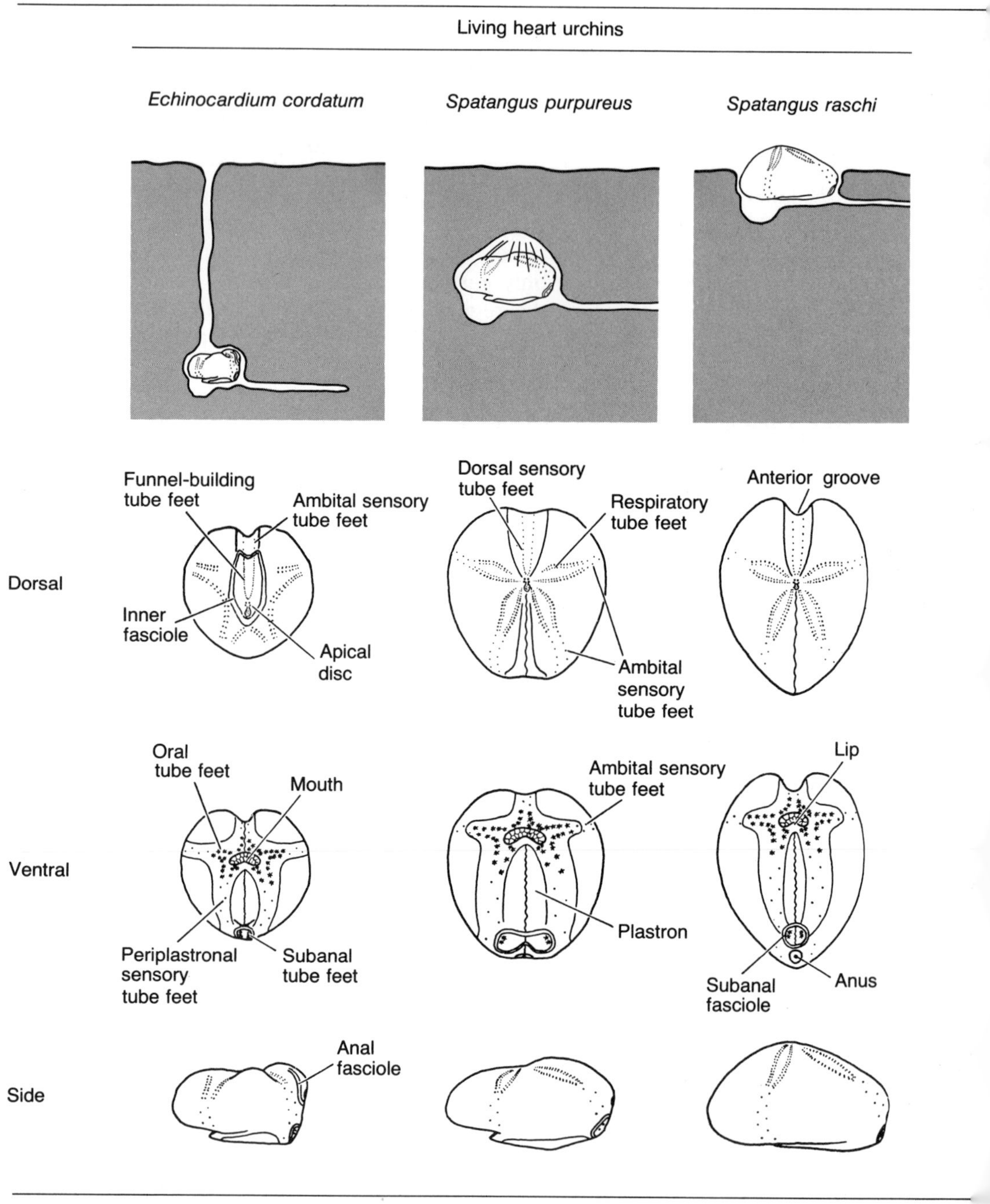

FIGURE 10-20
Morphology and habits of living and fossil heart urchins.

Cretaceous heart urchins

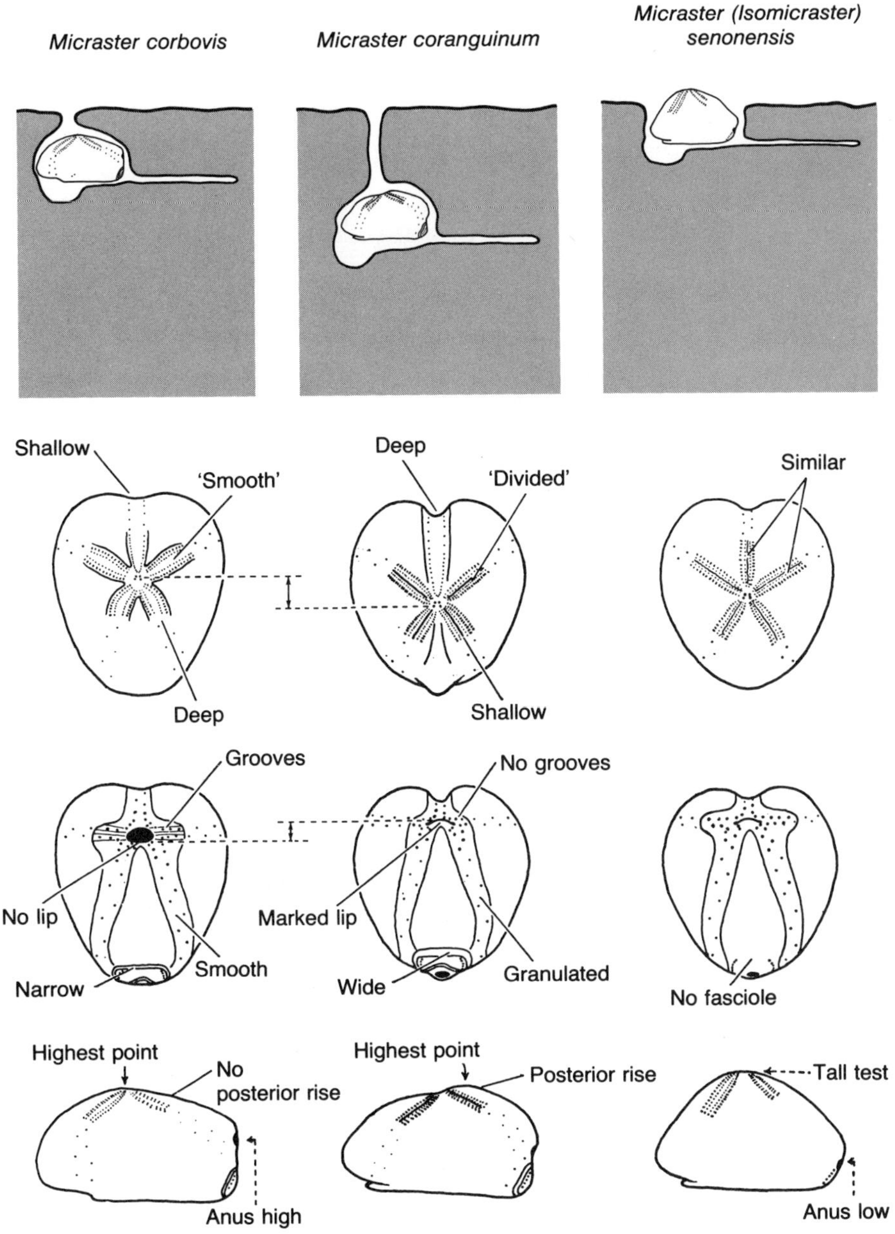

Micraster corbovis

Micraster coranguinum

Micraster (Isomicraster) senonensis

Shallow · 'Smooth' · Deep
Deep · 'Divided' · Shallow
Similar

Grooves · No grooves
No lip · Marked lip
Narrow · Smooth · Wide · Granulated
No fasciole

Highest point · No posterior rise · Anus high
Highest point · Posterior rise
Tall test · Anus low

(From Nichols, 1959b.)

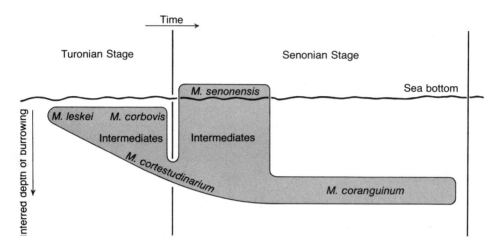

FIGURE 10-21
Life habit changes in the evolution of *Micraster* in the Cretaceous
of the south of England. (From Nichols, 1959b.)

The smaller the unit, the more complex and finely branched will be the
pattern; the finest taxonomic resolution is provided by observation at the
species level. Ultimately, every branching pattern can be reduced to a succession of individual ontogenies (Figure 6-3).

Large-scale diversification and divergence of lineages is known as **adaptive radiation.** An adaptive radiation commonly occurs immediately following
an evolutionary breakthrough, or adaptive innovation, and leads to the origin
of what, in retrospect, may be recognized as a new higher category. An example is provided by the bivalve molluscs (Stanley, 1968). The vast majority
of Paleozoic bivalve genera belong to epifaunal and primitive infaunal families. In the Mesozoic and Cenozoic Eras, a large number of new infaunal
taxa appeared. Their diversification has led to the great proponderance of
burrowing clams in modern seas. Fundamental unifying features, including
shell microstructure, hinge mechanism, gill type, and siphons, indicate that
the majority of post-Paleozoic burrowing taxa may well have radiated from
a basic ancestral stock. Most are united taxonomically within the subclass
Heterodonta. The reason for the adaptive radiation of the heterodont clams
is that they made a major adaptive breakthrough: they developed siphons
and related features through fusion of the sheet-like fleshy mantle that under-
lies the shell halves. Many are much more rapid burrowers than more primitive groups, and many live much deeper within the sediment, where they
find protection, especially in the intertidal realm. Siphon-feeding heterodont
bivalves have surpassed all other Recent marine taxa in the occupation of
niches in which both burrowing and suspension-feeding are employed.

Many other modifications have led to the origin and diversification of
major higher categories. Some of the most notable are: the shell-covered
amniote egg of reptiles, which freed early reptiles from dependence on the

aqueous environment for reproduction and permitted them to exploit late Paleozoic terrestrial habitats unavailable to their precursors, the amphibians; the pollen-and-seed reproductive mechanism of gymnosperm plants (pines, spruces, etc.) and angiosperm (flowering) plants, which freed these groups from dependence on moist surroundings for union of egg and sperm and enabled them to invade many terrestrial habitats inaccessible to their seed-less predecessors; and the opposable thumb of primates, which – in con-junction with stereoscopic vision – gave early primates dexterity, the most important prerequisite for manipulation and tool making, which in turn facilitated the development of a large, highly-developed brain.

Adaptive radiations tend to follow a characteristic pattern. The higher taxonomic categories comprised by a new group tend to appear relatively early and their rate of origin tends to decline as the radiation progresses. Lower taxonomic categories originate most rapidly later in the radiation. In Figure 10-22, this pattern is illustrated for the advanced siphon-feeding bivalves just described. This pattern, which is also displayed by scleractinian corals (Figure 10-4), reflects the fact that the major adaptive possibilities made available to any group by one or more new features are very rapidly exploited. Once the major ways of life are adopted, they tend to be sub-divided more and more finely among lower categories. The pattern may be altered or terminated by extinction at any time.

FIGURE 10-22
Rates of superfamily and genus evolution of siphon-feeding bivalves. Compare with Figure 10-4. (From Stanley, 1968.)

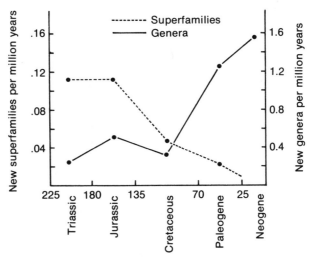

Time, in millions of years before the present

Occasionally in the geologic past, a group has made such an important adaptive breakthrough that it has won in competition with another group with similar life habits and habitat preferences, thus causing the other group's extinction. This phenomenon is termed *ecologic displacement*. Obviously, displacement can be documented only by temporal overlap in the histories of two groups and a decline of the group considered to be adaptively inferior. The geologic history of the fishes is marked by examples that seem to meet these conditions. The primitive ostracoderms, most of which were apparently scavengers or deposit feeders, declined in the Devonian during the ascendency of the placoderms (Figure 10-23). The placoderms were the first jawed vertebrates and their jaw mechanism was the adaptive innovation that led to their radiation in the Devonian, and perhaps to their displacement of many ostracoderms. The remaining ostracoderms and the placoderms both gave way in the late Paleozoic to shark-like fishes with cartilaginous skeletons and bony fishes. The chief advantage of the two new groups seems to have been their possession of an advanced and efficient set of fins, which made them far more adept swimmers than their predecessors.

The subsequent adaptive radiation of the bony fish took place in three major steps; sketches of specimens from the three taxonomic groups that were each dominant in turn are reproduced here as Figure 10-24. Each group displayed adaptive improvements over the preceding group. These improvements were achieved primarily through modification of the scales, jaws, tail, and fins, and through alteration of primitive lungs into a hydrostatic air bladder.

FIGURE 10-23
Stratigraphic ranges and relative abundances of major fish groups. (From Colbert, 1955.)

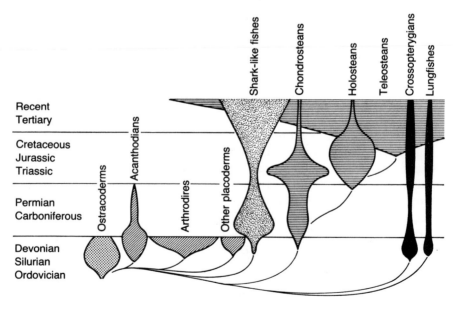

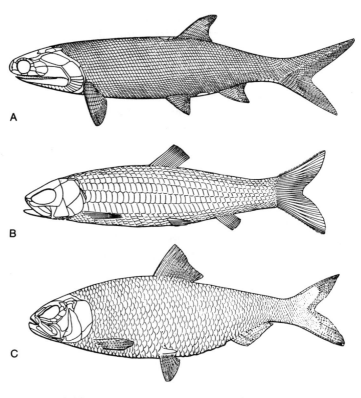

FIGURE 10-24
Representatives of the three major bony fish groups shown in
Figure 10-23, which show stages in evolution toward thinner scales,
a more symmetrical tail, forward movement of pelvic fins, and
shortening of the jaws. A: *Palaeoniscus* (a Permian chondrostean).
B: *Pholidophorus* (a Jurassic holostean). C: *Clupea* (a Cenozoic
teleost). (From Colbert, 1955.)

A new group of organisms cannot easily evolve to occupy niches already
filled by well-adapted members of another group, even if the new group has
the potential to produce more efficient and more diverse adaptations than
the entrenched group. The earliest members of the new group may possess
rudimentary features offering great evolutionary potential, but in their
primitive state these features may be inferior to the advanced features of
the entrenched group. A new group may, however, diversify in a region
that is geographically separated from the one occupied by the group it is
to displace, and then gain access to the region after attaining adaptive superiority. If the older group is geographically widespread, however, this is
unlikely to happen. During the Mesozoic Era, diversification of the earliest
mammals, which were small and unspecialized, was suppressed by the domi-
nation of terrestrial habitats by reptiles, especially dinosaurs, throughout

the world. As is well known, following the mass extinction of dinosaurs in the Late Cretaceous, mammals have radiated to attain greater adaptive diversity than Mesozoic reptiles. The Cenozoic mammalian diversification (Figure 10-25) provides a striking illustration of the evolutionary pattern described earlier as being typical of adaptive radiation. As the ecologic successors of the dinosaurs, the mammals exemplify what is known as *ecologic replacement.* (This phrase does not imply competitive displacement, but most workers consider displacement to be a variety of replacement.)

In a special type of repetitive ecologic replacement, a basic parent stock has given rise to successive groups of higher taxa, each replacing the former. The resulting pattern is called *iterative evolution.* Iterative evolution was apparently characteristic of Mesozoic ammonite groups, although it is difficult to determine exact relationships between ancestral and descendant groups. Salfeld (1913) established the generalization that two slowly evolving suborders (the Phylloceratina and Lytoceratina) lived predominantly in deep-water regions and gave rise to successive groups of short-lived, shallow-water taxa that are grouped together as the Ammonitina. The Ammonitina is thus considered to be a polyphyletic suborder (Figure 10-26).

Cambrian trilobites are also known to have undergone iterative evolution. The best-documented example has been provided by Kaufmann (1933, 1935) and discussed by Simpson (1953). Kaufmann recognized four successive lineages of the genus *Olenus* in a Upper Cambrian rock sequence in Sweden. He divided each lineage into three species or subspecies (Figure 10-27). Three of the lineages exhibit continuous morphologic gradation. All four show certain morphologic trends, the most striking being relative elongation of the pygidium (tail region). Furthermore, the earliest representatives of lineages 1 and 2 are nearly identical. The four lineages are believed to have evolved in succession from a more slowly evolving, geographically removed parental stock. Each lineage invaded the region of study, followed the recognized evolutionary trends, and then became extinct, to be replaced by another invasion of taxa derived from the parent stock.

EVOLUTIONARY CONVERGENCE AND PARALLELISM

In the course of evolution close morphologic similarity may arise between two unrelated groups as they take on similar life habits. This is termed *adaptive convergence,* and the similar taxa are referred to as *homeomorphs.* Some biologists reserve the term homeomorph for one of a pair of species whose members are indistinguishable to the untrained observer. Most paleontologists, however, apply the term to one of a pair of species whose members simply exhibit a strong resemblance.

Homeomorphy can also result from what is referred to as *parallel evolution,* or *parallelism,* in which two closely related stocks with minor morphologic differences undergo a series of similar evolutionary changes through time. Parallelism and convergence are distinguished on the basis of the

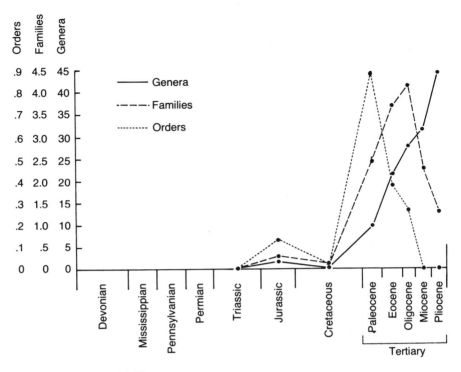

FIGURE 10-25

Rates of appearance (in taxa per million years) of mammalian orders, families and genera. (From Simpson, 1953.)

amount of similarity between the ancestral groups giving rise to the lineages to which the two species belong, but there is no clearcut separation between the two concepts.

The terms parallelism and parallel evolution have frequently been applied to crudely similar evolutionary trends in only one or a few characters, rather than whole organisms. There is some question whether this use of the terms is advisable.

It should be evident that parallelism is similar to iterative evolution, except that it occurs in contemporaneous, rather than successive, lineages. Similarly, two converging lineages may undergo the same changes simultaneously or at different times, which has led Cloud (1949) to label the two types of adaptive convergence as isochronous convergence and heterochronous convergence. Iterative evolution, parallelism, isochronous convergence, and heterochronous convergence are compared diagrammatically in Figure 10-28. True parallelism, which has probably been common in evolution, is not well documented in the fossil record. In contrast, many examples of convergence are known from the study of fossils because we can recognize adaptive convergence simply by discovering two similar forms that belong to widely separate higher taxa. To recognize parallelism, however, at least four taxa are required (two in each lineage).

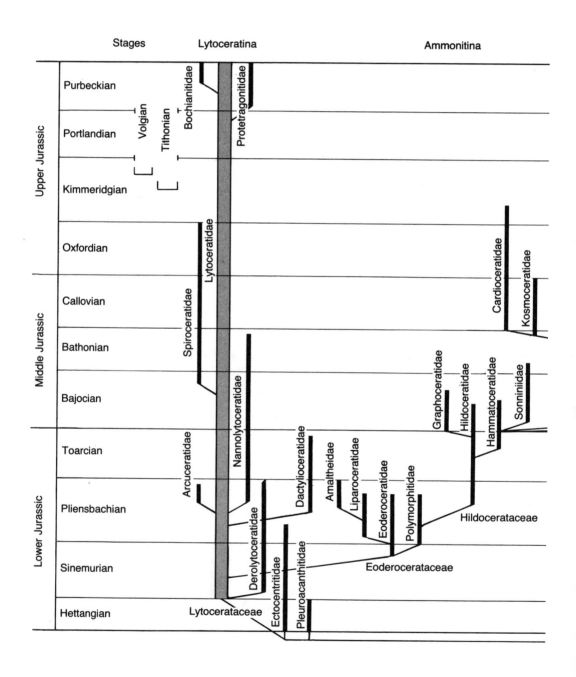

FIGURE 10-26
Iterative evolution of Jurassic ammonite families derived from the suborders
Phylloceratina and Lytoceratina. The suffix -aceae denotes a superfamily;
the suffix -idae denotes a family. (From Arkell, 1957.)

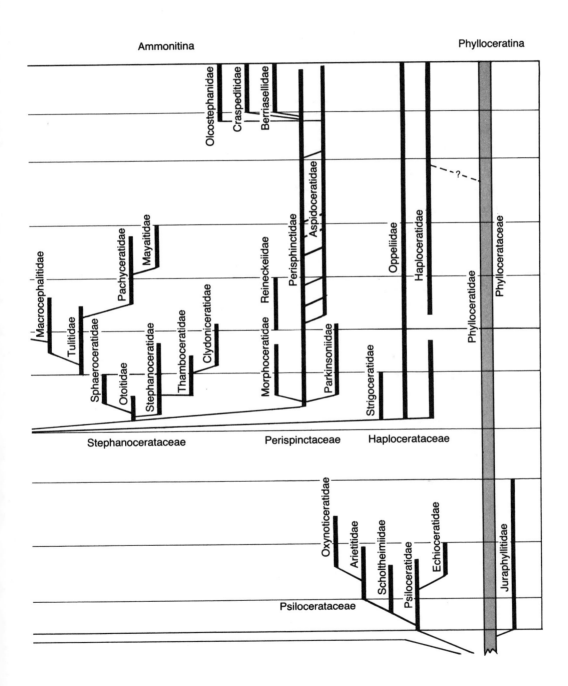

Paleozoic brachiopods are noted for having contributed many examples of homeomorphy deriving from adaptive convergence. One example, illustrated in Figure 8-20, was discussed earlier with regard to functional morphology. A well-known example of adaptive convergence on a large scale has been provided by the marsupial and placental mammals. A similar type of convergence on a smaller scale is illustrated by bivalve mollusc genera of the superfamily Mactracea (Figure 10-29).

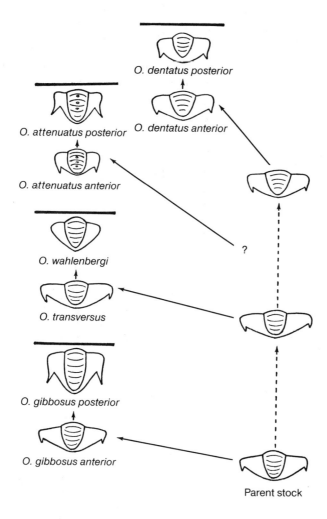

O. dentatus posterior

O. attenuatus posterior   *O. dentatus anterior*

O. attenuatus anterior

O. wahlenbergi

?

O. transversus

O. gibbosus posterior

O. gibbosus anterior

Parent stock

FIGURE 10-27
Iterative evolution in the Cambrian trilobite genus *Olenus,* represented by outlines of pygidia (tail regions). Notice that the four descendent lineages underwent similar morphologic changes. (From Simpson, 1953; after Kaufmann, 1933.)

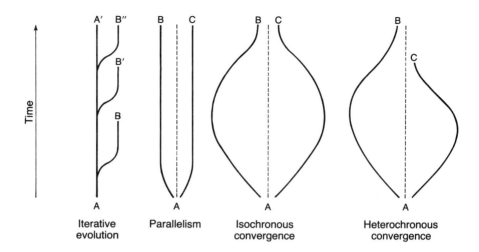

FIGURE 10-28

Ways in which evolving lineages may be related. The horizontal distance separating two species at any point is proportional to the differences between them relative to the differences between them at other points in their evolutionary histories. (Partly from Cloud, 1949.)

Extinction

A dominant feature of life history is extinction, which, though not always a part of evolution *per se*, exerts a strong influence on evolutionary patterns.

The success of an interbreeding population requires a good adaptive relationship between its genetic composition and the environment. Since the genetic composition of a population is, except under unusual circumstances, altered primarily through natural selection, disturbance of a good adaptive balance is usually caused only by environmental change. When the environment becomes unfavorable a population may evolve to become adapted to the changed environment, migrate to a more favorable area, or die out. When all populations of a species fail to meet unfavorable environmental changes by evolving or migrating, the species becomes extinct, and the extinction is said to have come about through *termination of a lineage*. Extinction of a species may also occur when phyletic changes have so accumulated that the organism is judged to be a new species; this type of extinction is called **phyletic extinction.**

Environmental changes causing extinction may be physical, chemical, or biologic. We have already seen how one taxonomic group may displace another through adaptive superiority. One group may also bring about extinction of another through excessive predation.

The time at which fossil species became extinct can seldom be established accurately because of gaps in the stratigraphic record. A large percentage of recognized fossil species are known from rocks that represent a length of time less than half that during which they lived. (It should be borne in

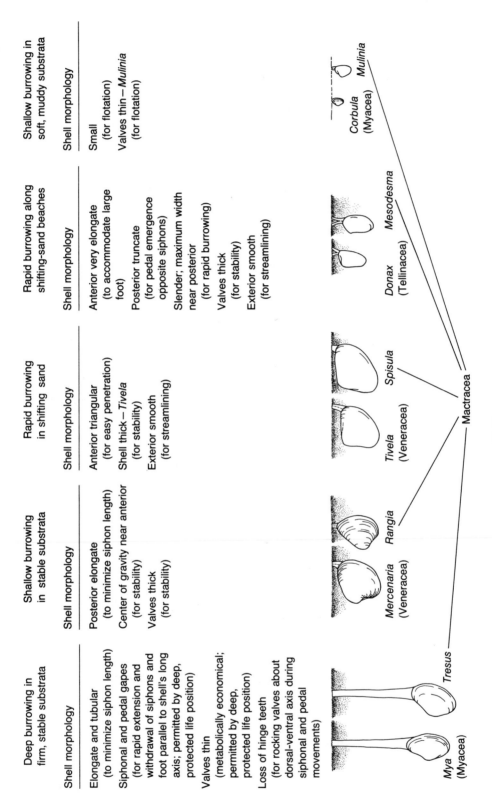

Deep burrowing in firm, stable substrata

Shell morphology

Elongate and tubular (to minimize siphon length)

Siphonal and pedal gapes (for rapid extension and withdrawal of siphons and foot parallel to shell's long axis; permitted by deep, protected life position)

Valves thin (metabolically economical; permitted by deep, protected life position)

Loss of hinge teeth (for rocking valves about dorsal-ventral axis during siphonal and pedal movements)

Mya (Myacea) — *Tresus*

Shallow burrowing in stable substrata

Shell morphology

Posterior elongate (to minimize siphon length)

Center of gravity near anterior (for stability)

Valves thick (for stability)

Mercenaria (Veneracea) — *Rangia*

Rapid burrowing in shifting sand

Shell morphology

Anterior triangular (for easy penetration)

Shell thick—*Tivela* (for stability)

Exterior smooth (for streamlining)

Tivela (Veneracea) — *Spisula*

Rapid burrowing along shifting-sand beaches

Shell morphology

Anterior very elongate (to accommodate large foot)

Posterior truncate (for pedal emergence opposite siphons)

Slender; maximum width near posterior (for rapid burrowing)

Valves thick (for stability)

Exterior smooth (for streamlining)

Donax (Tellinacea) — *Mesodesma*

Shallow burrowing in soft, muddy substrata

Shell morphology

Small (for flotation)

Valves thin—*Mulinia* (for flotation)

Corbula (Myacea) — *Mulinia*

Mactracea

FIGURE 10-29
Adaptive convergence between pairs of bivalve genera; one member of each pair belongs to the superfamily Mactracea, and the other, to some other superfamily (whose name is given in parentheses). All of these genera are extant. (From Stanley, 1970.)

mind that even if the fossil record were complete, division of lineages into species would involve both subjective and arbitrary decisions.) There is an additional problem: populations of a species do not necessarily die out simultaneously throughout its geographic range.

GEOGRAPHIC FACTORS

The most striking testimony to the importance of geography in extinction is provided by **relict** groups — taxa that persist in limited regions long after their counterparts have died out elsewhere. One of the most notable examples is the persistence of blastoids in the eastern hemisphere during the late Paleozoic. Blastoids were stalked, suspension-feeding echinoderms whose earliest known representatives are found in Silurian rocks. The group abounded during the Mississippian Period, but only three or four genera have been recognized in rocks of Pennsylvanian age. During the Permian Period, however, the group apparently underwent a renewed expansion, centering on the vicinity of the island of Timor, Indonesia, where 16 genera have been recognized. A small fraction of the Timor genera, and a few others, have also been found in Permian rocks of Australia and Asia. Only one Permian species is known from the western hemisphere, and it is found in the Canadian Arctic.

The marsupial mammals of Australia are a significant relict group living today. At the close of the Mesozoic, marsupials enjoyed a global distribution. Throughout most of the world, placental mammals displaced marsupials in the early Cenozoic, as the mammals began their spectacular adaptive radiation. Two isolated islands, South America and Australia, served as refuges where marsupials persisted without competition or predation from placentals. Near the close of the Pliocene Epoch, however, North and South America became connected by a narrow land bridge, the Isthmus of Panama. Northern placental species migrated southward, displacing and preying on marsupial species and causing their extinction. The only survivors were the opossums, primitive forms that had given rise to diverse South American marsupial groups during the earlier period of isolation. One unusually successful opossum species has extended its range northward into a large area of North America. Only in isolated Australia did other marsupials continue to flourish.

Other islands have also harbored important relict groups. The lemurs, forerunners of the higher primates, are well represented on Madagascar while only a few species inhabit the mainland of Africa. *Sphenodon,* a genus that is restricted to a few small islands off the coast of New Zealand, is the sole survivor of a primitive reptilian group, the rhynchocephalians; it originated in the Triassic Period and is not far from the ancestral line of the true lizards and dinosaurs. It is an interesting fact that many other relict groups, including many articulate brachiopod taxa, *Nautilus,* marsupials, and the primitive bivalve *Neotrigonia,* also persist in the Australia-New Zealand region. In both the marine and terrestrial realms, this region has apparently maintained some degree of biogeographic isolation for many millions of years.

While certain representatives of a taxonomic group may be protected as relicts by their geographic isolation, the group as a whole may suffer. An extremely small, isolated geographic region offers little opportunity for development of many semi-isolated populations in which new adaptations may arise. Diversity of environments also tends to be limited in small regions and major environmental changes may adversely affect an entire fauna.

An excellent example of decline in diversity through isolation is provided by the modern reef-building corals of the Caribbean region (Vaughan and Wells, 1943). During the early Cenozoic, the Tethyan Sea served to connect coral faunas of the Mediterranean region with those of the Indian Ocean and East Indian region. By the late Oligocene Epoch, most European elements of the Caribbean fauna were dying out. The remaining genera formed almost the same group that survives in the Caribbean region today. In the Miocene, the marine passage between the Mediterranean and the Indian Ocean became blocked. This barrier and subsequent climatic cooling virtually eliminated reef growth in the Mediterranean and left two separate coral faunas that persist to the present day: the West Indian fauna and the Indo-Pacific fauna. There was little connection between the two faunas before the late Pliocene emergence of the Isthmus of Panama, and there has been none since. The isolated Caribbean fauna, occupying a small geographic region, has dwindled since the Miocene to about 36 species, while the widespread Indo-Pacific fauna has diversified to more than 500.

MASS EXTINCTION

Among the greatest enigmas of the fossil record is the tendency for groups of higher categories to become extinct within relatively brief time intervals. Major extinctions, as we have seen, are commonly followed by major adaptive radiations involving ecologic replacement. Many major episodes of extinction approximately coincide with the boundaries that we recognize between our geologic series and systems. The reason for this coincidence is simply that many formal divisions have been placed where major breaks have been recognized in the fossil record. The most publicized episodes of mass extinction for animals occurred at the close of the Paleozoic and Mesozoic Eras.

The ends of the Cambrian, Ordovician, and Devonian Periods were times of mass extinction for marine life. The marine "faunal crisis" of the late Permian was far more dramatic, however, in terms of numbers of disappearing taxa. The fusulinid Foraminifera, productoid brachiopods, and many bryozoan and ammonoid taxa were among the most important groups to disappear. Higher taxa believed to have become extinct in the late Permian are listed in Figure 10-30. In marked contrast to the marine invertebrates,

FIGURE 10-30
Late Permian extinctions and Early Triassic appearances of higher taxa of animals. (From Schindewolf, 1962.)

Carboniferous	Permian	Triassic	Jurassic		
				Fusulinida	Foraminifera
				Conulariida	
				Tabulata	
				Streptelasmatina	
				Columnariina	Coelenterates
				Astrocoeniina	
				Fungiina	
				Faviina	
				Trilobita	
				Eurypterida	
				Beyrichiida	
				Leperditiida	
				Thysanura	Arthropods
				Palaeodictyoptera	
				Megasecoptera	
				Protohemiptera	
				Orthoptera	
				Unionacea	
				Carditacea	
				Cardiacea	Bivalve molluscs
				Myacea	
				Ostreacea	
				Bellerophontacea	
				Platyceratacea	
				Subulitacea	
				Patellacea	
				Trochacea	Gastropods
				Littorinacea	
				Cerithiacea	
				Naticacea	
				Solenochilida	
				Goniatitina	
				Ceratitina	Cephalopods
				Phylloceratina	
				Trepostomata	Bryozoans
				Cryptostomata	
				Dalmanellacea	Brachiopods
				Productacea	
				Blastoidea	
				Inadunata	
				Flexibilia	Stalked echinoderms
				Camerata	
				Articulata	
				Rhachitomi	Amphibians
				Stereospondyli	
				Ichthyosauria	
				Sauropterygia	
				Rhynchocephalia	
				Squamata	Reptiles
				Archosauria	
				Pelycosauria	
				Ictidosauria	

terrestrial vertebrates made the Permo-Triassic transition with relatively few higher-category extinctions. The diverse amphibian faunas of the early Permian had largely given way to reptilian faunas in the middle Permian. The Age of Reptiles thus really began in the middle Permian and continued across the Permo-Triassic boundary into the Mesozoic.

In the Mesozoic Era, the close of the Triassic Period marked the first major episode of extinction. As shown in Figure 10-31, eight of the 19 Upper Triassic orders and suborders of reptiles became extinct. Both terrestrial groups and marine groups were affected. At the same time, the previously abundant labyrinthodont amphibians disappeared. Among the invertebrates, the ammonites, which had diversified tremendously during the Triassic, nearly died out. Only a small stock of phylloceratinids survived into the Jurassic Period, during which they gave rise to the lytoceratinids. The two

FIGURE 10-31

Major amphibian and reptile groups at the Triassic-Jurassic transition. Asterisks mark those groups whose fossilized remains have not actually been found in Lower Jurassic rocks; since they have been found in the Upper Triassic and Upper Jurassic, it is obvious that they lived also during the Early Jurassic. (From Colbert, 1965.)

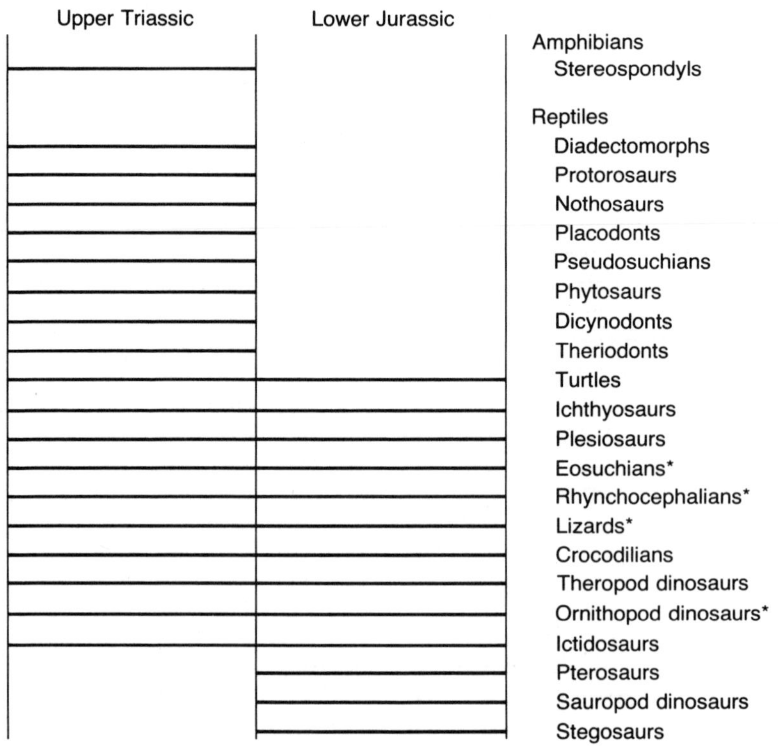

groups, together, gave rise to many other superfamilies in the Jurassic and Cretaceous (as discussed on page 288 and illustrated in Figure 10-26). Marine and terrestrial vertebrates also diversified extensively during the Jurassic and Cretaceous Periods. Then, in the Late Cretaceous, a major wave of extinction eradicated the dinosaurs. At the same time the ammonites and other previously marine groups (including the belemnites and several important gastropod and bivalve taxa) disappeared. The taxonomic effects of the Cretaceous faunal crisis are summarized in Figure 10-32. The taxonomic impact of the Cretaceous faunal crisis rivalled that of the Permo-Triassic crisis.

During the Cenozoic Era the only large-scale episode of mass extinction occurred during the Pleistocene Epoch, when many terrestrial mammal groups disappeared. The Pleistocene climatic fluctuations, which produced four major glaciations in the northern hemisphere, may not have caused the extinctions, however, because other ecologic groups, such as terrestrial plants and shallow-water marine invertebrates, did not undergo major extinctions. Many workers now believe that a newly evolved predator called man brought about the extinction of most of the terrestrial animal groups known to have disappeared.

Major extinctions among higher plant categories have not generally coincided with those described for animal groups. A striking aspect of evolution is that the two major changes in terrestrial floras have preceded major changes in terrestrial faunas. The general evolutionary history of higher plants is divisible into three major floras that have successively dominated nonmarine environments: the Paleozoic spore-plant flora, the Mesozoic gymnosperm flora, and the Cenozoic angiosperm flora. Transitions between these floras did not straddle boundaries between geologic eras but preceded them by about half a geologic period. It is probable that each transition involved ecologic displacement. The gymnosperm flora was dominant by the late Permian and the angiosperm flora by the Late Cretaceous. Dramatic mass extinctions have not been so characteristic of higher plant evolution as of animal evolution. Gradual decline of major plant groups has tended to occur during the rise of replacement groups.

Many hypotheses, some preposterous, have been advanced to explain mass extinctions recorded in the geologic record. Some have been applied to a single episode of mass extinction, others to all major episodes. We cannot review all mass-extinction hypotheses here, but will mention a few. Many are virtually untestable. For example, variation in solar radiation and epidemics of microbial disease are possible agents of extinction that would probably have left no direct geologic evidence.

Perhaps through wishful thinking, much attention has been given to a few hypotheses of mass extinction involving destructive agents that might have left tangible imprints in the rock record, independent of fossil occurrences. Foremost among these are changes in the earth's magnetic field, climatic conditions, and relative sea level.

It has been suggested that periodic reversals of the earth's magnetic polarity during the geologic past may have temporarily eliminated the shielding against cosmic radiation that the magnetic field provides. The resulting

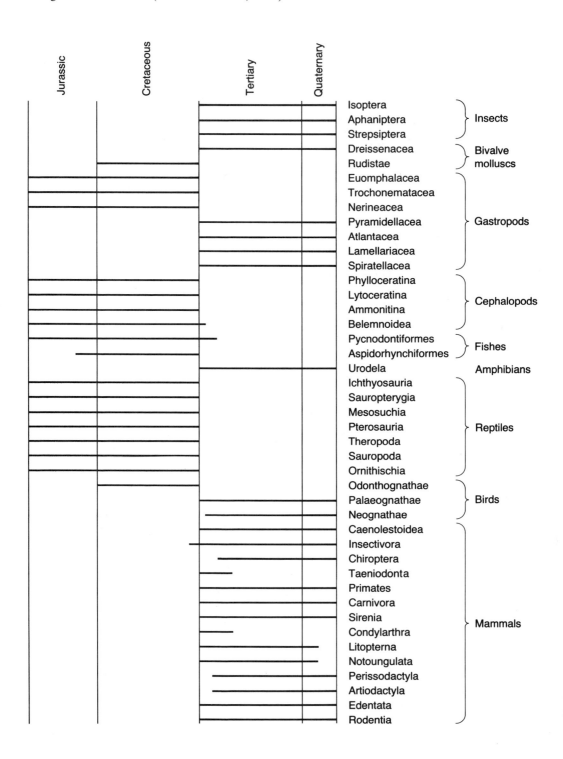

FIGURE 10-32
Late Cretaceous extinctions and early Tertiary appearances
of higher taxa of animals. (From Schindewolf, 1962.)

increase in cosmic radiation reaching the earth's surface may have caused marked temporary increases in mutation rates and led to higher rates of evolution and extinction. Studies of the directional properties of magnetism in rocks of many ages testify to the occurrence of polar reversals, but these reversals have not yet been shown to correspond to periods of rapid evolutionary turnover or mass extinction. Furthermore, mass extinctions have generally spanned greater time intervals than those thought to have been required for magnetic reversal.

Climatic conditions and relative sea-level changes are not unrelated. They have commonly been cited separately and in combination in mass-extinction hypotheses. The Permian geologic record shows that large, markedly arid land areas existed in many regions, in contrast to the widespread coal-swamp environments of the Pennsylvanian Period. In addition, the Permian Period was apparently a time of marked climatic contrasts, and perhaps fluctuations. The Permian record offers evidence of glaciation, as well as aridity. The Pleistocene record, however, provides evidence opposing climatic change as a primary factor in the Permian mass extinction. While many mammal groups disappeared, few major extinctions of marine invertebrate groups occurred in the Pleistocene. The Permo-Triassic faunal crisis, in contrast, primarily involved marine invertebrates.

Newell (1962), especially, has stressed the importance of sea-level change in causing extinction. Because the continental land masses show relatively little surface relief in many areas, relatively small sea-level fluctuations may rapidly immerse or drain large geographic areas and alter environmental conditions. The chief source of evidence favoring sea-level change as a major agent of extinction is the apparent limitation of many fossil species to time intervals bounded by periods of major withdrawal of ancient seas. Many of the major recognized stratigraphic boundaries coincide with both unconformities in rocks and mass extinctions of fossil groups. Still, it is difficult to make a case for the complete extinction of such highly diversified and widespread taxonomic groups as dinosaurs, trilobites, or ammonites on this basis. Certainly, attempts to relate rates of evolution and extinction to mountain-building episodes have largely failed.

Changes in the salinity of the oceans may, in the past, have occurred in association with climatic or sea-level changes. Beurlen (1956) suggested that the Permo-Triassic crisis, in particular, was a consequence of reduced salinity in the oceans. The groups most adversely affected by the crisis were predominantly stenohaline marine groups (bryozoans, corals, brachiopods, crinoids, ammonoids, trilobites); the least affected groups have many brackish-water representatives today (bivalves, gastropods, fishes). Beurlen assumed a decreased rate of salt addition to the oceans to have been the most likely cause of a worldwide decrease in ocean salinity. Fischer (1960a) has presented an alternative hypothesis, which calls for evaporation of sea water in restricted marginal marine basins and outflow of dense brine across basin thresholds into the deep sea. There it would have settled and only slowly mixed with the overlying waters, which would have become progressively more brackish. Cessation or reduction in rate of brine production at the close of the Permian would have permitted deep brines to mix completely with oceanic water, to reestablish "normal" salinity in the ocean basins. This idea, or some variation of it, gains support from our knowledge of Permian evaporite basins, notably the Delaware Basin of Texas. Still, it is only a hypothesis.

It seems most reasonable not to appeal to any single factor in attempting to account for mass extinction. Not only may different agents have caused different mass extinctions, but interplay between agents may have caused extinctions. One relevant factor is species interdependence, not only through food webs but also through such mechanisms as substratum formation (as in coral reefs) and symbiosis.

In this light, studies of marine phytoplankton have been brought to bear on the problem of major extinctions. The evolutionary history of marine phytoplankton, like that of the vertebrates, invertebrates, and land plants, is divisible into three primary biotas whose stratigraphic ranges *approximately* coincide with the three geologic eras. The history of the major marine phytoplankton groups is summarized in Figure 10-33. The acritarchs and other groups were diverse and abundant in the early Paleozoic. The fossil record shows a decline in early Paleozoic phytoplankton floras beginning in the Upper Devonian and extending through the Triassic. In the stratigraphic interval from the Carboniferous through the Triassic, phytoplankton assemblages are very rare; no major taxonomic group is abundant and widespread.

The Lower Jurassic record shows an adaptive radiation of new groups, including new acritarch taxa, coccolithophorids, and dinoflagellates. This radiation culminated in the Late Cretaceous, but was then terminated suddenly by major extinctions within the new groups.

The early Paleocene record, like that of the Carboniferous through Triassic, shows a meager phytoplankton flora. Another adaptive radiation, beginning in the late Paleocene, produced the modern phytoplankton flora. Extinctions, especially in the Oligocene, have reduced the relative abundance of certain groups that flourished in the early Tertiary, and diatoms and dinoflagellates have come to predominate in modern seas.

Interpretation of the phytoplankton fossil record is, in part, speculative. We cannot be certain that many groups of unpreservable organisms did not exist at times of apparent low diversity. That the acritarchs declined during the late Paleozoic and that they underwent a new adaptive radiation during the Jurassic (Figure 10-33) are, however, almost indisputable facts. That there were major phytoplankton extinctions in the Late Cretaceous is also beyond dispute.

Tappan (1968) has suggested that phytoplankton have exerted a strong influence on animal life throughout Phanerozoic evolutionary history. Their most direct influence is through their role in primary food production and oxygen liberation. Tappan has speculated that major periods of phytoplankton extinction and low productivity may have resulted from reduced physiographic relief, which lowered rates of nutrient supply from continental erosion. These periods may have caused major extinction of consumer organisms and oxygen-dependent groups.

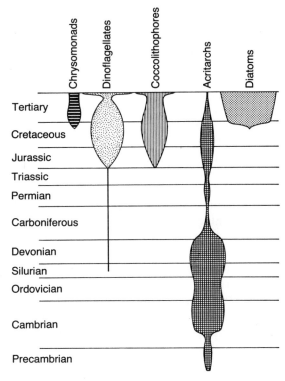

FIGURE 10-33
Geologic ranges of major marine phytoplankton groups.
(Modified from Downie, 1967.)

The fact that major floral changes on land have slightly preceded mass extinction of terrestrial animals may be of significance in a similar way. At least major floral changes in the marine and terrestrial realms may have compounded problems for animals subjected to other adverse conditions and thus have contributed to widespread animal extinctions.

One of the most puzzling features that must be explained by any reasonable mass-extinction hypothesis is the apparent tendency of mass extinction to follow *taxonomic* rather than *ecologic* lines. Fusulinid Foraminifera, productoid brachiopods, ammonites, and dinosaurs are among the major groups that appear to have developed a wide range of ecologic adaptations, yet all groups disappeared suddenly as taxonomic units. This situation suggests that agents of mass extinction may have operated on fundamental genetic and physiological characters rather than on less fundamental ecologic adaptations within groups. Whether or not their solutions are within reach of scientific inquiry, problems of mass extinction will continue to spark the imaginations of workers for years to come. Certainly the past decade has seen the emergence of many new lines of evidence shedding light on these problems.

Determining Phylogenetic Relationships

Phylogenetic relationships among taxonomic groups are determined from both biologic information and the fossil record, either separately or together. In studying extinct groups whose taxonomic relationships are uncertain, there has been a common tendency to attempt to establish affinities with phyla that have Recent representatives. This bias, often based on wishful thinking, has led to many erroneous conclusions. The Archaeocyatha (Figure 10-34) of the Cambrian System offer an interesting example. The cup-shaped skeletons of archaeocyathids formed reef-like structures, especially in the vicinity of Australia. The morphology and sessile habit of archaeocyathids suggests that they were suspension feeders. The holes in the archaeocyathid wall suggest the presence of a current system similar to that of sponges, in which water is channeled inward through small openings and out through one or more larger openings. A condition in which the current direction was reversed would be hydrodynamically inefficient. For many years some workers classified archaeocyathids with sponges. Lack of similarity between the archaeocyathid skeleton and the sponge skeleton, however, has convinced most workers that archaeocyathids were not closely related to sponges. Groups that defy classification with recognized living groups are commonly termed "problematica." We must accept the possibility that we may never ascertain their affinities.

There has been greater success in attempts to establish the taxonomic affinities of the extinct graptolites. Although universal agreement is still lacking, most workers accept the argument of Kozlowski (1966 and earlier studies) that the graptolites were hemichordates, closely resembling living pterobranchs (Figure 10-35). The detailed structure and chemical composition of the graptolite skeleton, as well as the pattern of colonial budding, all support such a relationship. Kozlowski's interpretations stem from his work with exceptionally well-preserved colonies etched from chert with hydrofluoric acid. Kozlowski imbedded free specimens in paraffin and cut serial sections with a microtome in the same way that biologists section

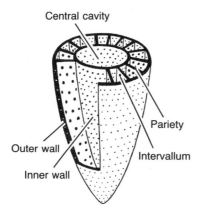

Central cavity

Outer wall

Inner wall

Pariety

Intervallum

FIGURE 10-34
Schematic drawing of an archaeocyathid. (From Okulitch, 1955.)

Recent organisms. In contrast, most graptolite skeletons are found flattened along bedding planes of black shale. Kozlowski's success demonstrates the importance of seeking unusually well-preserved fossil material for taxonomic study. It also appears to represent a fruitful attempt to establish affinities between a problematic fossil group and a living group.

Looking at the opposite side of the coin, there are certain living taxonomic groups whose affinities are understood largely through study of the fossil record. The subungulate mammals, for example, include modern-day conies (small rodent-like forms), elephants, and sea cows. It is highly unlikely that these three strikingly different groups would be lumped together were it not for their fossil records, which show basic similarities among their primitive Tertiary ancestors on the continent of Africa, where they apparently shared a common ancestry.

In general, it is much easier to establish phylogenies for major vertebrate groups than for major invertebrate and plant groups because all recognized classes and orders of the Vertebrata have originated since the Cambrian. Although all higher vascular plant taxa have apparently originated since the early Paleozoic, the fossil record of plants is less complete. Furthermore, fossil plant remains usually reveal less about whole-organism morphology than do vertebrate remains. Invertebrate animals fall into several phyla whose late Precambrian and Cambrian origins are almost universally undocumented by the known fossil record. In some instances, however, we have a moderately good knowledge of the post-Paleozoic phylogenies within invertebrate phyla.

FIGURE 10-35
Morphologic comparison of pterobranchs and graptolites.
A and B: Wall structure, consisting of "fusellar" bands,
of *Cephalodiscus* and *Rhabdopleura*.
C: Stolon (stippled) of *Rhabdopleura*.
D and E: *Mastigograptus* and *Bulmanicrusta*,
showing similar structures. (From Kozlowski, 1966.)

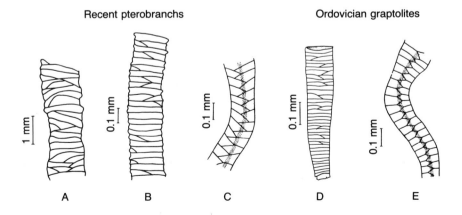

Recent pterobranchs   Ordovician graptolites

A   B   C   D   E

Unfortunately, the origins of most higher categories are shrouded in mystery; commonly new higher categories appear abruptly in the fossil record without evidence of transitional ancestral forms. Simpson (1953) has listed several reasons for this situation. Among them are the following:

1. Appearance of a new higher category has usually marked a major adaptive breakthrough, often accompanying inhabitation of previously unoccupied niches; evolution under such conditions has tended to be *very rapid.*
2. Any lineages of the ancestral group that were similar enough to enter into competition with the new group are likely to have been rapidly displaced.
3. Often, times of higher category appearance are represented by gaps in the geologic record. (In some instances, rapid evolutionary turnover and unconformities may have resulted from the same widespread environmental change.)
4. Change in habitat during the adaptive breakthrough has made discovery of certain transitional forms unlikely.
5. Major adaptive breakthroughs have commonly occurred in relatively small populations or taxonomic groups.
6. Transitions have commonly been made in taxa whose members were small relative to average size in both the ancestral and descendant higher categories.
7. Transitions have commonly taken place in restricted geographic areas, and possibly the same transitions occurred at different times in different areas.

The fossil record does occasionally provide what might be termed as a "missing link," a species that appears to represent a transitional stage between higher taxa. One such form is the reptile-like bird *Archaeopteryx,* of the Middle Jurassic, already discussed in other connections and pictured in Figure 1-1. *Archaeopteryx* possessed both reptilian and avian characters. Its possession of feathers suggests that it was warm-blooded, like modern birds, but it also had large teeth, solid bones, and other reptilian skeletal features.

A spectacular transition, described by Erben (1966), is shown by a graded morphologic series of cephalopods in the Lower Devonian Hunsrück Shale, of Germany. In the 1930's Schindewolf postulated the origin of the ammonoids from the bactritid nautiloids. His arguments stressed the similarity of the initial chambers of bactritid and early ammonoid shells. The siphuncle (see page 174) of the bactritids was also marginal, like that of most ammonoid shells. The morphologic series reported by Erben apparently confirms Schindewolf's hypothesis. The series contains bactritids, ammonoids, and intermediate forms (Figure 10-36). The various taxa constituting the series have not been studied with respect to relative stratigraphic position, but their existence together in a local rock sequence strongly supports the idea of a bactritid origin for the ammonoids.

FIGURE 10-36
Adult shells of species from the Devonian Hunsrück Shale of Germany showing the apparent evolutionary sequence leading to the ammonoids. A-E: Bactritid nautiloids. F-L: Early ammonoids. (From Erben, 1966.)

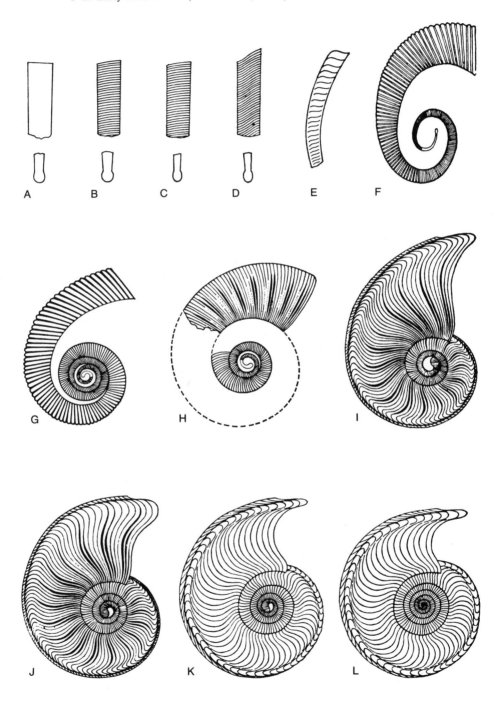

Just as the fossil record may provide transitional links between major taxa, so may discoveries of hitherto unknown Recent organisms. The deep-sea mollusc *Neopilina* provides a possible illustration. In 1952, workers of the Danish Galathea Expedition dredged members of this unusual genus from the ocean bottom off the Pacific Coast of Central America at a depth of about 3,500 meters. Lemche and Wingstrand (1959) have described the anatomy of the genus, which has many primitive characters. The most unusual feature is the general body plan, which displays paired series of pedal muscles, kidney-like structures, and gills (Figure 10-37). Other molluscan groups possess only one pair of each organ type, or multiple pairs that appear to have had a secondary (nonprimitive) origin.

During the two decades preceding the discovery of *Neopilina,* some paleontologists, including Wenz and Knight, had suggested that early Paleozoic fossils belonging to the molluscan class Monoplacophora (see Figure 10-37) were the cap-shaped ancestors of gastropods, and perhaps of other higher molluscan classes. Upon its discovery, *Neopilina* was recognized as a monoplacophoran, though fossil representatives of the group were known only from the lower and middle Paleozoic. *Neopilina* is a "living fossil," representing a group that has apparently found refuge for many millions of years in the deep sea.

The evolutionary position of the Monoplacophora, whose anatomy is known only through *Neopilina,* is still a matter of debate. One issue is whether the semisegmented body plan of the primitive Monoplacophora indicates that the Mollusca arose from segmented annelid worms. Molluscs and annelids have long been considered to be related because of similarities in their early ontogenetic development. Another question is whether *Neopilina*-like monoplacophorans did, indeed, give rise to higher molluscan classes. Most hypothetical models of the "ancestral mollusc" have portrayed an organism with single pairs of organs. If *Neopilina* is in the main line of early molluscan evolution, loss of the semisegmented structure that it possesses probably preceded the origin of higher mollusc classes. It has not been demonstrated conclusively that the group to which *Neopilina* belongs was ancestral to any of the other molluscan classes. Despite its uncertain phylogenetic position, *Neopilina* is one of the most significant living taxa to have been discovered in recent decades.

Most progress in the establishment of phylogenies is made through consideration of both fossil and Recent evidence. Used alone, either type of evidence may lead to misinterpretation. There is no need here to cite specific examples, but many unreasonable hypotheses linking taxonomic groups have been proposed by both paleontologists and neontologists whose approaches included data from only the fossil record or only the Recent.

FIGURE 10-37

Monoplacophora. A-F: The living species *Neopilina galatheae* Lemche. A and C: Ventral views. B and D: Dorsal views. E: Right lateral view. F: Anterior view.
G: The Silurian fossil *Pilina unguis* (Lindstrom); shell interior, showing paired muscles by which the body and foot were attached. (A-F from Lemche and Wingstrand, 1959; G courtesy of the Swedish Museum of Natural History.)

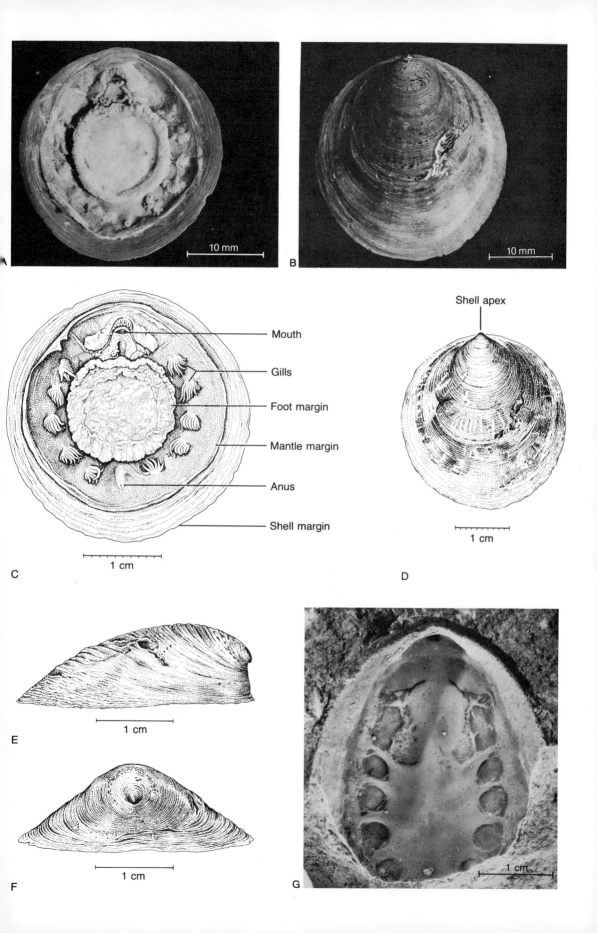

B

Mouth

Gills

Foot margin

Mantle margin

Anus

Shell margin

Shell apex

C

D

E

F

G

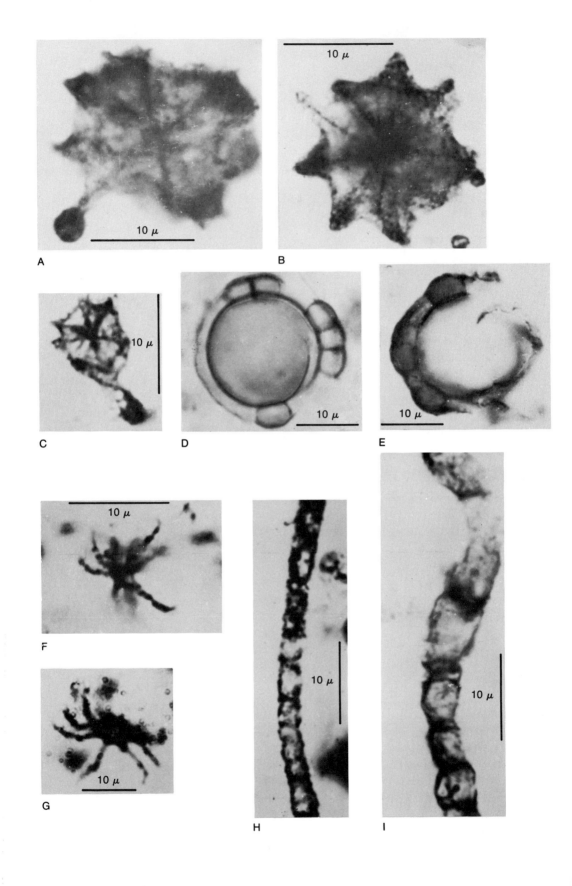

A

B

C

D

E

F

G

H

I

Evolution of Ecosystems

The history of life has been highlighted by adaptive innovations. Many of these have been of minor significance in the overall ecologic history of the earth; others have had special significance in leading directly or indirectly to new ways of life and the exploitation of new habitats. New ecosystems have thus evolved, and the biosphere has changed markedly since life first arose.

Perhaps the first major event in life history was development of the capacity for self-replication using nucleic acids. Unfortunately, the fossil record can tell us little about the timing of this event, which has been followed by hundreds of millions of years of alteration and contamination of sediments.

A second major event in organic evolution was the origin of the cell. The earliest fossils of organisms having an apparent cellular structure reported to date are from the 3.2-billion-year-old Onverwacht Series of South Africa. They consist of carbonaceous rods and spheroids preserved in cherts and other fine-grained sedimentary rocks. Better preserved bacterium-like rods and alga-like spheroids have been described by Barghoorn and Schopf (1966) from the Fig Tree Series of South Africa (minimum age 3.1 billion years). Stromatolites (Figure 9-15) are also known from the early Precambrian, but apparently did not become common until the middle Precambrian.

The most famous and best-studied middle Precambrian deposit yielding fossil microorganisms is the Gunflint Iron Formation of Ontario, Canada, which is approximately 1.9 billion years old. All early and middle Precambrian microorganisms that have been described from this deposit are of two basic types: blue-green alga-like structures (some of which form stromatolites) and bacterium-like structures (Figure 10-38). Blue-green algae and bacteria differ from all other living cellular organisms in possessing certain primitive cell features. The most important of these is that their genetic material is not neatly arranged in discrete chromosomes within a nucleus: their cells are thus said to be *procaryote*. Because many sets of genes may be present in procaryote cells, it is possible that single mutations do not easily affect individual cells, and, because reproduction is not sexual, mutations may not be easily spread throughout populations. These features may explain the unusually slow evolution of the bacteria and blue-green algae and the resulting similarity between their Precambrian and Recent representatives (Schopf, 1968).

FIGURE 10-38
Precambrian microflora of the Gunflint Chert, Lake Superior region. A-C: *Kakebekia umbellata*. D and E: *Eosphaera tyleri*. F and G: *Eoastrion simplex*. All three of these species are of uncertain affinities. H and I: *Gunflintia grandis* (alga-like filaments). (From Barghoorn and Tyler, 1965.)

Chromosomes and a discrete nucleus characterize *eucaryote* cellular organization, which typifies all higher organisms (including higher algae and protozoans). Attainment of the eucaryote organization was one of the most significant steps in organic evolution, apparently opening the way for the rapid evolution and differentiation of unicellular and multicellular taxa. Schopf (1968) has described certain eucaryote fossils, which appear to represent higher algae and fungi, from the Bitter Springs Formation of Australia, which is approximately 1 billion years old (Figure 10-40). The apparent absence of similar forms in the comparable Gunflint deposits and other middle Precambrian sequences that have been studied has led Schopf to suggest that the procaryote-eucaryote transition may not have occurred until the late Precambrian.

The fossil record has thus far revealed little concerning the mode of appearance of higher (eucaryote) algae and primitive multicellular animals. The oldest known multicellular animal faunas are the Ediacara fauna of Australia (Figure 10-39) and similar faunas from other continents. The Ediacara fauna appears to consist entirely of soft-bodied animals, some of which are, or closely resemble, coelenterates, annelids, and echinoderms. The precise age of the Ediacara fauna is uncertain, but the rocks containing it are older than those in which are found fossils of the first known trilobites and younger than rock radiometrically dated as being 700 million years old.

FIGURE   10-39
Representatives of the Ediacara Fauna. A: *Dickensonia*, a problematical segmented form (×1).
B: *Spriggina*, which appears to be an annelid worm (×1.7). C: *Parvincorina*, a genus of unknown affinities (×1.35). D: *Beltanella*, which is thought to be a jellyfish (×0.7). E: *Rangea*, which resembles certain sessile, colonial coelenterates (×0.6). F: *Tribrachidium*, which may belong to the Echinodermata (×1.2).
G: *Arborea*, a genus of unknown affinities (×1.1). (From Glaessner, 1962.)

A   B

C

D

E

F

G

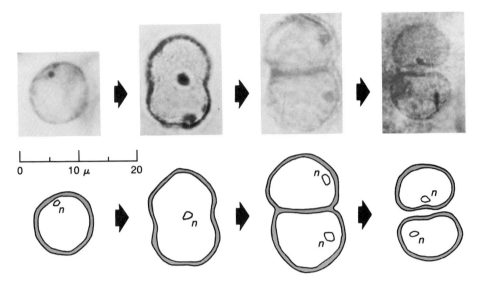

FIGURE 10-40

Glenobotrydion aenigmatis Schopf from the Precambrian Bitter Springs Formation of Australia, apparently showing stages in cell division. Dark bodies (*n*) are believed to be the residues of cell nuclei. (From Schopf, 1968.)

The virtual absence of trails, burrows, or other Lebensspüren in Precambrian rocks suggests that multicellular life did not exist in abundance before the late Precambrian. It apparently arose rapidly and in great diversity soon after the procaryote-eucaryote transition.

The earth's primitive atmosphere is thought to have lacked oxygen, which has been added chiefly through plant photosynthesis. Berkner and Marshall (1964) have suggested that buildup of atmospheric oxygen to a level capable of supporting cellular animal life (about one percent of the present oxygen level) did not occur until about the beginning of the Cambrian. Fischer (1965) has suggested that local "oxygen oases" may have permitted evolution of multicellular marine animals in areas of high plant concentration during the latter part of the Precambrian, even if the atmospheric oxygen pressure at the earth's surface had not yet increased to the amount required for cellular animal life.

The sudden appearance of diverse shelled invertebrate organisms in the Cambrian has inspired many imaginative hypotheses. Brooks and Hutchinson have independently suggested that the chief adaptive advantages of exoskeleton formation were in thwarting newly evolved predators (see Hutchinson, 1958). Others believe that the rapid evolution of marine organisms was made possible by major changes in ocean chemistry. Berkner and Marshall, and Fischer, have suggested that early exoskeletons protected organisms against ultraviolet radiation. The ozone (O_3) layer that shields the earth's surface from intense ultraviolet radiation formed as the earth's atmosphere became enriched in oxygen. Even after the critical oxygen level for animal life was attained, amounts of ultraviolet radiation dangerous for animal life may have reached the earth's surface.

Whatever the reason for the sudden appearance of many shelled marine animals in the Cambrian, the appearance of exoskeletons opened many new doors for evolution. Exoskeletons ultimately served both for protection against predation and for support of muscular systems in taxa that could not have existed as soft-bodied forms.

Phanerozoic adaptive advances that have had widespread ecologic significance include:

1. Origin of the vertebrate skeleton (Ordovician?).
2. Development of jaws and fins in fishes, which led to extensive exploitation of nektonic marine and freshwater habitats (Silurian and Devonian).
3. The rise of land plants, which opened the way for vertebrate invasion of the land (Siluro-Devonian).
4. Origin of legs and lungs in amphibian precursors, which made possible the invasion of the terrestrial realm by vertebrates (Devonian).
5. Evolution of the amniote egg in reptiles, which freed them from dependence on aqueous habitats for reproduction and permitted their extensive adaptation to terrestrial niches (Carboniferous).
6. Development of breathing structures and wings in insects, which led to their rapid diversification (Carboniferous).
7. Development of the pollen-and-seed method of reproduction by gymnosperms, which freed them from restriction to moist surroundings for reproduction and permitted their extensive exploitation of terrestrial habitats (late Paleozoic).
8. Extensive diversification of shelled marine invertebrates and their development of special devices for infaunal life, including the siphons of bivalve and gastropod molluscs and the petaloid tube feet of echinoids; all of these infaunal groups diversified extensively in the Mesozoic Era.
9. Evolution of dinosaur hip structures, which gave dinosaurs a remarkable capacity for locomotion and led to their unprecedented success in terrestrial habitats (Triassic).
10. Origin of feathered, warm-blooded birds, which greatly exceeded their flying reptile ancestors in their range of adaptation (Jurassic).
11. Appearance of the flower and enclosed seed in angiosperms, which opened the way for their tremendous adaptive radiation into previously unoccupied niches (Early Cretaceous).
12. Appearance of basic mammalian characters, especially placental embryonic development, which preceded the many advanced mammalian adaptations of the Cenozoic Era (Triassic-Cretaceous).
13. Appearance of angiosperm grasses, which created many new grassland niches for mammal groups and marine eel-grass niches for benthonic invertebrates (early Tertiary).
14. Origin of grasping hands and stereoscopic vision in primates, which preceded their remarkable increase in brain capacity (early Tertiary).

The precise times of origin of most major adaptive innovations remain uncertain. Inasmuch as the innovations marked major ecologic advances and gave rise to new higher categories, however, we would expect them to have been followed rapidly by conspicuous adaptive radiations. The fossil record, therefore, is probably reasonably reliable in documenting the general time of origin of major innovations and in providing a chronology of the exploitation of major earth-surface habitats.

Supplementary Reading

DeBeer, G. R. (1958) *Embryos and Ancestors*. London, Oxford University Press, 197 p. (A thorough review of heterochrony, with many examples.)

Olson, E. C. (1965) *The Evolution of Life*. London, Weidenfeld and Nicolson, 300 p. (A discussion of the major aspects of evolution, especially those evident in the vertebrate fossil record.)

Rensch, B. (1960) *Evolution above the Species Level*. New York, Columbia University Press, 419 p. (A treatise on the evolution of higher taxa with references to many German articles.)

Simpson, G. G. (1953) *The Major Features of Evolution*. New York, Columbia University Press, 434 p. (A classic discussion of rates, trends, and patterns of evolution.)

CHAPTER 11

Biostratigraphy

Stratigraphy is the study of the geometry, composition, and time relations of stratified rocks. Its special concern is with the history of these rocks, including their fossil components. Biostratigraphy is simply that branch of stratigraphy that is primarily concerned with fossils. In its broadest sense, biostratigraphy can be considered to include most of the topics discussed in this volume. In its narrower sense, as used here, it concerns the spatial distribution and temporal relations of fossils and fossil-bearing rocks.

In 1961, the American Commission on Stratigraphic Nomenclature published its "Code on Stratigraphic Nomenclature" in the Bulletin of the American Association of Petroleum Geologists. To a much greater extent than the "International Code of Zoological Nomenclature" (page 107), this stratigraphic code embraces scientific concepts rather than simply rules of nomenclature. It therefore contains a larger number of controversial statements than the zoological code. Still, it is a valuable formulation of prevailing stratigraphic opinions and has tended to promote uniformity in classification and naming of stratigraphic units. It will certainly be revised to accommodate changes in prevailing ideas and techniques in the years to come. In general, stratigraphy is a highly subjective field of study for which few unique principles serve as an underpinning. Perhaps the only basic stratigraphic principle is the "law" of superposition, which states that in any undisturbed sedimentary sequence the oldest bed is at the base and the youngest bed at the top. Many principles of paleoecology, evolution, sedimentology, and geomorphology are commonly applied to stratigraphic problems. For example, the so-called "law of biotic succession," often cited in elementary or historical discussions of biostratigraphy, states that biotas have followed one another in an orderly succession through geologic time. This succession is simply the product of organic evolution and might properly be considered to be a part of evolutionary paleontology borrowed by biostratigraphy.

Rock-stratigraphic Units

Stratigraphy deals with units, primarily units of rock and of time that are used to divide stratigraphic sequences and geologic history. The stratigraphic record is divided into three-dimensional rock bodies on the basis of lithology, the physical character of rock. Rock units, often called ***rock-stratigraphic units,*** are recognized without regard to time relations or fossil content, except insofar as fossils contribute to their physical and chemical composition. Units are recognized on the basis of lithologic uniformity, which is inevitably judged with a great deal of subjectivity. The basic rock-stratigraphic unit is the ***formation.*** Formations may be lumped into higher units called ***groups,*** or divided into smaller units called ***members.***

It is commonly stipulated that to be designated a formation a unit should be "mappable." Because of the great variety of scales used for geologic maps, this stipulation is of questionable value. Much like an arbitrary taxonomic subdivision, such as a "genus" or "family," a formation is a category that defies quantitative description. As with higher taxonomic units, most workers develop a general concept of the traditional magnitude of rock units. Still, in both taxonomy and stratigraphy, opinions differ among various workers as to the amount of subdivision desirable in particular instances. In general, increased detail of knowledge tends to promote increased subdivision.

One of the first facts learned by the introductory geology student is that the stratigraphic record is far from complete. In all regions major unconformities in rocks and erosional surfaces at the earth-air interface reveal the existence of large gaps in the rock record.

The student also learns that most rock units seeming to show continuous deposition were actually formed by discontinuous deposition. Chemical precipitation in certain restricted environments, such as closed evaporite basins, may produce nearly continuous deposition, but in most other settings sediment deposition is sporadic. Nearly all marine detrital sediments undergo some transportation on the sea floor before reaching their site of final deposition. Many, of course, are carried into marine depositional basins from the land. While sediment may be supplied to a depositional basin nearly continuously (usually with fluctuations in *rate* of supply), at any given time *final* deposition of sediment layers will occur only in certain areas. Non-deposition, or erosion, will occur simultaneously in others. Periodic "instantaneous" agents such as storms may produce waves and currents that erode thick sediment layers that took many years to accumulate.

A single bed, or lamina, of sediment represents a single depositional event. Surfaces between beds or laminae then indicate gaps in the rock record, just as unconformities do on a larger scale. Early in the twentieth century, Barrell recognized the discontinuousness of deposition and for gaps due to short time periods of nondeposition or erosion he introduced the term *diastems*.

One of the most striking illustrations of the discontinuousness of sediment deposition comes from study of preserved animal burrows. Figure 11-1, from a report by Rhoads (1966), shows a slab of Silurian rock sectioned perpendicular to bedding. The rock contains three distinct types of burrows, which tended to destroy some of the depositional bedding. Burrowing extended downward from the sediment-water interface to a depth of several centimeters. If deposition had been continuous and burrowing activity had kept pace with it, we would expect the burrowing animals to have maintained their position in the upper few centimeters during sediment accumulation and to have largely destroyed primary bedding throughout.

In fact, the rock section shows four layers (A-D) in which bedding was not entirely destroyed. The upper beds in these layers (especially in the second layer from the bottom) are partly obliterated by burrows, producing irregular upper surfaces. Rhoads has interpreted the rock slab as representing four distinct depositional events. Careful observation reveals that the four distinct bedded layers, each several centimeters thick, were deposited in succession. Each layer was suddenly "dumped" on the one below. The sediment-water interface on which each was deposited is shown in Figure 11-1. Some of the lowermost beds of each layer are preserved because burrowing organisms established themselves only in its upper part, near the sediment-water interface.

Vertical "escape" burrows are represented by heavy arrows in Figure 11-1. They cut across the normal burrows, most of which are oriented horizontally. The escape burrows strongly support the idea of sudden deposition, for apparently they were upward migration routes of animals to a new sediment-water interface. Some animals may not have been so fortunate, perhaps being unable to make their way upward through a new blanket of sediment. The idea of sudden deposition also gains favor from the fact that layer B, whose lower beds have been least disturbed by burrowing, was the thickest of the four.

Similar analysis by Seilacher (1962) has demonstrated that many of the famous deep-water flysch deposits of Europe were formed by successive events of deposition, in which thick beds were laid down suddenly from time to time. Most burrows and trails in flysch deposits are preserved on the bottoms of sandstone beds underlain by shale. For years there was disagreement as to whether burrowing took place only at the interface between sand and shale or occurred throughout homogeneous sandstone units but was

FIGURE 11-1
Section of rock from the Silurian Stonehouse Formation, Nova Scotia, with preserved animal burrows. (From Rhoads, 1966; sample, courtesy of R. K. Bambach.)

10 cm

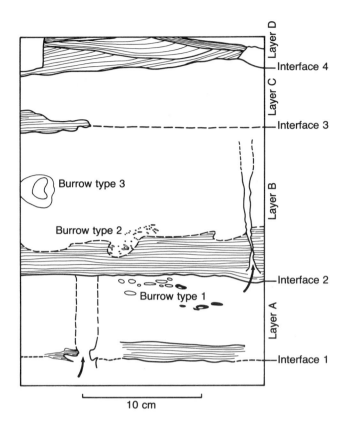

10 cm

only well preserved at the lower surface (which individual burrows disrupted). Using various lines of evidence, Seilacher concluded that some animals were surface burrowers and others penetrated the sediment to considerable depth. He next restricted his interpretation to the latter group, and reasoned that each species must have burrowed to a characteristic sediment depth. Therefore, if deposition of each bed was sudden, there should be a maximum thickness for all sandstone units on whose undersurfaces each burrow type is found. Figure 11-2 shows that data compiled for various burrow types bear out this prediction and support the idea of sudden deposition.

Even though these examples offer unusually favorable opportunities for demonstration of discontinuous deposition, there is reason to believe that most sedimentary rocks have had similar histories of discontinuous deposition, whether deposition of the included beds was separated by gaps of days, years, or thousands of years.

Just as environments change laterally over the earth's surface, sediments being deposited simultaneously in different places differ in chemical and physical characteristics. The general aspect of a body of rock deposited over a wide area during a certain time interval may, therefore, show considerable variation. Portions of a body of rock showing lateral changes in

Predepositional burrows · Post depositional burrows

Thickness of beds, in cm	Total number of beds in section A	B	C	Granularia A	B	C	Fucusopsis A	B	C	Phycosiphon A	B	C	Scolicia A	B	C	Nereites A	B	C	Lorenzinia A	B	C	[fan] A	B	C	Palaeodictyon Regular A	B	C	Palaeodictyon Irregular A	B	C	Scolicia A	B	C	Ceratophycus A	B	C
0–1	21	–	3	6	–	–	–	–	–	4	–	1	10	–	2	3	–	–	3	–	–	2	–	–	4	–	–	–	–	–	2	–	–	–	–	–
1–2	41	12	16	19	3	5	–	–	–	13	2	8	12	7	6	5	1	2	9	–	1	3	–	–	4	1	5	3	1	–	1	–	1	1	2	1
2–3	33	14	18	18	7	8	–	–	–	12	1	7	4	5	1	4	2	3	5	–	1	2	–	–	7	2	6	1	2	–	5	2	–	12	1	–
3–4	19	17	9	15	9	7	1	–	–	9	4	1	–	4	–	2	1	2	5	2	–	1	–	–	2	3	7	2	–	–	2	1	–	8	–	–
–5	18	14	7	15	7	4	1	–	–	9	3	4	1	3	1	1	–	1	3	–	1	1	–	–	4	1	6	3	3	–	1	3	–	4	3	–
–6	17	15	8	10	14	4	–	–	–	4	4	6	–	–	–	–	–	–	4	4	1	–	1	–	1	2	3	2	2	1	4	1	–	1	2	–
–7	12	11	10	11	11	9	1	–	–	5	1	6	1	–	1	1	–	1	–	–	–	–	–	–	4	1	5	1	1	–	5	2	–	2	5	–
–8	10	11	8	7	11	4	–	–	–	6	–	2	–	–	–				–	1	–	–	–	–	2	–	6	–	–	–	1	2	–	3	–	–
–9	12	14	6	10	13	4	–	–	–	2	–	2	1	–	1				1	–	2	1	–	–	1	–	3	1	1	–	4	1	–	2	3	–
–10	12	5	2	12	4	2	1	–	–	6	1	–							2	–	–	1	–	–	–	–	–	2	1	–	2	1	–	3	2	–
–12	13	17	5	11	16	3	–	–	–	3	1	3							5	1	–	1	–	–	6	7	3	2	1	–	4	4	–	–	1	–
–14	7	16	4	7	16	4	1	–	–	1	–	–							–	–	–	1	–	–	2	7	3	–	3	–	1	3	–	1	1	–
–16	5	5	5	6	5	5	–	–	–										2	–	–	–	–	–	2	–	–	2	–	–	3	1	2	–	3	–
–18	7	6	1	7	6	1	–	1	–										2	1	–	1	–	–	–	–	–	2	1	–	3	1	–	1	–	–
–20	8	8	–	6	8	–													2	–	–	1	–	–	4	1	–	–	–	–	2	1	–	2	–	–
–25	14	10	–	12	10	–													5	–	–				–	–	–	2	–	–	5	1	–	–	–	–
–30	9	10	1	7	9	1													3	1	–	1	1	–	1	1	–	–	–	–	3	1	–	3	–	–
–35	10	6	–	10	6	–													–	1	–	–	–	–	1	1	–	1	1	–	2	1	–	1	1	–
–40	4	3	–	4	3	–													–	1	–	1	–	–	–	–	–	–	–	–	2	–	–	–	–	–
–45	7	7	–	7	7	–													–	–	–	1	–	–	–	–	–	1	1	–	–	–	–	–	1	–
–50	4	7	–	3	7	–													1	1	–	1	–	–	1	1	–	–	–	–	1	1	–	–	1	–
–60	8	3	–	8	3	–													1	1	–	2	1	–	–	–	–	2	–	–	3	–	–	3	1	–
–80	7	5	–	7	5	–													–	–	–				–	–	–	1	–	–	1	–	–	–	–	–
–100	2	3	–	2	3	–													–	–	–				1	–	–	–	–	–	–	–	–	–	–	–
–150	12	8	–	12	7	–													2	1	–				1	–	–	1	1	–	2	–	–	–	–	–
–200	4	7	–	3	7	–													1	1	–				–	–	–	1	1	–	–	2	–	1	–	–
Sum	655			493			6			130			60			28			72			24			121			49			89			75		

aspect are known as *facies.* In recognizing facies, emphasis may be placed on *lithofacies,* based on prominent lithologic features, or *biofacies,* based on fossil features. Facies may be found within rock units that are formally classed as groups, formations, or members. Facies may be used to define lateral boundaries between formal rock units, though the boundaries are then usually gradational rather than abrupt.

Biostratigraphic Units — The Biozone

Superimposed on the rock-unit subdivision of the stratigraphic record is a system of subdivision based on fossil occurrences. The units of this subdivision, called *biostratigraphic units,* are also tangible rock bodies, but their boundaries are defined by various paleontologic criteria, such as the appearance, maximum abundance, and disappearance of fossil species or genera in various local rock sequences.

The fundamental biostratigraphic unit is the zone. Many sorts of zones have been recognized. We can envisage an abstract kind of zone known as a *biozone,* which represents all rocks throughout the world that were deposited during the time interval in which a species lived. A biozone is an abstraction because we can never delineate it physically. No species is present in all rocks of its biozone.

It is instructive to examine the reasons that species have not filled their biozones. We can assume that many species have originated through evolutionary divergence of geographically isolated populations of a preexisting species. The extent to which a new species arising in this manner has come to fill its biozone has depended partly on how rapidly it spread throughout areas of the earth's surface that it could potentially inhabit. In a sense, we are comparing rates of evolution and rates of geographic dispersal.

If there were no *effective barriers,* most species would have spread over large geographic areas quite rapidly relative to rates of species evolution. Many examples of the rapid dispersal of species given access to new regions are presented in the excellent book by Elton (1958) on the ecology of animal invasions. In fact, however, geographic restrictions apply to all species. Biogeographic ranges, particularly their latitudinal extents, are limited by environmental conditions, especially temperature. In addition oceans are barriers to terrestrial animals and land masses are barriers to marine animals. Large abrupt changes in altitude on land or of depth in the ocean may also form effective barriers.

It is difficult to formulate general rules to describe the effectiveness of barriers in preventing species dispersal. In part, dispersal across barriers

FIGURE 11-2
Burrows and trails preserved on the undersurfaces of beds in three stratigraphic sections (A, B, and C) in the Flysch of northern Spain. Postdepositional burrow types tend to be restricted to thinner beds; predepositional burrows show no such relationship to thickness of bed. (From Seilacher, 1962.)

is a matter of chance. Such agents as floating logs, bird droppings, and strong winds have undoubtedly transported many species across major barriers. Simpson (1940a) has referred to chance mechanisms of dispersal as "sweepstakes routes." Chains of islands often form stepping stones for "sweepstakes" migration of both terrestrial species and shallow-water marine species from one continent, or continental margin, to another.

More continuous, but narrow, dispersal routes are sometimes referred to as "corridors" of migration. Because of their narrowness, corridors of migration are subject to opening and closing, perhaps repeatedly, in the course of geologic history. The Isthmus of Panama was a land corridor to South America for North American placental mammals that enabled them to displace the previously isolated South American marsupial groups near the close of the Pliocene Epoch (page 295).

Many plant and animal groups have developed special mechanisms for migration across barriers. For example, planktonic larvae may be viewed as specially adapted dispersal stages in the life histories of certain benthonic marine invertebrates (page 201). It has been found that most planktonic larvae can temporarily delay metamorphosis if they fail to locate a suitable substratum. For many years it was believed that transport of larvae of shallow-water invertebrates across the Atlantic Ocean was virtually impossible. Scheltema (1968), however, has found that larvae of shallow-water benthonic species found on both sides of the Atlantic are commonly transported across the Atlantic from west to east and east to west by major ocean currents (Figure 11-3). Seeds and spores are important wind-transported dispersal agents of many terrestrial plant groups. Under unfavorable environmental conditions, many aquatic and terrestrial unicellular organisms alter to inactive, cyst-like resting stages that, like spores, may be blown by the wind to transport species across otherwise effective barriers.

It is not uncommon for a species to disappear from most parts of its range but persist in a limited region as one or more relict (page 295) populations.

There are additional reasons that no fossil evidence of a species is found in many parts of its biozone. Even within their geographic ranges, species can generally inhabit only certain types of local environments. Consequently, many fossil species are noted for their restriction to certain rock types. Such fossils may be known as "facies fossils." In most instances, their local distribution patterns reflect primary ecologic distribution. Furthermore, no species is preserved in all rock units deposited in environments in which it lived. Finally, a species may actually occur in a rock unit but may be misidentified because of inadequate preservation or confusion with another species.

Prevailing winds and ocean currents disrupt latitudinal temperature gradients in many parts of the world. Sharp temperature changes and abrupt physiographic barriers tend to confine *groups* of species to certain regions, which are called **biogeographic provinces.** Species limited to any given region are termed **endemic.** Most recognized biogeographic provinces contain some endemic species, but also share species with other provinces. Hence, provincial boundaries are not always clear-cut; differences of opinion as to

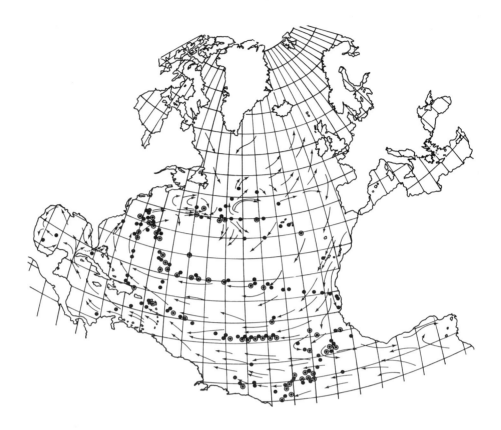

FIGURE 11-3.
Distribution of planktonic larvae of the gastropod family Architectonicidae.
Circled dots indicate stations where larvae of *Philippia krebsii*, a member of
the family, have been found. Adults belonging to the Architectonicidae
occur in shallow-water areas throughout the tropics. Arrows show major
water currents at the ocean surface. (From Scheltema, 1968.)

precise boundary designations and degrees of subdivision desirable have
led to a variety of provincial classifications in certain parts of the world.

Several provincial classifications for the coastal waters of eastern North
America are shown in Figure 11-4. Eastern and western boundaries are the
continental margin and shoreline. The Gulf Stream swings away from the
coast in the vicinity of Cape Hatteras, forming a sharp temperature discontinuity. Likewise, the Labrador Current, sweeping south along the north-
eastern coast, is diverted seaward at Cape Cod. Northern and southern
boundaries in other parts of the world do not necessarily coincide with those
of Figure 11-4 (in either latitude or mean annual temperature).

In the paleoecology chapter we discussed the relationship between temperature gradients and taxonomic-diversity gradients. We would expect
widespread warm climates of the past to have produced broad biogeographic
provinces and steep temperature gradients to have produced narrower ones.
One of the most interesting and best-studied geologic periods from this

standpoint is the Jurassic. Arkell (1956) and other authors have noted that many Lower Jurassic taxa (especially of cephalopods) were cosmopolitan; for every recognized North American ammonite species there is a similar European counterpart, for example. In contrast, Middle and Upper Jurassic faunas are markedly provincial. Arkell recognized three major Upper Jurassic faunal provinces, which he termed "realms": the Boreal Realm (of northern latitudes), the Pacific Realm (of areas bordering the Pacific Ocean), and the Tethyan Realm (including the rest of the world, but especially well represented by fossil faunas of the Mediterranean region). Origin of the realms was by no means instantaneous, and their faunas shifted and overlapped from time to time. But the realms tended to persist and are recognized in the Cretaceous as well. Most workers have considered their distinctness to have been caused by the development of steep temperature gradients and physiographic barriers to dispersal. Hallam (1969), however, has suggested that Arkell's Pacific Realm should be considered a subdivision of the

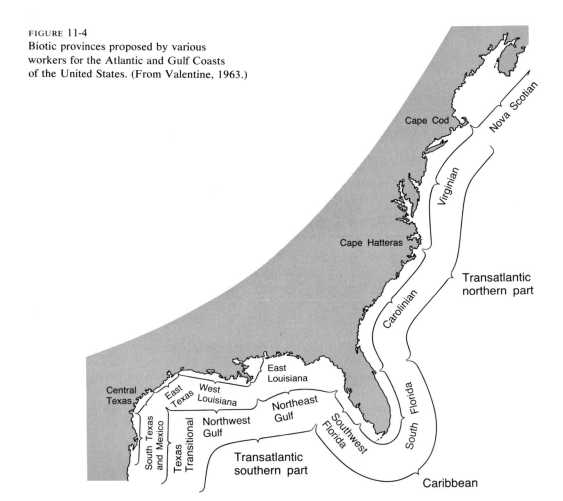

FIGURE 11-4
Biotic provinces proposed by various
workers for the Atlantic and Gulf Coasts
of the United States. (From Valentine, 1963.)

Tethyan Realm. He has also proposed that the Boreal Realm was not established and maintained by cold temperatures or physiographic barriers, but was formed by a vast inland sea of slightly reduced salinity. Hallam has compared its faunas, which were characterized by low taxonomic diversity, to those of the modern Baltic Sea. Whatever the cause of Jurassic faunal differentiation, it is easy to see why intercontinental correlation is generally simpler and more accurate for Lower Jurassic sequences than for Upper Jurassic sequences.

Valentine (1963) has pointed out that the geologic record of a biogeographic province represents a biostratigraphic unit according to the Code of Stratigraphic Nomenclature. He has recommended the adoption of a formal system of biostratigraphic nomenclature for provincial units. Such a system might be very useful.

Correlation with Fossils

Given the partial representation of fossil species in their biozones, to what extent can the stratigraphic distributions of fossils be used profitably to determine relative geologic time? From our previous discussions, it should be evident that biostratigraphic units can only be used in an approximate way to determine time relationships. Establishing the time equivalence of two spatially separate stratigraphic units is known as **correlation**. Unfortunately, the term "correlation" is also used by some workers to imply equivalence of rock type or fossil content without regard to time. It is less confusing to use terms like "lithologic equivalence" and "biostratigraphic equivalence" for the latter sorts of relationships.

Time correlation, though usually only approximate, can be undertaken by various methods, only some of which make use of fossils. Nevertheless, fossils represent by far the most important tools for time correlation. Establishment of biostratigraphic equivalence is the first step. This equivalence may then be interpreted as demonstrating approximate time equivalence, or correlation. We will review several types of biostratigraphic equivalence in the following sections.

STRATIGRAPHIC RANGES AND ZONES

Most stratigraphic work is based on studies of vertical fossil distribution in local stratigraphic sections (rock sequences exposed for study). A stratigraphic section is really a three-dimensional exposure of rock strata, often on the side of a hill or along a creek bed, where there is considerable relief. For correlation purposes, however, it is commonly treated as if it were one-dimensional. The two-dimensional columns used to represent stratigraphic sections in geologic literature merely serve to permit diagrammatic portrayal of rock types at various levels (Figure 11-5). Techniques used to study faunal distribution in a stratigraphic section vary with structural patterns of the rock units, terrain, lithology, and fossil content. In common

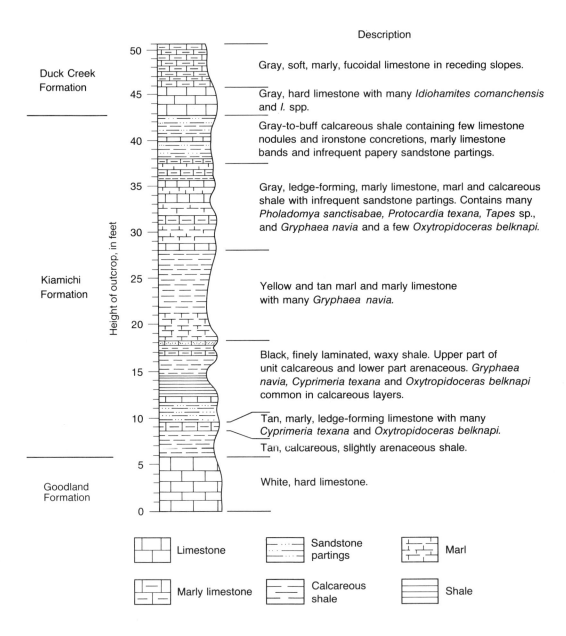

Description

Gray, soft, marly, fucoidal limestone in receding slopes.

Gray, hard limestone with many *Idiohamites comanchensis* and *I.* spp.

Gray-to-buff calcareous shale containing few limestone nodules and ironstone concretions, marly limestone bands and infrequent papery sandstone partings.

Gray, ledge-forming, marly limestone, marl and calcareous shale with infrequent sandstone partings. Contains many *Pholadomya sanctisabae*, *Protocardia texana*, *Tapes* sp., and *Gryphaea navia* and a few *Oxytropidoceras belknapi*.

Yellow and tan marl and marly limestone with many *Gryphaea navia*.

Black, finely laminated, waxy shale. Upper part of unit calcareous and lower part arenaceous. *Gryphaea navia*, *Cyprimeria texana* and *Oxytropidoceras belknapi* common in calcareous layers.

Tan, marly, ledge-forming limestone with many *Cyprimeria texana* and *Oxytropidoceras belknapi*.

Tan, calcareous, slightly arenaceous shale.

White, hard limestone.

Duck Creek Formation

Kiamichi Formation

Goodland Formation

Height of outcrop, in feet

	Limestone
	Marly limestone
	Sandstone partings
	Calcareous shale
	Marl
	Shale

FIGURE 11-5
Stratigraphic column for the Cretaceous Kiamichi Formation, at a bluff southwest of Meacham Field, Tarrant County, Texas. The scale is in feet above the base of the outcrop, which is located within the Goodland Formation. (From Perkins, 1960.)

stratigraphic practice, a section is measured in feet or meters, and significant changes in lithology are then plotted on the vertical scale. Fossil species are carefully collected throughout the section and their vertical positions are recorded. It is then possible to plot the ***stratigraphic range*** of each fossil species on the scale. This is simply the stratigraphic interval bounded by its lowermost and uppermost occurrence. Species ranges of a typical stratigraphic column are shown in Figure 11-6.

A common method of correlation uses upper and lower boundaries of the stratigraphic range of a single species in two or more stratigraphic sections. Under certain circumstances, such "earliest" or "latest" appearances indicate approximate time equivalence. They may, however, represent different times because of species migration or differing time of extinction from place to place. Palmer (1965), for example, has demonstrated large-scale migration of late Cambrian trilobite faunas of the Great Basin region of the United States. Each of the three faunas shown in Figure 11-7 apparently arose from an offshore stock. Detailed correlation reveals that, after its appearance, the second fauna migrated from deep water toward the shoreline, apparently displacing the preceding fauna as it moved eastward. After undergoing considerable evolution, it was displaced by eastward migration of the third fauna. Thus, the biostratigraphic unit defined by each of the three faunas (such as the one defined by the pterocephaliid fauna) has non-synchronous boundaries. The probable reason that migration was not an "instantaneous" event in geologic time was that displacement took place gradually and as the result of ecologic competition. There is no evidence that faunas changed in response to environmental changes, for major faunal transitions do not coincide with major changes in sediment character.

A correlation method that has historical importance is based on the maximum abundance, or "acme," of a species in its stratigraphic range. The importance of local environment in determining abundance in life and extent of preservation after death invalidate this approach for correlation over large geographic distances. In a particular place, however, where uniform environmental conditions may have prevailed, correlations judiciously based on maximum abundance may be reasonably accurate.

Obviously, reliance on more than one species would tend to improve correlation accuracy. Most recognized biostratigraphic zones are based on the occurrence together of two or more species. Such a zone, known as an ***assemblage zone,*** is customarily named for a single characteristic genus or species, (which need not necessarily be present throughout the entire horizontal or vertical extent of the zone).

PERCENTAGE OF COMMON TAXA

Some correlations have been based on comparison of a local fossil assemblage with "standard" assemblages from regions in which stratigraphic relationships are better understood. Similarities are usually evaluated in terms of numbers of taxa (usually species) found in both the assemblage in question and a standard assemblage. For example, the percentage of tax?

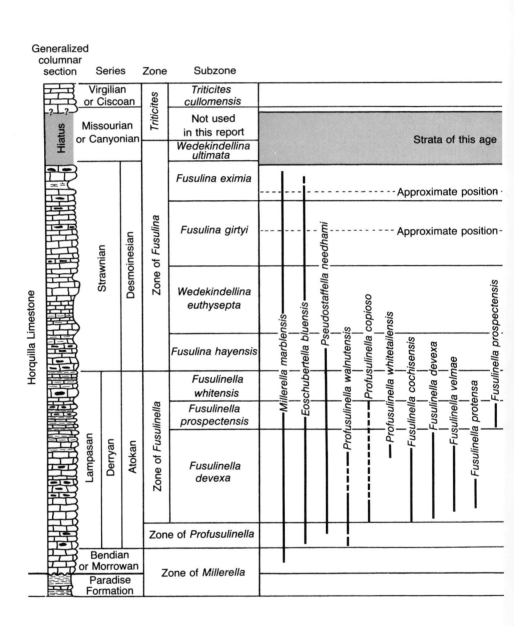

FIGURE 11-6
Generalized section of the lower and middle parts of the Pennsylvanian Horquilla Limestone of Arizona. (From Ross and Sabins, 1965.)

Stratigraphic range of Fusulinid species

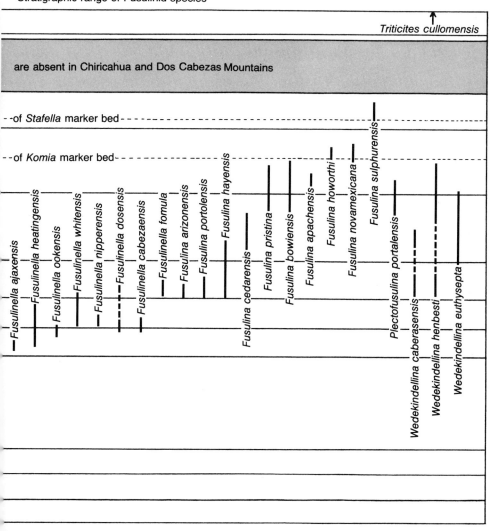

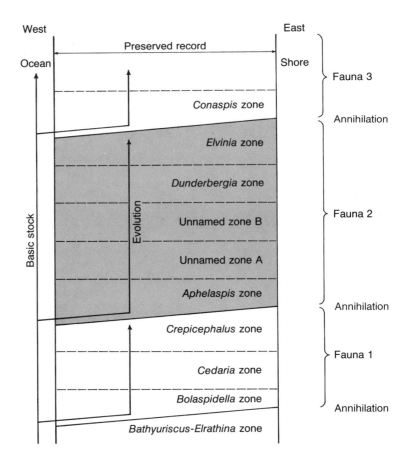

FIGURE 11-7
Schematic diagram showing successive invasions of shallow
Late Cambrian seas of the western United States by
trilobite faunas evolving from a basic parent stock in deeper
water to the west. Horizontal dashed lines represent time.
(From Palmer, 1965.)

of either assemblage also found in the other assemblage may be calculated
and used as an index of comparison. An alternative is to use the ratio between number of common taxa and total number of unique taxa in the two
assemblages.

Simpson (1960) has advocated use of the following index for comparison
of two fossil faunas:

$$\frac{\text{number of common taxa}}{\text{number of taxa in the smaller fauna}} \times 100.$$

This index is especially meaningful when one fauna contains many more
taxa than the other. (The sparser fauna usually represents less complete

preservation of a community.) Simpson has also discussed methods for inclusion of relative abundance data, which are especially important in comparison of closely similar faunas.

Correlations based on percentages of common species can be made without concern for stratigraphic ranges of taxa, but accuracy can be greatly reduced by ecologic biases. It is therefore usually necessary to qualify purely statistical conclusions, depending upon knowledge of biozone ranges and species ecology.

INDEX FOSSILS

Taxa found to be especially useful in correlation are commonly referred to as *index fossils,* or *guide fossils.* The attributes of an ideal index fossil are: wide geographic distribution, ecologic tolerance, abundance, rapid evolutionary rate, and distinct morphologic features. Index fossils are especially useful for interregional and intercontinental correlation.

Planktonic and nektonic species are generally the best marine index fossils because their occurrence in life is usually widespread and independent of local benthonic conditions. Good examples are provided by many Ordovician and Silurian graptolite taxa and Mesozoic ammonite taxa. Benthonic groups, such as Cambrian trilobites and late Paleozoic fusulinid Foraminifera, have also produced important index fossils.

Unfortunately, a second, spurious usage of the term "index fossil" has become common in some geologic circles. In this usage, which has value in geologic mapping, a taxon is described as an "index fossil" if it is especially common in a local rock-stratigraphic unit and can therefore be used in identifying isolated outcrops of the rock unit when lithology alone is inadequate. This use of the term is confusing and should be avoided.

MORPHOLOGIC FEATURES

Sudden morphologic changes within species and genera can be used profitably for correlation, especially of rocks and sediments of Cenozoic age. Among the most spectacular examples are those involving shell-coiling direction (dextral versus sinistral) in planktonic foraminifera. Coiling reversal in *Globorotalia menardii* has been widely used for recognition of the Pliocene-Pleistocene boundary in deep-sea sediment cores (Figure 11-8). Shell-coiling direction of planktonic foraminifera has also been used for correlation of older Cenozoic deposits. Coiling direction appears to be affected by temperature, and widespread reversal of coiling direction is apparently a response to a major change in climate. It seems likely that coiling direction itself has no adaptive significance. It is probably a nonadaptive manifestation of a single gene or group of genes that also controls temperature adaptation (Mayr, 1963, p. 236).

The great utility of coiling-direction reversals for correlation stems from the fact that reversal is apparently triggered suddenly and occurs over broad geographic areas.

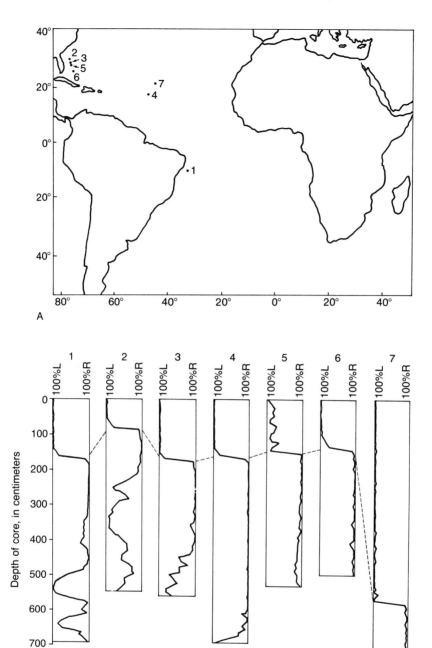

FIGURE 11-8
Changes in direction of shell coiling among fossil Foraminifera which are assigned to *Globorotalia menardii,* but which may actually belong to more than one species. A: Places at which sediment cores were taken.
B: Coiling direction of fossils in the sediment cores. L is coiling to the left; R, to the right. The Pliocene-Pleistocene boundary, as delineated by the reversal in coiling direction, is indicated by the dashed lines between cores. (From Ericson, Ewing, and Wollin, 1963.)

ECOLOGIC PATTERNS

Under certain circumstances, facies and their characteristic fossils exhibit stratigraphic patterns that permit correlation of two or more sections. A simple example is illustrated in Figure 11-9 in which a hypothetical stratigraphic cross section reveals a period of *transgression* followed by a period of *regression*. During deposition of the hypothetical rock units, land existed in the west and the ocean lay to the east. The shoreline shifted first to the west, as the sea encroached on the land, and later to the east, as the sea receded from the land. Each of the marine rock units represents a distinct facies, with its own characteristic lithology and fossil content. The marine facies represent adjacent environments that trended parallel to the shoreline. The point marking the westernmost extent of each facies represents the time of maximum westward transgression. A line drawn through these points for all facies is a correlation line, representing a single time in the history of development of the sequence. In most geologic settings only isolated outcrops or well cuttings are available for study. Still, the single-time surface can be approximately determined from the vertical position in all sections of the facies deposited in the deepest water.

FIGURE 11-9

Cross section of a hypothetical depositional sequence produced by a transgression and regression. The time line (line connecting points of deepest deposition for each column, *A* through *G*) represents the time of maximum westward transgression. (From Israelsky, 1949.)

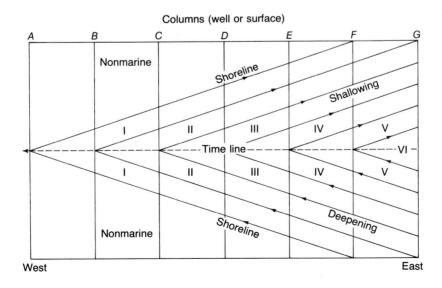

QUANTITATIVE CORRELATION METHODS

Shaw (1964) has presented a statistical method for correlation based on detailed knowledge of groups of local sections containing numerous fossil species in common.

Each of two local sections to be correlated is first plotted on a different axis of a graph. The units of the graph axes represent distances, usually measured in feet, from the bases of the sections to their tops. Each species occurring in the two sections can then be represented on the graph by points representing the base of its range, the top of its range, or both, in the two sections (Figure 11-10). If species ranges in the two sections (X and Y) are identical, the line connecting all points will be straight and will form an angle of 45° with each axis.

Let us postulate that the *times* of appearance and disappearance of each of the nine species common to both sections were identical at the two locations. We will retain this assumption throughout our discussion. The line connecting points will then be an approximate correlation line. The reason that it is not necessarily an exact correlation line, even though the sections are identical, is that our hypothetical situation indicates only that each of the plotted *points* is a point of correlation. If the rates of rock accumulation differ during a time interval between two points, then the real correlation curve, if drawn, would not follow a straight line between the two points. The limiting cases in an analysis like this would be those in which all rock accumulation of the interval between two points occurred at one section before there was any accumulation in the other section. These cases, which are shown in Figure 11-11, would enable the investigator to determine the *maximum* error that could exist by simply connecting the two points with a straight line. Evaluation of the *actual* amount of error is a matter of judgment, based on geologic information. Shaw believes that use of a straight line is reasonable for most sections containing marine sediments deposited in what are now the central regions of continents.

The phrase "rock accumulation" is used instead of "sediment accumulation" because different types of sediment are compacted to different extents after deposition. Rock thickness in the stratigraphic record is therefore not proportional to original sediment thickness in a constant way.

Consider the case in which rates of accumulation are approximately constant for each of two sections, but differ between the sections. We are still assuming that the ranges of a species in the two sections represent a single time interval. The species ranges will be spatially longer in one section than in the other and the line connecting points on the graph will no longer form an angle of 45° with the axes (Figure 11-12).

For any given time there is a ratio between the rates of rock accumulation in two sections being studied. Often this ratio may change. Shaw has found that a "dog leg" pattern (Figure 11-13) is typical of graphs used to compare ranges of species common to two sections. In other words, changes in relative rates of accumulation between sections tend to be abrupt. Otherwise, curved graphic patterns would be more common.

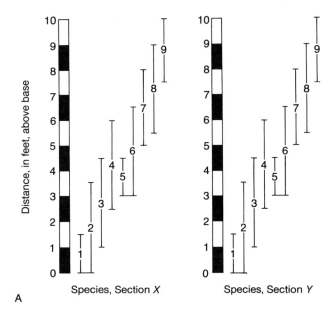

Species, Section X

A

FIGURE 11-10
Confirmation, through use of a graph, that ranges of nine fossil species are identical in two hypothetical sections. A: Diagrams showing vertical ranges of the species in sections. B: Graph on which has been plotted the top and base of the range of each species in the two sections. Connecting all the points plotted gives a straight line at 45°. (From Shaw, 1964.)

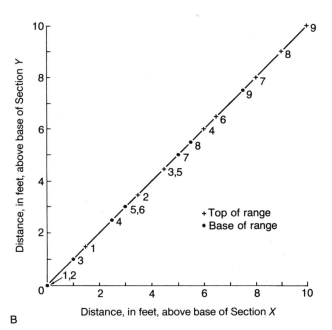

B

Statistical theory can be applied to the raw stratigraphic data to compute "best-fit" correlation lines for graphic representation. The statistical signifi-cance of a correlation line (an estimate of the probability of error) can also be calculated.

Having made correlation-line comparisons of several local sections, it is useful to construct what Shaw calls a "composite standard." Usually, the section that is most complete (having no major depositional breaks) and con-tains the largest biota is used as a base. Fossil ranges of all other local sec-tions can be projected onto the base section. This is accomplished for each of the other sections by applying the equation that represents its line of correlation with the base section. In effect, differences in rates of rock ac-cumulation among all the sections are cancelled out when the data are thus combined. In a composite standard, such as the one in Figure 11-14, the relative ranges of various species are estimated for the entire local area. In fact, relative ranges shown for species in a composite standard will nearly

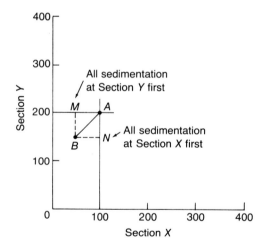

FIGURE 11-11
Portion of a graph like that in Figure 11-10, showing limiting cases of completely noncontemporaneous deposition in two sections during a certain time interval. *A* and *B* are two recognizable events, such as the appearance or disappearance of a species, that mark the beginning and end of a time interval in sections *X* and *Y*. Interpolation of a straight line between *A* and *B* is perfectly accurate only if rates of deposition were identical in the two sections throughout the interval. The distances *BM* and *BN* show the maximum possible error that could be introduced if the assumption of equal rates of deposition were wrong. This maximum possible error would exist if all sediment that accumulated during the interval in either section *X* or section *Y* was deposited before deposition began in the other section. In either case the correlation error would be 50 feet. (Modified from Shaw, 1964.)

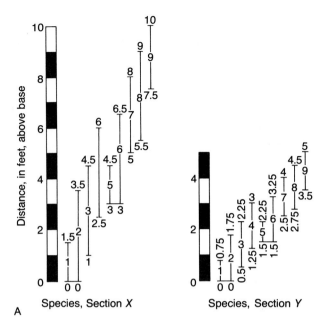

A

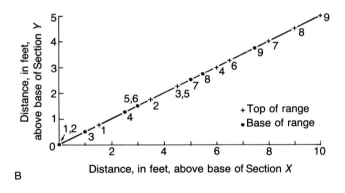

B

FIGURE 11-12
Diagrams and graphs like those in Figure 11-10, of two
sections representing the same time interval and having
identical fossil species, but different rates of rock accumulation.
(From Shaw, 1964.)

always be incomplete. Most will, however, approach the actual relative
ranges more closely than will the stratigraphic ranges of any single section.
The composite standard is an abstraction based on rates of rock accumulation in the base section. Its vertical scale is, therefore, not in units of feet
or time. Its units are undefined.

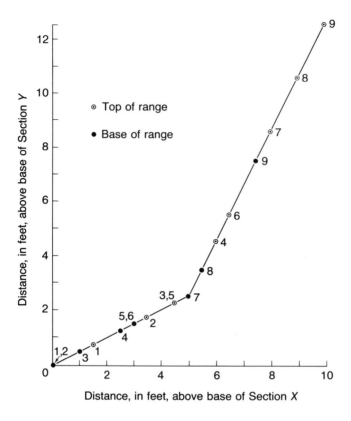

FIGURE 11-13
Graph of species ranges in two sections
illustrating a sudden change in relative rate of
rock accumulation. (From Shaw, 1964.)

If constructed from adequate data, the composite standard can be extremely important to stratigraphic, paleoecologic, and evolutionary studies. It can be correlated with a single section or distant composite standard by the same technique used for comparing local sections. The composite standard makes quantitative correlation between widely separated regions possible. Presumably, it can also be used for intercontinental correlation, using ranges of genera rather than species. By comparison of a series of sections with the composite standard, geographic migration of a species could potentially be demonstrated. In a similar manner, relict populations might be recognized in certain regions. In addition, a composite standard nearly always provides a much clearer picture of relative times of origin and extinction of species than can be provided by any single section. Its use, therefore, makes recognition of evolutionary lineages and phylogenetic relationships less speculative than they would otherwise be.

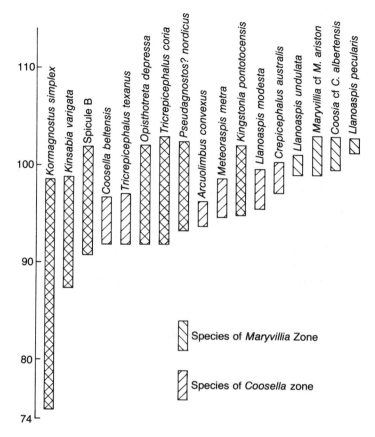

FIGURE 11-14
Graphic representation of composite standard ranges for the seventeen most common species of the *Coosella* and *Maryvillia* (trilobite) zones of the Cambrian Riley Formation, based on eight local sections. (From Shaw, 1964; data for local sections from Palmer, 1955.)

Time and Time-rock Units

Very early in the history of modern geology it became convenient to divide regional stratigraphic sequences into large units that could be distinguished from one another on the basis of fossil content. Thus, the stratigraphic *systems,* from Cambrian through Quaternary, came to be recognized over the course of several decades during the nineteenth century. Delineation of these did not proceed according to any well-planned scheme, nor were they established in order of their temporal relationships. There was a natural tendency for early workers to establish system boundaries at major faunal and floral breaks. Many of the boundaries seem especially well placed, even today, for some breaks that were chosen represent distinct evolutionary changes or extinctions that occurred during short time intervals on a worldwide scale. Nevertheless, there is an unfortunate tendency among students to accept boundaries between geologic systems as being natural breaks carefully chosen at the most appropriate possible stratigraphic levels. In fact, other boundaries of equal or even greater utility could have been chosen. There is a large measure of arbitrariness in the stratigraphic procedures by which our geologic time scale has been established.

The geologic time scale, of course, remained a *relative* time scale for many years. Not until the advent of radiometric age dating (just after the turn of the century) could the scale be related reasonably accurately to *absolute* time, measured in years.

Geologic systems are still defined on the basis of **type sections,** which are local stratigraphic sections in the **type areas** in which the systems were first recognized. Most type areas are in Europe. One problem with this traditional practice is that not all type sections represent the most useful rock sequences for definition of their respective systems, as now recognized throughout the world. Many modern workers support the idea of redefining the systems by defining their *boundaries* in selected areas in which the boundaries can be placed within fossiliferous rock sequences formed by nearly continuous deposition.

In order to understand the problems of extending a system from its type section to other biotic provinces or continents, a diagram like the one in Figure 11-15 is useful. As long as one system is defined in one area, and an adjacent system in another area, the systems may in fact contain rocks that overlap in age. It is just as likely that an age gap may exist between two systems. In other words, even the type sections of geologic systems, as now defined, do not neatly divide the stratigraphic column. Correlation, by means of fossils, of rocks in a type section with rocks of other regions is not perfectly accurate, as we have seen, and further serves to reduce the degree of synchroneity of recognized system boundaries throughout the world.

Systems have been subdivided into **series,** and series into **stages.** All these subdivisions are most commonly recognized by using fossils. All three of these kinds of units—systems, series, and stages—have traditionally been called **time-rock units,** implying that they are **isochronous units,** that is, units bounded by surfaces that are everywhere the same age. This view is taken in the Code of Stratigraphic Nomenclature mentioned at the beginning of this chapter. "Time-rock unit" used in this way is a somewhat misleading phrase. Because systems, series, and stages are all recognized primarily on the basis of biostratigraphic equivalence (except in their type areas) the recognized subdivision boundaries are no more synchronous than are the biostratigraphic boundaries used to define them.

This problem is recognized by most modern workers. But the question is really one of accuracy. Some workers, such as Hedberg (1951) argue that, because systems and series represent larger time intervals than zones, the percentage of error in their recognized boundaries is smaller. They believe that most system boundaries throughout the world are so nearly synchronous that the phrase "time-rock unit" is appropriate. Weller (1960) advocates retention of the phrase "time-rock unit" to describe systems and series, but only as long as it is understood that most units to which the phrase is

FIGURE 11-15
Problems of correlation with type sections. *A*, *B*, and *C* are type sections of three geologic systems in different areas. The vertical scale is based on time, rather than on sediment thickness. On the right is a composite section showing the systems in another area. Gaps, due to nondeposition or erosion, are found in all these sections, and are diagrammed as empty boxes. Lines of fossil correlation (dashed) show varying degrees of accuracy, time lines being horizontal. Systems *A* and *B* overlap in time; there is a gap in time between systems *B* and *C*. Even if perfect correlation were possible, parts of the composite section on the right could not be assigned to a system.

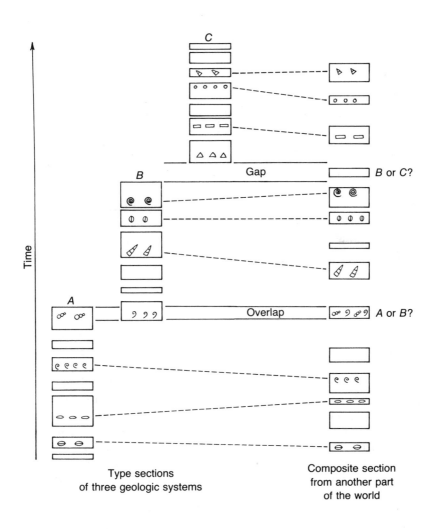

Type sections
of three geologic systems

Composite section
from another part
of the world

applied are only approximately isochronous. Dunbar and Rodgers (1957, page 293), however, reject the attempts of many workers to use the phrase "time-rock unit" in this way: "Historically, series and systems, as they have always been used by stratigraphers, are as dependent on fossils as zone, and if zone is a biostratigraphic term, so are they." The question of whether systems and series should be regarded as time-rock units is still widely debated.

The geologic time scale itself is an abstraction. The time units of the scale are defined by the boundaries between systems in their type areas. As we saw by looking at Figure 11-15, these time units may overlap or be separated by gaps. Thus, although time is a continuum, our geologic time scale, as now recognized, does not neatly subdivide this continuum.

It is easy to understand why many workers now desire that geologic systems be redefined in places other than the original type areas using *boundaries* within fossiliferous rocks, rather than the rock units themselves. Thus, a system could have its lower boundary defined in one place and its upper boundary defined elsewhere. The type section for each boundary could be so selected that the boundary would be defined by good index fossils contained within rocks formed by nearly continuous deposition. The new systemic boundaries could be so chosen as to approximate the traditional ones. Some of the old type sections might be retained, but used to define a boundary between two systems rather than a system itself. Series, which also have type areas, might be subjected to similar revision, although many are presently defined in areas in which they are represented by well-exposed, fossiliferous strata.

An alternative proposal has been presented by Bell and colleagues (1961) in a report dissenting from some of the ideas and rules set forth in the Code of Stratigraphic Nomenclature. This minority-view proposal points out that radiometric age dating has provided a time scale based on years that is essentially independent of the so-called "relative time scale" based on presence of fossils and physical stratigraphic evidence. It suggests that the term "chronologic" be restricted to the radiometric time scale and that the term *geochronologic* be applied to the traditional time scale of eras, periods, and epochs. This proposal also suggests that the rock bodies (systems and series) representing the geochronologic units be called *chronostratigraphic units.* Chronostratigraphic units might then be defined by using biostratigraphic units such as zones, as well as any other pertinent criteria. Their boundaries would not have to coincide with biostratigraphic boundaries, but commonly would. The time span of each geochronologic unit would be represented by widespread rocks (the chronostratigraphic unit), plus all the gaps in these rocks owing to nondeposition or erosion. The type section is maintained in this scheme, but its limitations are recognized. In a sense, the type section in this scheme is like the type specimen of a species, the chronostratigraphic unit is like a hypodigm (page 106), and the geochronologic unit is like a species. This analogy should be clear if the reader understands the taxonomic concepts presented on pages 106 and 117.

Accuracy Of Correlation

As implied by the controversy over designation of systems, series, and stages as time-rock units, there is considerable debate regarding the accuracy of correlation by fossils.

As we have seen, faunas and floras of past times tended to be segregated into somewhat isolated geographic realms, just as they are today. The relatively high degree of biotic uniformity within provinces makes correlation more accurate within provinces than between them. Correlation between provinces is commonly used to establish large-scale, approximately synchronous "surfaces" for reconstruction of paleogeography for brief intervals of geologic time. Often, two or more continents are included, as in the map that is reproduced here as Figure 11-16. Because paleogeographic maps typically portray large areas, they seldom provide local detail. Correlations used for their construction need not be nearly as accurate as those used for local environmental studies.

FIGURE 11-16
Paleogeographic map showing major seaways and locations of well-studied fusulinid faunas (dots) during the mid-Carboniferous. (From Ross, 1967.)

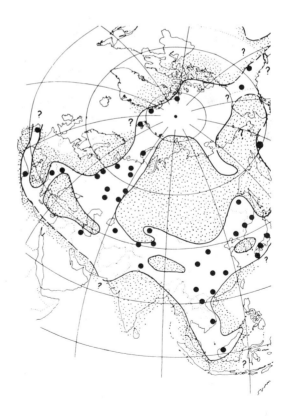

Most intercontinental correlations are based on index fossils, rather than on comparison of large fossil faunas or floras. Commonly, the index fossils in the two areas being compared belong to two species of the same genus; sometimes they belong to the same species. If Simpson's estimate of 0.5–5 million years for average species duration is correct (Simpson, 1952), this range of values establishes the approximate accuracy of correlations based on single fossil species, assuming no other evidence is applied to a given problem. Actually, many marine invertebrate species have lived for longer than 5 million years (Fig. 10-11). An idea of the accuracy of correlation based on genera, rather than species, is given by the estimate of 8 million years for duration of Cenozoic carnivore genera, and 78 million years for Cenozoic bivalve genera (Simpson, 1953). Certainly, there has been much variation in the duration of genera and species throughout geologic history. Some higher categories have been characterized by much longer mean duration of their component genera and species than others. It is obvious why one prime requisite of a good index fossil is a relatively short stratigraphic range. The statistical value of using several taxa for a single correlation should also be readily apparent.

Many intercontinental correlations are accomplished in a series of two or more steps, one biota being judged to have existed contemporaneously with a second, the second with a third, and so on. Thus, the third biota is approximately correlated with the first, though the two may have few (or no) species or genera in common. Each added step in a sequence of equidistant correlations doubles the inaccuracy of the total correlation, all other factors being equal. A great deal of subjectivity enters into the making of most correlations in choice and weighting of both the taxa and the events (such as first appearances and last appearances) that are used.

The statements of many workers suggest that system boundaries traced around the world are nearly synchronous surfaces. Other workers, such as Weller (1960, page 562) are of a different persuasion:

> Because of more or less effective isolation, the organic relations between different faunal or floral provinces are likely to be more remote than those connecting different parts of a single province. Consequently, comparisons and correlations must be made on the basis of more general evolutionary changes, and the results are less precise. Just what allowance should be made on account of these less direct relationships is uncertain. . . . On the whole, interprovincial correlations based on paleontology are probably not accurate to much less than one-quarter of a geologic period.

For most correlations within a single biogeographic province, stratigraphic sections are separated by intervals of a few kilometers (seldom exceeding 100 kilometers) and the total area of study has dimensions measured in tens or hundreds of kilometers (seldom exceeding 2,000 kilometers). Correlation on this scale can often be based on biotas possessing many common faunal or floral elements. The required accuracy of correlation is, however, also much greater than for correlation between provinces. Within a single province, extremely accurate correlation is often desired for local paleoecologic studies. Rate of species evolution is too slow for changes in fossils to be of value in most such local studies. Other stratigraphic features, such as nearly isochronous rock units, may prove to be more useful in correlation. Beds formed by volcanic ash falls or by other apparently widespread deposition over short lengths of time may be especially valuable and are commonly referred to as **key beds.**

In fact, under certain circumstances, the presence of similar biotas in rocks of separate areas may indicate *different* ages for the rocks. For example, suppose a worker identifies a certain fossil fauna as a marine intertidal community that lived in a narrow belt along the ancient shoreline. Discovery of this biota in two isolated outcrops 300 kilometers apart along an axis perpendicular to the known shoreline orientation would strongly suggest different ages for the two outcrops.

Some workers have expressed hope that improved radioactive dating techniques and increased numbers of absolute age determinations may provide much better correlation accuracy than has been attained with the use of fossils. Unfortunately, this sanguine outlook is largely unjustified except for very young rocks.

The estimated *precision* of a radiometrically-determined date is commonly listed as a standard error (a "plus-or-minus" value following the date). Lack of precision is the result of both statistical variation among samples of the same origin and errors in measurement. In addition, most measured dates are subject to considerable errors in *accuracy,* primarily because of loss of the measured daughter product of radioactive decay through weathering, heating, or solution alteration, for example.

The absolute geologic time scale (Figure 11-17) that was provided by Kulp (1961) is based on selected age determinations made from samples that are believed to have remained as virtually closed systems. The standard errors attached to his selected values are therefore the approximate total errors for most determinations. The age determinations used to construct the post-Cambrian part of the time scale are plotted against their standard errors in Figure 11-18, which shows a considerable increase in error with increasing age. We can assume that fossil correlation is also poorer for older rock units because of poorer preservation and less widespread rock exposures. Whether correlation based on fossil evidence or radiometric age dating is more reliable varies from study to study.

There are also difficulties in attempting to relate radiometric age determinations to established biostratigraphic subdivisions. One major problem is that fossils are seldom directly datable by radiometric methods. Many measured dates are for minerals formed at the time of deposition of sediments containing fossils, but others are for igneous rock units whose stratigraphic relationships only indicate their approximate time relationships to fossil-bearing sedimentary rocks. To construct the time scale of Figure 11-17, Kulp chose only measured dates that represent minerals whose origin can be closely fitted into a biostratigraphic framework.

Additional problems arise through poor biostratigraphic correlation of those fossils that are closely associated with radiometrically dated minerals.

Era	Period	Epoch	Approximate beginning of interval	Number of years ago, in millions
	Quaternary	Pleistocene	1	0
Cenozoic	Tertiary	Pliocene	13	
		Miocene	25	
		Oligocene	36	
		Eocene Upper	45	50
		Middle	52	
		Lower	58	
		Paleocene	63	
Mesozoic	Cretaceous	Upper Maestrichtian	72	
		Campanian	84	
		Santonian	90	
		Coniacian		
		Turonian		100
		Cenomanian	110	
		Lower Albian	120	
		Aptian		
		Neocomian	135	
	Jurassic	Upper		150
		Middle Bathonian	166	
		Lower Bajocian	181	
	Triassic	Upper		
		Middle	200	200
		Lower	230	
Paleozoic	Permian	Upper		250
		Middle	260	
		Lower	280	
	Carboniferous — Pennsylvanian			300
	Carboniferous — Mississippian	Visean	320	
		Tournaisian	345	
	Devonian	Upper	365	350
		Middle	390	
		Lower	405	400
	Silurian		425	
	Ordovician	Upper Trenton	445	
		Middle		450
		Lower		
			500	500
	Cambrian	Upper	530	
		Middle		550
		Lower		
	?	?	?	600

FIGURE 11-17
Absolute geologic time scale.
(Data from Kulp, 1961.)

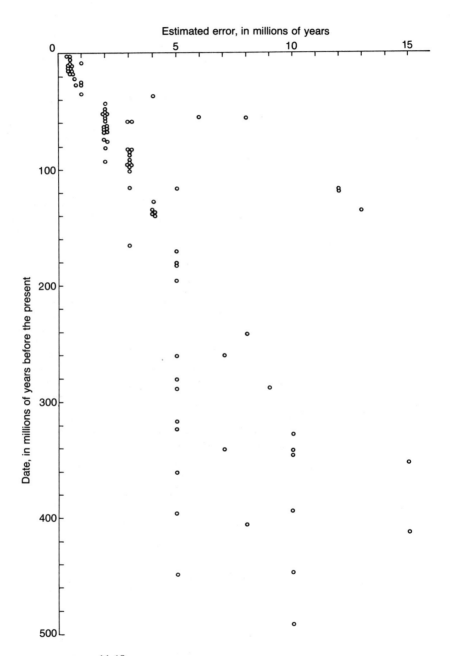

FIGURE 11-18
A plot of post-Cambrian dates used by Kulp (1961) to construct the
absolute time scale (Figure 11-17) versus their estimated analytical errors.

Ideally, it would be desirable to obtain the absolute ages of all system and stage boundaries in type areas. Because absolute age determinations are restricted to certain mineral occurrences and geologic conditions, however, this goal can seldom be attained. The dates of most system and stage boundaries of Kulp's time scale have been estimated from age determinations made far from the type locations of the various systems and stages.

The problem of determining absolute ages for biostratigraphic system boundaries is illustrated by current opinions regarding the Silurian System. The time interval given by Kulp (1961) for the Silurian Period is 405–425 million years before the present. Kulp indicates that the age of the lower boundary of the Silurian System is especially uncertain. In the time scale of Holmes (1960) the Silurian boundaries are dated at 400 and 440 million years before the present. In other words, the time interval represented by the system in the two time scales differs by a factor of 2, and the absolute age assigned to the lower boundary differs by 15 million years.

Radiometric age determinations are of little value in local correlations, especially for rocks well back in the stratigraphic record. Most local environmental and paleoecologic studies require much greater precision than is provided by radiometric measurement, even in sediments of Tertiary age, where nearly all measurements are subject to errors of 0.5 million years or more.

There is some hope that quantitative biostratigraphic techniques of the type introduced by Shaw (1964) may improve correlation accuracy in the future. The primary roadblock is the extremely detailed stratigraphic measurement and fossil collection required. An astronomical number of man-hours would be required for extensive application of Shaw's method, especially for the establishment and correlation of many composite standard sections.

Furthermore, some workers refuse to accept the validity of quantitative approaches to correlation. Their arguments are similar to those of taxonomists who attack the validity of numerical taxonomy (page 138). The question raised is whether it is valid to give equal statistical weight to all fossil taxa in correlation (or to all measured morphologic characters in taxonomy). Just as some taxonomists argue for the special utility of certain "key characters" in the classification of taxonomic groups, some stratigraphers argue for the special utility of certain index fossils in correlation (see, for example, Jeletsky, 1965). The opposing argument claims that if enough measurements are used, the less meaningful data merely dilute, but do not invalidate, the statistical conclusions. Perhaps the real question is whether the necessary quantities of data can be gathered economically enough to justify widespread use of statistical approaches.

Biostratigraphers are constantly striving to improve correlations and correlation techniques. The assessment of correlation error in this chapter is not meant to belittle the validity or utility of our existing stratigraphic framework, but to alert the student to the potential pitfalls of biostratigraphic correlation and, perhaps, to stimulate him to contribute improvements to existing correlations and correlation techniques.

Supplementary Reading

American Commission on Stratigraphic Nomenclature (1961) Code of stratigraphic nomenclature. *Bull. Amer. Assoc. Petrol. Geol.,* **45**:645–665. (Outlines principles and practices for consistency in classifying and naming stratigraphic units.)

Berry, W. B. N. (1968) *Growth of a Prehistoric Time Scale.* San Francisco, W. H. Freeman and Company, 158 p. (History of the development of the relative geologic time scale.)

Donovan, D. T. (1966) *Stratigraphy.* Chicago, Rand McNally, 199 p. (A concise summary of stratigraphic principles with reference to their historical development.)

Dunbar, C. O., and Rodgers, J. (1958) *Principles of Stratigraphy.* New York, Wiley, 356 p. (Contains chapters encompassing biostratigraphy, with many examples from geologic literature.)

Eicher, D. L. (1968) *Geologic Time.* Englewood Cliffs, Prentice-Hall, 150 p. (A highly readable capsule summary of stratigraphic concepts.)

Krumbein, W. C., and Sloss, L. L. (1963) *Stratigraphy and Sedimentation,* 2nd Ed. San Francisco, W. H. Freeman and Company, 660 p. (A popular text emphasizing principles and incorporating the Code of Stratigraphic Nomenclature as an appendix.)

Weller, J. M. (1960) *Stratigraphic Principles and Practice.* New York, Harper, 725 p. (Includes useful discussions of practical stratigraphic techniques).

CHAPTER 12

Uses of Paleontologic Data in Geophysics and Geochemistry

Most research problems using fossil data relate directly to the biologic history of the earth. As we have seen, fossil data also contribute information to many areas of research not normally considered biologic. In biostratigraphy, paleontologic data are used to help develop a chronology of the physical history of the earth. In the chapter on paleoecology, we saw how fossil data may be used to reconstruct physical environments (in terms of distribution of land and sea, land elevations and water depths, behavior of sedimentary processes, etc.). Such information has direct application to purely geologic problems such as studying continental drift, rates of subsidence, and the development of sedimentary basins.

In this chapter we introduce examples of the direct application of fossil data to problems in geophysics and geochemistry. Interdisciplinary thought was necessary in each example, and each demonstrates a type of research that has been instrumental in producing major scientific breakthroughs.

Growth Bands in Marine Invertebrates

In skeletons grown by accretion, it is often possible to observe a succession of growth bands, or lines. We saw an example in echinoids as part of our discussion of ontogenetic variation (Chapter 3). Growth banding is also evident among corals and molluscs. Figure 12-1 illustrates a common type of coral banding.

FIGURE 12-1
Growth banding displayed by specimens of the coral *Holophragma calceoloides* from the Devonian Visby Marl of Gotland. (Specimens provided by John W. Wells; photography by R. M. Eaton.)

Growth bands have been used in studies of both ontogenetic variation and the functional significance of ontogenetic change. They are valuable in this regard because they make it possible to study the ontogeny of shape using only one adult individual. Also, some assessment of the environment has been possible from comparisons of growth rates in the same or closely related species. It has been shown, for example, that present-day species of reef corals grow more rapidly in warmer water. This information has been used to help deduce what climates of the geologic past were like.

Until very recently, there was almost no rigorous interpretation of growth bands in terms of absolute time. Prominent growth banding has often been suggested to represent an annual growth cycle. By analogy to counting growth rings in trees, attempts have been made to deduce from its growth bands the age, in years, of a fossil at the time of its death. The contributions of this line of inquiry have not, however, been substantial, partly because the age of an organism at the time of death is only rarely of significance in interpreting the fossil record and partly because visible increments of growth on animal skeletons are not actually known to be annual. Growth banding in invertebrates often shows either no predictable periodicity or a set of cycles more complicated than that of a simple annual fluctuation.

In 1963, John W. Wells, an authority on fossil and living corals, looked into this problem more deeply than any of his predecessors. He observed that growth bands in some modern corals are much more closely spaced than they would be if they were the visible expression of an annual growth cycle: he found about 360 tiny growth increments in an amount of skeleton thought to have grown in one year, and suggested that the finer growth lines may actually be the marks of a *daily* growth cycle. From a physiologic viewpoint this is not an unreasonable interpretation because calcification in reef corals is apparently related to the photosynthetic activity of symbiotic algae, which differs markedly between night and day.

Wells then turned to a few well-preserved coral specimens from Devonian and Carboniferous rocks for which annual growth banding had already been demonstrated with some confidence. He studied details of the growth history recorded between the annual bands, and found that the number of smaller growth bands was consistently larger than expected. In Devonian material the number averaged about 400 (ranging between 385 and 410), and in Carboniferous material ranged between 385 and 390. To many investigators, this would have seemed to be a clear indication that the smaller growth increments were not diurnal or that the larger growth increments had been incorrectly interpreted as being annual. But Professor Wells was aware that entirely independent research in geophysics suggested that the number of days in the Devonian year should indeed have been greater than 365.

On purely theoretical grounds, geophysicists and astronomers have reasoned that the rate of rotation of the earth on its axis is decreasing due to tidal friction, which, of course, would decrease the number of days per year. The geophysical calculations give a specific rate of decrease in the earth's rotational speed and thus the decrease in the number of days per year can be calculated. The relationship is shown in Figure 12-2. Day length should be increasing by about 0.0016 second per century. Archeological discoveries of Egyptian astronomical data have made it possible to obtain tentative confirmation of this increase but the time that has elapsed between the Egyptian civilization and the present is so short that conclusive documentation is not possible. Wells' data on Devonian and Carboniferous corals are in remarkable accord with the theoretical predictions.

The coral data provide a tentative but very striking corroboration of archeological interpretation and geophysical theory. The way is now open for further verification and definition of the relationship between geologic time and the earth's rotational speed. If the change in the number of days per year can be learned for certain, reasoning can be done from the other direction, using the number of days per year observed in a fossil's growth increments to determine its geologic age, for example. This method of geochronology has obvious advantages over methods based on radioactive decay in that it does not depend on syngenetic minerals or associated igneous rocks. With geochronology based on growth banding, correspondence between the "relative" time scale developed in biostratigraphy and the "absolute" time scale developed by radioactive decay methods could be made much firmer. Also, geochronology based on growth banding is not subject to the biasing effects of chemical change (due to diagenesis and metamorphism) that plagues radioactive decay methods. To be sure, dating by growth banding depends upon excellent preservation but it is morphologic preservation rather than chemical preservation. The principal limiting factor in growth-band geochronology is that the number of days per year changes at the rate of only about one per 10 million years.

Since the pioneer work of Wells, similar studies have been carried out on corals and other organisms that grow by accretion. In 1965, for example, Scrutton published a detailed study of cyclic growth in some Devonian corals. Though he did not find convincing evidence of annual periodicity in growth, he did recognize small growth increments (morphologically

FIGURE 12-2
The number of days per year throughout geologic time, calculated in accordance with the idea that the rate of the earth's rotation on its axis is decreasing as a result of tidal friction. (From Wells, 1963.)

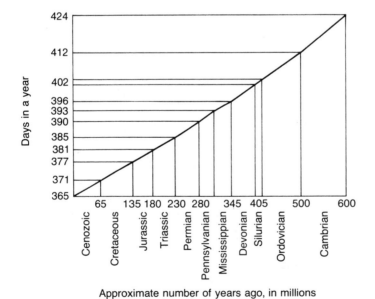

comparable to Wells' diurnal bands) that were grouped in clusters averaging 30.6 ridges. The similarity between this number and the present-day length of the lunar month led Scrutton to suggest that the Devonian lunar month was 30.6 days long. This is biologically plausible because many marine organisms, including corals, are known to be tied physiologically to lunar, or tidal, cycles.

Scrutton's month of 30.6 days is only slightly longer than the present-day period of one tidal cycle (29.5 days), but the difference appears to be significant. Some of Scrutton's data are shown in Table 12-1. The lowest average number of ridges per band for a single specimen is 29.9—the Devonian month length as determined from this specimen is almost a half a day longer than that of today. (It should be noted that the widely recognized length of the lunar month, 28 days, is only a convention for legal purposes and has no precise scientific basis.)

TABLE 12-1

Counts of Growth Ridges in Fossil Corals Used To Estimate the Length of the Devonian Lunar Month

Specimen Number	Number of Bands Analyzed	Average Number of Ridges per Band
BM R44851	12	30.9
OUM DT2	13	30.85
OUM DT3	10	30.7
OUM DT4	16	30.7
OUM DT5	12	30.4
OUM DT6	14	30.8
OUM DT7	12	31
OUM DT8	6	30.6
OUM DZ32	9	30
OUM DZ33	8	29.9
Average for ten specimens		30.59

Source: From Scrutton, 1965.

Scrutton's conclusions are tentative but clearly open the way for further geophysical application of growth-band data to studies of the evolution of the earth-moon system through geologic time.

Some of the most detailed growth-band work has been applied to bivalve shells. Work by Barker (1964) established recognizable cycles much more complex than any considered previously. Barker studied thin sections of four Recent bivalve species collected in shallow marine waters from Prince Edward Island, Canada, to Florida. Differences in texture, composition, and thickness of growth layers made possible recognition of five different cycles. The periods represented by these cycles are: one year, one-half year, one-half lunar month, one day, and six hours. Table 12-2 lists the cycles and relates each to shell growth.

358

TABLE 12-2

Periodic Growth Layers in Bivalves

Growth Layer	Thickness	Sublayers	Data Pertinent to Growth	Probable Growth Period
Fifth-order	1–16 μ	Calcium carbonate	Shell open or closed at high or low tide	6 hours
		Conchiolin	Shell in process of opening during tidal change	
Fourth-order	5–60 μ	Thick fifth-order	Day (fast growth)	24 hours
		Thin fifth-order	Night (slow growth)	
Third-order	0.1–0.9 mm	Thick fourth-order	Maximum tidal amplitude	15 days
		Thin fourth-order	Minimum tidal amplitude	
Second-order	0.9–11.0 mm	Conchiolin-rich	Normal conditions	½ year
		Conchiolin-poor	Equinoctial conditions	
First-order	2–22 mm	Thick second-order Conchiolin-rich	Summer	1 year
		Thin second-order Conchiolin-poor	Winter	

Source: Data from Barker, 1964.

Much of Barker's work was confirmed in 1968 by Pannella and Mac-Clintock in studies of comparable material from other Recent bivalves (see Figure 12-3). Also, Scrutton's work on the length of the lunar month was extended in 1968 by Pannella, MacClintock, and Thompson. Their data suggest (Figure 12-4) that the change in number of days per lunar month may not have been regular throughout geologic time.

The whole field of growth-band research is an extremely active one, and, clearly, it can make major contributions to our knowledge of the parts of geologic history that depend on the dynamics of the earth-sun-moon system.

Fossils in Deformation Studies

A second example of the application of fossil data to nonbiologic problems is the use of fossils in structural or geophysical studies of rock mechanics and rock deformation. The major directions of strain that have accompanied deformation are of particular interest. If the amount and character of strain are to be assessed, the rock must contain features or particles whose original size and shape is known. Few nonbiologic structures meet these requirements. Many studies have made use of objects assumed to have been originally spherical (ooids, mudballs, certain concretions, and some nodules), but such an assumption is often difficult to make. Undeformed ooids, for example, vary in shape even though the average shape is that of a sphere. Only by working with large samples is it possible to assume original sphericity. Furthermore, the morphology of a sphere is simple and there is a limit to what can be learned about patterns of stress and strain from its deformation. Also, original size is rarely known.

Fossils have considerably more potential. In the first place, they are more common (at least in sedimentary rocks). Secondly, their original shape is often better known and can be defined with greater precision (through reference to undeformed fossils of the same species). Thirdly, the initial shape is generally more complex and the complexity of the deformation interpretation is thus potentially higher. The original size or volume can be estimated with particular ease for animals such as trilobites that grow by molting and have successive growth stages that are morphologically distinct. Where

FIGURE 12-3
Daily growth bands in a section of the modern bivalve *Tridacna*. Numbers refer to days. Variation in form and thickness of bands defines a 14-day tidal cycle. (From Pannella and MacClintock, 1968.)

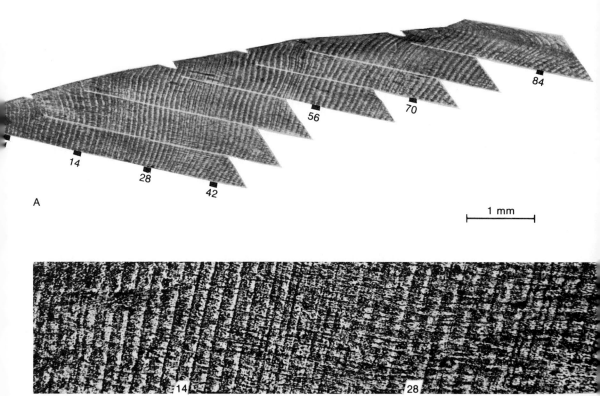

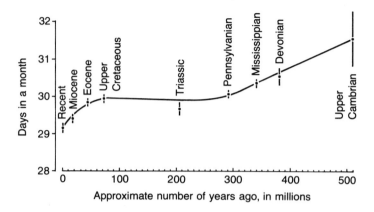

FIGURE 12-4
Variation in number of days per lunar month through
geologic time as estimated from fossil growth-band data.
(From Pannella, MacClintock, and Thompson, 1968.)

fossils are preserved as impressions in rock, they behave during deformation in the same fashion as the matrix (rather than being less or more resistant to stress). Very often, an inorganic particle, such as an ooid, is considerably more resistant to stress than the surrounding rock and its deformation
does not accurately measure the stress system applied to the rock as a whole.

We will consider here only deformation in two dimensions — the type commonly encountered in fossils lying on bedding planes. We will also limit our-
selves to what the geophysicist calls "homogeneous strain;" that is, change
in volume and shape but change in which straight lines remain straight,
parallel lines remain parallel, and in which all lines of the same original
orientation are lengthened or shortened by the same relative amount.

In each example we will be concerned with determining the shape of the
"strain ellipse" — that is, the shape that would be produced if a circle were
subjected to the same stresses. The most important characteristics of the
ellipse are the direction of maximum elongation, the amount of maximum
elongation, and the amount of minimum elongation. (The last two are the
major and minor axes of the strain ellipse, respectively.) This does not tell
us anything about the *stresses* causing the deformation because identical
strain patterns can be caused by a wide variety of stress systems. For example, a given strain ellipse may be produced by tension parallel to the direction of maximum elongation or by compression in a perpendicular direction.

Figure 12-5 shows three deformed trilobite pygidia. In *A* and *B*, the deformation is symmetrical with respect to the original bilateral symmetry of the
trilobite. In *A*, the direction of maximum elongation parallels the axis of
bilateral symmetry and in *B*, it is perpendicular to this axis. If the original

FIGURE 12-5
Tectonically deformed pygidia of trilobites. (From Ramsay, 1967.)

morphology of a deformed trilobite fossil is known, it is possible to determine the directions of maximum and minimum elongation. If the ratio of the principal dimensions in the undeformed pygidium is known, the ratio of maximum and minimum elongations can be calculated. In order to determine the absolute values of the elongations, it is necessary to know the original dimensions themselves. This can be done only if there is negligible variation in specimen size.

Figure 12-6 shows a sketch of a more complicated example of two-dimensional deformation of bilaterally symmetrical fossils. By comparing deformed and undeformed brachiopods in this illustration we can see that a fossil's orientation with respect to the stress system is crucial. Notice that specimens A and B are comparable to pygidia B and A, respectively, in Figure 12-5. Each is either parallel to or perpendicular to the direction of maximum elongation. All other specimens are at an oblique orientation and are thus comparable to C in Figure 12-5.

These are several methods for dealing with the deformation shown in Figure 12-6. In one, a morphologic dimension (length of hinge line, h, in this example) is measured on each specimen as is the angle (α) between the line along which the dimension is measured and an arbitrary line that has been drawn across the section for the purposes of this analysis (*PQ* in Figure 12-6). The results are plotted (Figure 12-7). The scatter of points conforms to a regular curve described by the following equation:

$$h = h_0 \left[\frac{\lambda_1}{1 + \left(\frac{\lambda_1}{\lambda_2} - 1\right) \sin^2 (\alpha + \theta)} \right]^{1/2},$$

where h_0 is the undeformed length of hinge line, h is its deformed length, α is the angle between the hinge line and the arbitrary reference line *PQ*, θ is the angle between the reference line and the direction of maximum elongation (the major axis of the strain ellipse), and λ_1 and λ_2 are the minimum and maximum elongations.

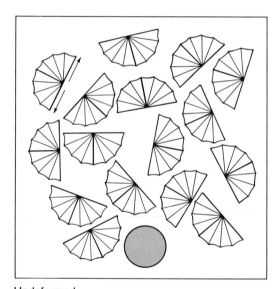

Undeformed

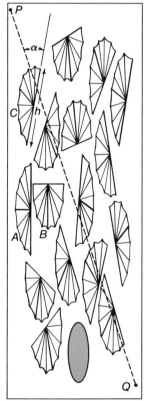

Deformed

FIGURE 12-6
Effect of deformation on the shape of bilaterally symmetrical fossil brachiopods. All fossils were originally the same size. (From Ramsay, 1967.)

In Figure 12-7, the points are close enough to the ideal curve that the important features of the strain ellipse may be determined graphically. The α value of the highest point on the curve is θ and the ratio of the length measurements at the highest and lowest points is the ratio of the two elongations (λ_1/λ_2). The strain ellipse can be defined in this manner *without* precise knowledge of the morphology of the undeformed fossils. The analysis does not even depend on the fossils having original bilateral symmetry. It is important, however, to be able to assume that the fossils were originally all the same size! Any variability in specimen size will produce increased scatter in plots such as Figure 12-7. If variability is too high and if the number of fossils is small, other methods of analysis must be used.

Figure 12-8 shows two deformed brachiopod shells of different species. To determine the strain ellipse, all that need be known is that the fossils, which we may refer to as A and B, were both bilaterally symmetrical before deformation. Three angular measurements on the deformed fossils are necessary: the angle (ψ) between the original morphologic mid-line and a perpendicular to the hinge line for each fossil and the angle (α) between the hinge lines of the two fossils. The following two equations may be expanded and solved for λ_2/λ_1 (yielding the shape of the strain ellipse) and for θ (the orientation of the long axis of the ellipse with respect to one of the fossils).

$$\sqrt{\frac{\lambda_2}{\lambda_1}} = \frac{\tan \theta}{\tan (\theta + \psi_A)}$$

$$\sqrt{\frac{\lambda_2}{\lambda_1}} = \frac{\tan (\theta + \alpha)}{\tan (\theta + \alpha + \psi_B)}$$

The analysis is even simpler if a graphical solution (employing a Mohr circle) is used.

FIGURE 12-7
Data on deformed hinge-line lengths from the brachiopods in Figure 12-6. (After Ramsay, 1967.)

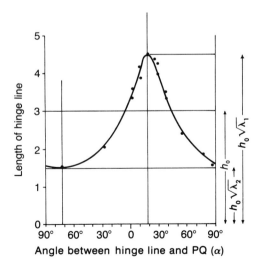

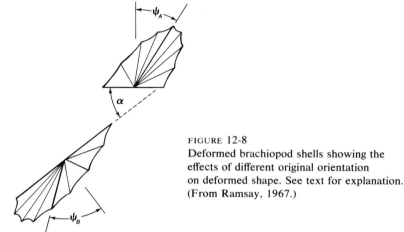

FIGURE 12-8
Deformed brachiopod shells showing the
effects of different original orientation
on deformed shape. See text for explanation.
(From Ramsay, 1967.)

Figure 12-9 shows sketches of a bivalve shell before and after deformation. Even though the shell lacks bilateral symmetry, the shape and orientation of the strain ellipse can be determined if the angles between three linear elements in the shell are known before and after deformation. The linear elements chosen are shown as heavy lines. They allow the measurement of two angles (α and β). The following equations can be solved simultaneously to yield R, which equals λ_2/λ_1, and θ, the orientation of the direction of maximum elongation with respect to the deformed fossil.

$$\frac{R^{1/2} \tan \theta + \tan \alpha}{1 - R^{1/2} \tan \theta \tan \alpha} = \frac{R^{1/2} (\tan \theta + \tan \alpha)}{1 - \tan \theta \tan \alpha}$$

$$\frac{R^{1/2} \tan \theta + \tan (\alpha + \beta)}{1 - R^{1/2} \tan \theta \tan (\alpha + \beta)} = \frac{R^{1/2} [\tan \theta + \tan (\alpha + \beta)]}{1 - \tan \theta \tan (\alpha + \beta)}$$

As before, the problem is amenable to graphical solution.

The student should refer to Ramsay (1967) for a more detailed treatment of the mathematical analysis of fossil deformation. The discussion we have given here is intended only as a general introduction to the subject and is not complete.

The examples just discussed illustrate the importance of fossils in studies of rock deformation and emphasize the importance of paleontologic training for geophysicists as well as the importance of geophysical training for paleontologists. The effect of rock deformation on fossil morphology means that morphologic variation displayed by a species cannot be understood without assessment of possible deformation effects.

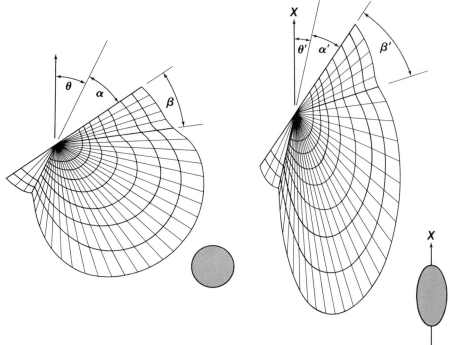

FIGURE 12-9
Undeformed (left) and deformed (right) bivalve
shells. (From Ramsay, 1967.)

Analysis of Diagenesis

Study of chemical diagenesis requires knowledge of the mineralogy, chemical composition, and fabric of original sediment. Without this knowledge,
the geochemist can see only the end products of diagenesis. Many limestones
are formed primarily from the skeletal remains of marine invertebrates. For
fossils of organisms that have close living relatives, it is usually possible to
specify quite accurately the original skeletal material; thus fossils often provide the key to the interpretation of diagenetic processes.

Most skeleton-bearing marine invertebrates secrete calcium carbonate in
one or more of the following forms: low-magnesium calcite, high-magnesium
calcite, and aragonite. The distinction between the first two is arbitrary, but
is conventionally placed at 4 percent magnesium (in substitution for calcium). Aragonite, because of its lattice structure, never contains a significant
amount of magnesium. Distribution of the three carbonate forms among
living marine invertebrates is shown in Table 12-3.

Both high-magnesium calcite and aragonite are unstable with respect to
low-magnesium calcite at the temperatures and pressures that prevail on the
earth's surface. In seawater, however, they tend to remain unaltered (in what

365

TABLE 12-3

Distribution of Skeletal Carbonates Among
Marine Invertebrates. (C = common; R = rare.)

Organism	Aragonite	Aragonite and Calcite	Calcite	High-magnesium-Calcite	High-magnesium-Calcite and Aragonite
Foraminifera					
Benthonic	R			C	
Pelagic			C		
Sponges				C	
Corals					
Madreporian	C				
Alcyonarian	R			C	
Bryozoa	C			R	C
Brachiopods			C	R	
Echinoderms				C	
Molluscs					
Gastropods	C	C			
Pelecypods	C	C	C		
Cephalopods	C			R	
Annelids	C			C	C
Arthropods					
Decapods				C	
Ostracods			C	R	
Cirripeds			C	R	
Algae					
Benthonic	C			C	
Pelagic			C		

Source: After Chave, 1964.

is sometimes called a "metastable" condition). Exposed to freshwater in subaerial environments, they tend to alter to low-magensium calcite and other minerals.

Aragonite skeletons sometimes alter to low-magnesium calcite with no apparent loss of microstructure. Commonly, however, they are entirely dissolved by acidic freshwater, leaving discrete cavities in the rock. The dissolved calcium carbonate is often deposited nearby as interlocking low-magnesium calcite crystals between primary sediment grains; these crystals cement, or lithify, the sediment into limestone. Presumably much of the calcite cement in sandstone is also derived from the solution of fossils. In limestone, crusts of secondary calcite often envelop aragonite grains before their solution, forming molds when the grains are later dissolved away (Figure 12-11).

High-magnesium calcite tends to alter to low-magnesium calcite in subaerial environments without large-scale solution. In one process, magnesium ions are simply removed by aqueous solutions and replaced by calcium ions with no apparent loss of skeletal microstructure. Alternatively, high-magnesium calcite may stabilize by "exsolution" to form dolomite (which contains about equal amounts of calcium and magnesium) and low-magnesium calcite. Evidence of the latter process is seen in some rocks by selective dolomitization of high-magnesium calcite fossils (Figure 12-10).

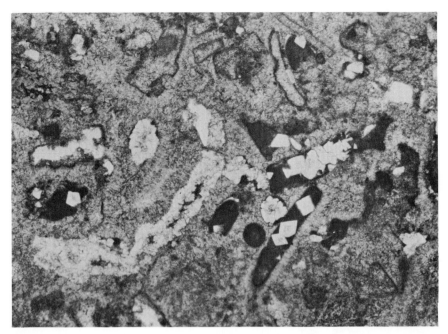

A

B
figure 12-10
Selective dolomitization of high-magnesium calcite coralline algae in Tertiary limestone of Eniwetok Atoll. A: Light-colored dolomite rhombs restricted to dark, rod-shaped segments of *Corallina* (×30). B: Enlargement of a single algal segment seen in lower right-hand area of A. (From Schlanger, 1963.)

FIGURE 12-11
Fossils of the aragonitic alga *Halimeda,* seen in thin sections of the Pleistocene Key Largo Limestone, Florida (×23). A: An intact aragonite segment of the alga. B: A mold of calcite cement that had encrusted a segment before the segment was dissolved by acidic fresh water. (From Stanley, 1966.)

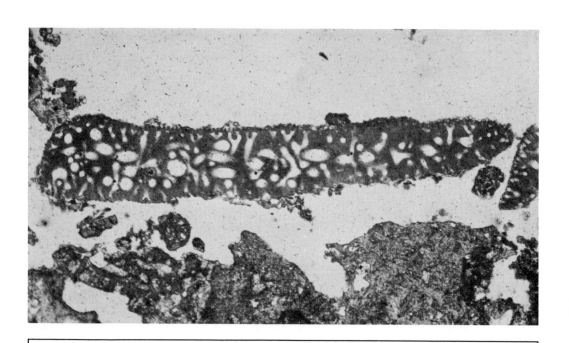

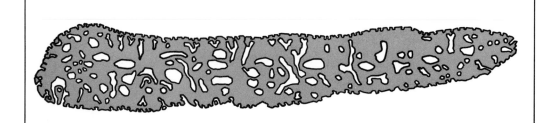

Primary aragonite

A

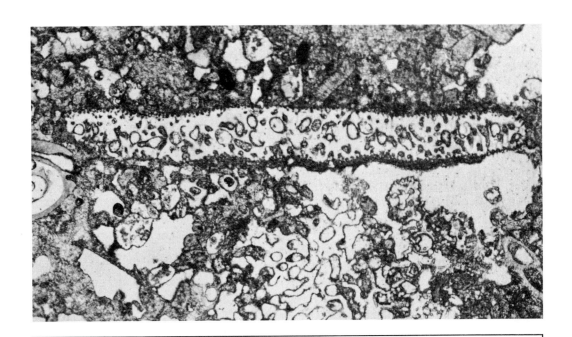

Sparry calcite cement

B

The use of fossil carbonates by the geochemist to interpret diagenesis is analogous to the use of deformed fossils by the structural geologist to monitor a geologic process. In many cases, however, fossil skeletal debris is sufficiently abundant to influence the course of diagenesis. Then the sedimentary geochemist cannot make complete or accurate interpretations without considering biologic factors. It is likely that evolutionary changes in the type of skeletal material secreted by organisms have produced parallel changes in diagenesis. For example, the percentage of invertebrate taxa secreting aragonite was apparently much smaller in the Paleozoic than it is today. The greater percentage of calcite skeletons probably had an influence on the mode of lithification of Paleozoic limestones.

Summary and Perspective

As the training of paleontologists becomes more interdisciplinary, the kind of work discussed in this chapter will become particularly attractive. This does not mean that the interdisciplinary research is independent of past and future descriptive work. In fact, such research is possible only because a large backlog of descriptive information is available. We may thus look upon the studies discussed in this chapter as by-products of the biologic study of fossils. In future years we will undoubtedly see many major advances, advances that will depend upon the sort of insight and interdisciplinary knowledge exemplified by Wells when he recognized that the presence of 400 growth bands in the annual accretion of the skeletons of Devonian corals was not an aberration or an error in his analysis, but was evidence that could be used to document an astronomical process.

Supplementary Reading

Land, L. S. (1967) Diagenesis of skeletal carbonates. *Jour. Sedim. Petrology,* **37**:914–930. (Summarizes both geochemical and petrographic data.)

Pannella, G., and MacClintock, C. (1968) Biologic and environmental rhythms reflected in molluscan shell growth. *Paleont. Soc. Mem. 2,* p. 64–80.

Pannella, G., MacClintock, C., and Thompson, M. N. (1968) Paleontologic evidence of variations in length of synodic month since Late Cambrian. *Science,* **162**:792–796. (Suggests that deceleration of the earth's rotation may not have been constant.)

Ramsay, J. G. (1967) *Folding and Fracturing of Rocks.* New York, McGraw-Hill, 568 p. (Contains an excellent treatment of the use of fossils in rock deformation studies.)

Wells, J. W. (1963) Coral growth and geochronometry. *Nature,* **197**:948–950.

References Cited

Ager, D. V. (1963) *Principles of Paleoecology*. New York, McGraw-Hill, 371 p.

American Commission on Stratigraphic Nomenclature (1961) Code of stratigraphic nomenclature. *Bull. Amer. Assoc. Petrol. Geol.*, **45**:645–665.

Arkell, W. J. (1956) *Jurassic Geology of the World*. London, Oliver and Boyd, 806 p.

Arkell, W. J. (1957) Introduction to Mesozoic Ammonoidea. *In* Moore, R. C., ed. *Treatise on Invertebrate Paleontology*. Boulder, Colo., Geological Society of America and University Press of Kansas, Part L, Mollusca 4, p. 81–129.

Bakker, R. T. (1968) The superiority of dinosaurs. *Discovery* (Peabody Mus. Nat. Hist.), **3**:11–22.

Barghoorn, E. S., and Schopf, J. W. (1965) Microorganisms from the Late Precambrian of central Australia. *Science*, **150**:337–339.

Barghoorn, E. S., and Schopf, J. W. (1966) Microorganisms three billion years old from the Precambrian of South Africa. *Science*, **152**:758–763.

Barghoorn, E. S., and Tyler, S. A. (1965) Microorganisms from the Gunflint Chert. *Science*, **147**:563–577.

Barker, R. M. (1964) Microtextural variation in pelecypod shells. *Malacologia*, **2**:69–86.

Bell, W. C., Kay, M., Murray, G. E., Wheeler, H. E., and Wilson, J. A. (1961) Note 25 — Geochronologic and chronostratigraphic units. *Bull. Amer. Assoc. Petrol. Geol.*, **45**:666–670.

Berkner, L. V., and Marshall, L. C. (1964) The history of growth of oxygen in the earth's atmosphere. *In* Brancazio, P. J., and Cameron, A. G. W., eds., *The Origin and Evolution of Atmospheres and Oceans*. New York, Wiley, p. 102–126.

Berry, W. B. N. (1968) *Growth of a Prehistoric Time Scale*. San Francisco, W. H. Freeman and Company, 158 p.

Best, R. V. (1961) Intraspecific variation in *Encrinurus ornatus. Jour. Paleont.*, **35**:1029–1040.

Beurlen, K. (1956) Der Faunenschnitt an der Perm-Trias-Grenze. *Z. Deut. Geol. Ges.*, **108**:88–99.

Blackwelder, R. E. (1967) *Taxonomy: A Text and Reference Book*. New York, Wiley, 698 p.

Bonner, J. T. (1952) *Morphogenesis*. Princeton, Princeton University Press, 296 p.

Boucot, A. J., Brace, W., and deMar, R. (1958) Distribution of brachiopod and pelecypod shells by currents. *Jour. Sedim. Petrology*, **28**:321–332.

Brinkmann, R. (1929) Statistisch-biostratigraphische Untersuchungen an mitteljurassischen Ammoniten über Artbegriff und Stammesentwicklung. *Göttingen, Abhandlung Gesell. Wiss.*, vol. 13, no. 3.

Brown, C. A. (1960) *Palynological Techniques*. Baton Rouge, La., C. A. Brown, 188 p.

Bulman, O. M. B. (1933) Programme-evolution in the graptolites. *Biol. Rev.*, **8**:311–334.

Callomon, J. H. (1963) Sexual dimorphism in Jurassic ammonites. *Trans. Leicester Lit. Philos. Soc.*, **57**:21–56.

Camp, C. L., and Hanna, G. D. (1937) *Methods in Paleontology*. Berkeley, University of California Press, 153 p.

Campbell, K. S. W. (1957) A Lower Carboniferous brachiopod-coral fauna from New South Wales. *Jour. Paleont.*, **31**:34–98.

Carpenter, F. M. (1953) The geological history and evolution of insects. *Amer. Scientist*, **41**:256–270.

Chave, K. E. (1964) Skeletal durability and preservation. *In* Imbrie, J., and Newell, N. D., eds. *Approaches to Paleoecology*. New York, Wiley, p. 377–387.

Clark, M. G., and Shaw, A. B. (1968) Paleontology of northwestern Vermont. XV. *Jour. Paleont.*, **42**:382–396.

Clark, W. E. Le Gros, ed. (1945) *Essays on Growth and Form Presented to D'Arcy Wentworth Thompson*. Oxford, The Clarendon Press, 408 p.

Clarke, G. L. (1965) *Elements of Ecology*. New York, Wiley, 560 p.

Clarkson, E. N. K. (1966) Schizochroal eyes and vision of some Silurian acastid trilobites. *Palaeontology*, **9**:1–29.

Cloud, P. E. (1949) Some problems and patterns of evolution exemplified by fossil invertebrates. *Evolution*, **2**:322–350.

Colbert, E. H. (1948) Evolution of the horned dinosaurs. *Evolution*, **2**:145–163.

Colbert, E. H. (1955) *Evolution of the Vertebrates*. New York, Wiley, 479 p.

Colbert, E. H. (1965) *The Age of Reptiles*. New York, Norton, 228 p.

Corliss, J. O. (1959) Comments on the systematics and phylogeny of the Protozoa. *Syst. Zool.*, **8**:169–190.

Craig, G. Y., and Hallam, A. (1963) Size-frequency and growth-ring analyses of *Mytilus edulis* and *Cardium edule*, and their paleoecological significance. *Palaeontology*, **6**:731–750.

Dacqué, E. (1921) Vergleichende biologische Formenkunde der Fossilen niederen Tiere. Berlin, Gebrüder Bornträger, 777 p.

DeBeer, G. R. (1958) *Embryos and Ancestors.* London, Oxford University Press, 197 p.

Denton, E. J., and Gilpin-Brown, J. B. (1961) The buoyancy of the cuttlefish *Sepia officinalis* (L.) *Jour. Marine Biol. Assoc. United Kingdom,* **41:** 319–342.

Denton, E. J., and Gilpin-Brown, J. B. (1966) On the buoyancy of the pearly nautilus. *Jour. Marine Biol. Assoc. United Kingdom,* **46:**723–759.

Denton, E. J., and Gilpin-Brown, J. B. (1967) On the buoyancy of *Spirula spirula. Jour. Marine Biol. Assoc. United Kingdom,* **47:**181–191.

Dobzhansky, T. (1951) *Genetics and the Origin of Species.* New York, Columbia University Press, 364 p.

Donovan, D. T. (1966) *Stratigraphy.* Chicago, Rand McNally, 199 p.

Dorf, E. (1960) Climatic changes of the past and present. *Amer. Scientist,* **48:**341–364.

Downie, C. (1967) The geological history of the microplankton. *Rev. Palaeobotany and Palynology,* **1:**269–281.

Dunbar, C. O., and Rodgers, J.(1957) *Principles of Stratigraphy.* New York, Wiley, 356 p.

Durham, J. W. (1950) Cenozoic marine climates of the Pacific Coast. *Geol. Soc. Amer. Bull.,* **61:**1243–1264.

Durham, J. W. (1955) Classification of clypeasteroid echinoids. *Univ. Calif. Publ. Geol. Sci.,* **31:**73–198.

Durham, J. W. (1966) Classification. *In* Moore, R. C., ed. *Treatise on Invertebrate Paleontology.* Boulder, Colo., Geological Society of America and University Press of Kansas, Part U, Echinodermata 3, p. 270–295.

Durham, J. W. (1967) The incompleteness of our knowledge of the fossil record. *Jour. Paleont.,* **41:**559–565.

Eagar, R. M. C. (1947) A study of a non-marine lamellibranch succession in the *Athraconaia lenisulcata* zone of the Yorkshire Coal Measures. *Philos. Trans. Roy. Soc. London* ser. B, **233:**1–54.

Easton, W. H. (1960) *Invertebrate Paleontology.* New York, Harper, 701 p.

Eicher, D. L. (1968) *Geologic Time.* Englewood Cliffs, Prentice-Hall, 150 p.

Elton, C. S. (1958) *The Ecology of Invasions by Animals and Plants.* London, Methuen, 181 p.

Engel, A. E. J., et al. (1968) Alga-like forms in Onverwacht Series, South Africa: oldest recognized lifelike forms on earth: *Science,* **161:**1005–1008.

Erben, H. K. (1966) Über den Ursprung der Ammonoidea. *Biol. Rev.* **41:** 641–658.

Ericson, D. B., Ewing, M., and Wollin, G. (1963) Pliocene-Pleistocene boundary in deep-sea sediments. *Science,* **139:**727–737.

Fenton, C. L., and Fenton, M. A. (1958) *The Fossil Book.* Garden City, Doubleday, 482 p.

Finks, R. M. (1960) Late Paleozoic sponge faunas of the Texas region. *Bull. Amer. Mus. Nat. Hist.,* **120:**1–160.

Fischer, A. G. (1960a) Brackish oceans as the cause of the Permo-Triassic marine faunal crisis. *In* Nairn, A. E. M., ed. *Problems In Palaeoclimatology.* London, Interscience, p. 566–580.

Fischer, A. G. (1960b) Latitudinal variations in organic diversity. *Evolution,* **14:**64–81.

Fischer, A. G. (1965) Fossils, early life, and atmospheric history. *Proc. National Acad. Sci.,* **53:**1205–1215.

Gilluly, J. (1949) Distribution of mountain building in geologic time. *Geol. Soc. Amer. Bull.,* **60:**561–590.

Ginsburg, R. N. (1957) Early diagenesis and lithification of shallow-water carbonate sediments in south Florida. *In* LeBlanc, R. J., and Breeding, J. G., eds. *Regional Aspects of Carbonate Deposition,* Soc. Econ. Paleont. Mineral. Spec. Publ. 5, p. 80–100.

Ginsburg, R. N. and Lowenstam, H. A. (1958) The influence of marine bottom communities on the depositional environment of sediments. *Jour. Geol.,* **66:**310–318.

Glaessner, M. F. (1962) Pre-Cambrian fossils. *Biol. Rev.* **37:**467–494.

Gould, S. J. (1966) Allometry and size in ontogeny and phylogeny. *Biol. Rev.* **41:**587–640.

Gould, S. J. (1970) Evolutionary paleontology and the science of form. *Earth-Science Rev.* **6:**77–119.

Grant, R. E. (1966) Spine arrangement and life habits of the productoid brachiopod *Waagenoconcha. Jour. Paleont.,* **40:**1063–1069.

Grant, R. E. (1968) Structural adaptation in two Permian brachiopod genera, Salt Range, West Pakistan. *Jour. Paleont.,* **42:**1–32.

Grant, V. (1963) *The Origin of Adaptations.* New York, Columbia University Press, 606 p.

Greiner, H. (1957) *"Spirifer disjunctus":* Its evolution and paleoecology in the Catskill Delta. *Bull. Peabody Mus. Nat. Hist., Yale Univ.,* 11, 75 p.

Hallam, A. (1969) Faunal realms and facies in the Jurassic. *Palaeontology,* **12:**1–18.

Harland, W. B., et al. (1967) *The Fossil Record.* London, Geol. Soc. London, 827 p.

Heaslip, W. G. (1968) Cenozoic evolution of the alticostate venericards in Gulf and East Coastal North America. *Palaeontographica Americana,* **6:**55–135.

Hecker, R. F. (1965) *Introduction to Paleoecology.* New York, Elsevier, 166 p.

Hedberg, H. D. (1951) Nature of time-stratigraphic units and geologic-time units. *Bull. Amer. Assoc. Petrol. Geol.,* **35:**1077–1081.

Hedgpeth, J. W., ed. (1957) *Treatise on Marine Ecology and Paleoecology, 1, Ecology.* Geol. Soc. Amer. Mem. 67, 1296 p.

Hessler, R. R., and Sanders, H. L. (1967) Faunal diversity in the deep-sea. *Deep-Sea Research,* **14:**65–78.

Heywood, V. H., and McNeil, J., eds. (1964) *Phenetic and Phylogenetic Classification.* Systematics Assoc. Publ. 6, 164 p.

Holmes, A. (1960) A revised geological time-scale. *Edinb. Geol. Soc. Trans.,* **17:**183–216.

Hudson, J. D. (1963) The recognition of salinity-controlled mollusc assemblages in the Great Estuarine Series (Middle Jurassic) of the Inner Hebrides. *Palaeontology,* **6:**318–326.

Hudson, J. D. (1968) The microstructure and mineralogy of the shell of a Jurassic mytilid (Bivalvia). *Palaeontology,* **11:**163–182.

Hunt, A. S. (1967) Growth, variation, and instar development of an agnostid trilobite. *Jour. Paleont.,* **41:**203–208.

Hunt, O. D. (1925) The food of the bottom fauna of the Plymouth fishing grounds. *Jour. Marine Biol. Assoc. United Kingdom,* **13:**560–599.

Hutchinson, G. E. (1958) The biologist poses some problems. *Amer. Assoc. Adv. Sci. Publ.* 67, p. 85–94.

Huxley, J. S. (1931) The relative size of antlers in deer. *Proc. Zool. Soc. London,* **1931**:819–864.

Huxley, J. S. (1932) *Problems of Relative Growth.* New York, Dial Press, 276 p.

Hyman (1959) *The Invertebrates,* IV, *Smaller Coelomate Groups.* New York, McGraw-Hill, 738 p.

Imbrie, J. (1956) Biometrical methods in the study of invertebrate fossils. *Bull. Amer. Mus. Nat. Hist.,* **108**:211–252.

Imbrie, J. (1959) Brachiopoda of the Traverse Group (Devonian) of Michigan, Part I. *Bull. Amer. Mus. Nat. Hist.* **116**:345–410.

Imbrie, J., and Newell, N. D. (1964) *Approaches to Paleoecology.* New York, Wiley, 432 p.

Israelsky, M. (1949) Oscillation chart. *Bull. Amer. Assoc. Petrol. Geol.,* **33**:92–98.

Jackson, R. T. (1912) Phylogeny of the Echini with a revision of Paleozoic species. *Boston Soc. Nat. Hist., Mem.* 7, 443 p.

Jeletsky, J. A. (1965) Is it possible to quantify biochronological correlation? *Jour. Paleont.,* **39**:135–140.

Johnson, R. G. (1960) Models and methods for the analysis of the mode of formation of fossil assemblages. *Geol. Soc. Amer. Bull.,* **71**:1075–1086.

Jurva, R. (1952) Seas. In *A General Handbook on the Geography of Finland. Fennia,* **72**:136–160.

Kaesler, R. L. (1967) Numerical taxonomy in invertebrate paleontology. *In* Teichert, C., and Yochelson, E. L., eds. *Essays in Paleontology and Stratigraphy.* Lawrence, University Press of Kansas, p. 63–81.

Kaufmann, R. (1933) Variationsstatistische Untersuchungen über die "Artabwandlung" und "Artumbildung" an der oberkambrischen Trilobitengattung *Olenus* Dalm. *Abh. Geol. Pal. Inst. Univ. Greifswald,* **10**: 1–54.

Kaufmann, R. (1935) Exakt-statistische Biostratigraphie der Olenus-Arten von Sudöland. *Geol. Foren. Stokholm Förhandl.,* **1935**:19–28.

Kermack, K. A. (1954) A biometrical study of *Micraster coranguinum* and *M. (Isomicraster) senonensis. Philos. Trans. Roy. Soc. London,* B, **237**:375–428.

Kinne, O. (1964) The effects of temperature and salinity on marine and brackish water animals, II. Salinity and temperature-salinity combinations. *Oceanogr. Mar. Biol. Ann. Rev.,* **2**:281–339.

Koch, D. L., and Strimple, H. L. (1968) A new Upper Devonian cystoid attached to a discontinuity surface. *Iowa Geological Survey Report of Investigations 5,* 49 p.

Kozlowski, R. (1966) On the structure and relationship of graptolites. *Jour. Paleont.,* **40**: 489–501.

Kraft, P. (1932) Die Mikrophotographie mit infraroten Strahlen. *Internat. Cong. of Photog. 8, Dresden, 1931, Berichte,* p. 341–345.

Krumbein, W. C., and Sloss, L. L. (1963) *Stratigraphy and Sedimentation,* 2nd Ed. San Francisco, W. H. Freeman and Company, 660 p.

Kulp, J. L. (1961) Geologic time scale. *Science,* **133**:1105–1114.

Kummel, B., and Raup, D. M., eds. (1965) *Handbook of Paleontological Techniques.* San Francisco, W. H. Freeman and Company, 852 p.

Kummel, B., and Steele, G. (1962) Ammonites from the *Meekoceras gracilitatus* zone at Crittenden Spring, Elko County, Nevada. *Jour. Paleont.,* **36:**638–703.

Kurtén, B. (1953) On the variation and population dynamics of fossil and recent mammal populations. *Acta Zool. Fennica,* **76:**1–22.

Kurtén, B. (1964) Population structure in paleoecology. *In* Imbrie, J., and Newell, N. D., eds. *Approaches to Paleoecology.* New York, Wiley, p. 91–106.

Ladd, H. S., ed. (1957) *Treatise on Marine Ecology and Paleoecology, 2, Paleoecology.* Geol. Soc. Amer. Mem. 67, 1077 p.

Land, L. S. (1967) Diagenesis of skeletal carbonates. *Jour. Sedim. Petrology,* **37:** 914–930.

Laporte, L. F. (1968) *Ancient Environments.* Englewood Cliffs, Prentice-Hall, 116 p.

Lawrence, D. R. (1968) Taphonomy and information losses in fossil communities. *Geol. Soc. Amer. Bull.,* **79:**1315–1330.

Ledley, R. S. (1964) High-speed automatic analysis of biomedical pictures. *Science,* **146:**216–223.

Leich, H. (1965) Eine neue Lebensspur von *Mesolimulus walchi* und ihre Deutung. *Aufschluss,* **1:**5–7.

Lemche, H., and Wingstrand, K. G. (1959) The anatomy of *Neopilina galatheae* Lemche, 1957 (*Mollusca, Tryblidiacea*). *Galathea Rept.* (Copenhagen), **3:**1–71.

Leon, R. (1933) Ultraviolettes Licht endeckt Versteinerungen auf "leeren" Platten. Ein Pantpod im Jura-Kalk. *Senckenberg Naturf. Gesell., Natur. Mus.* **63:**361–364.

Lerman, A. (1965) On rates of evolution of unit characters and character complexes. *Evolution* **19:**16–25.

Lowenstam, H. A. (1950) Niagaran reefs of the Great Lakes area. *Jour. Geol.,* **58:**430–487.

McLean, J. D., Jr. (1959–date) *Manual of Micropaleontological Techniques.* Alexandria, Va., McLean Paleont. Lab.

Macurda, D. B., ed. (1968) *Paleobiological Aspects of Growth and Development, a Symposium.* Paleont. Soc. Mem. 2, 119 p.

Marks, J. G. (1952) Especies vivientes de moluscos que se encuentran en las formaciones Terciarias de Venezuela. *Acta Cientifica Venezolana,* **3:**135–136.

Martin-Kaye, P. (1951) Sorting of lamellibranch valves on beaches in Trinidad, B.W.I. *Geol. Magazine,* **88:**432–434.

Mayr, E. (1942) *Systematics and the Origin of Species.* New York, Columbia University Press, 334 p.

Mayr, E. (1963) *Animal Species and Evolution.* Cambridge, Harvard University Press, 797 p.

Mayr, E. (1969) *Principles of Systematic Zoology.* New York, McGraw-Hill, 428 p.

Mayr, E., Linsley, E. G., and Usinger, R. L. (1953) *Methods and Principles of Systematic Zoology.* New York, McGraw-Hill, 328 p.

Medawar, P. B. (1945) Size, shape, and age. *In* Clark, W. E. Le Gros, ed. *Essays on Growth and Form presented to D'Arcy Wentworth Thompson.* Oxford, The Clarendon Press, p. 157–187.

Moore, H. B. (1958) *Marine Ecology.* New York, Wiley, 493 p.

Moore, R. C., Lalicker, C. G., and Fischer, A. G. (1952) *Invertebrate Fossils*. New York, McGraw-Hill, 766 p.

Muir-Wood, H. M. (1965) Chonetidina. *In* Moore, R. C., ed. *Treatise on Invertebrate Paleontology*. Boulder, Colo., Geological Society of America and University Press of Kansas, Part H, Brachiopoda, p. 412–439.

Newell, N. D. (1949) Phyletic size increase—an important trend illustrated by fossil invertebrates. *Evolution, 3:*103–124.

Newell, N. D. (1956) Fossil populations. *System. Assoc. Publ. 2,* p. 63–82.

Newell, N. D. (1959) Adequacy of the fossil record. *Jour. Paleont., 33:*488–499.

Newell, N. D. (1962) Paleontological gaps and geochronology. *Jour. Paleont., 36:*592–610.

Nichols, D. (1959a) Changes in the Chalk heart-urchin *Micraster* interpreted in relation to living forms. *Philos. Trans. Roy. Soc. London, B, 242:*347–437.

Nichols, D. (1959b) Mode of life and taxonomy in irregular sea-urchins. *System. Assoc., 3:*61–80.

Nicol, D. (1962) The biotic development of some Niagaran reefs—an example of an ecological succession or sere. *Jour. Paleont., 36:*172–176.

Odum, E. P. (1959) *Fundamentals of Ecology*. Philadelphia, Saunders, 546 p.

Odum, H. T., and Odum, E. P. (1955) Trophic structure and productivity of a windward coral reef community on Eniwetok Atoll. *Ecol. Monogr., 25:*291–320.

Okulitch, V. J. (1955) Archaeocyatha. *In* Moore, R. C., ed. *Treatise on Invertebrate Paleontology*. Boulder, Colo., Geological Society of America and University Press of Kansas, Part E, Archaeocyatha and Porifera, p. 1–20.

Olson, E. C. (1965) *The Evolution of Life*. London, Weidenfeld and Nicolson, 300 p.

Ostrom, J. H. (1964) A functional analysis of jaw mechanics in the dinosaur *Triceratops*. *Postilla* (Peabody Mus. Nat. Hist., Yale Univ.), *88:*1–35.

Palframan, D. F. B. (1966) Variation and ontogeny of some Oxfordian ammonites: *Taramelliceras richei* (de Loriol) and *Creniceras renggeri* (Oppel), from Woodham, Buckinghamshire. *Palaeontology, 9:*290–311.

Palmer, A. R. (1955) The faunas of the Riley Formation in central Texas. *Jour. Paleont., 28:*709–786.

Palmer, A. R. (1957) Ontogenetic development of two olenellid trilobites. *Jour. Paleont., 31:*105–128.

Palmer, A. R. (1965) Biomere—a new kind of stratigraphic unit. *Jour. Paleont., 39:*149–153.

Pannella, G., and MacClintock, C. (1968) Biological and environmental rhythms reflected in molluscan shell growth. *Paleont. Soc. Mem. 2,* p. 64–80.

Pannella, G., MacClintock, C., and Thompson, M. N. (1968) Paleontological evidence of variations in length of synodic month since Late Cambrian. *Science, 162:*792–796.

Pavlow, A. P. (1901) *Le Crétacé inférieur de la Russie et sa fauna*. Nouv. Soc. Imp. Nat. Moscou, livr. 21, 87 p.

Perkins, B. F. (1960) *Biostratigraphic Studies in the Comanche (Cretaceous) Series of Northern Mexico and Texas*. Geol. Soc. Amer. Mem. 83, 138 p.

Peterson, C. G. J. (1913) Valuation of the sea. II. The animal communities of the sea bottom and their importance for marine zoogeography. *Rep. Danish Biol. Sta.,* **21:**1–44.

Pianka, E. R. (1966) Latitudinal gradients in species diversity: a review of concepts. *Amer. Nat.,* **100:**33–46.

Ramsay, J. G. (1967) *Folding and Fracturing of Rocks.* New York, McGraw-Hill, 568 p.

Raup, D. M. (1966) Geometric analysis of shell coiling: general problems. *Jour. Paleont.,* **40:**1178–1190.

Raup, D. M. (1967) Geometric analysis of shell coiling: coiling in ammonoids. *Jour. Paleont.,* **41:**43–65.

Raup, D. M. (1968) Theoretical morphology of echinoid growth. *Paleont. Soc. Mem. 2,* p. 50–63.

Raup, D. M., and Michelson, A. (1965) Theoretical morphology of the coiled shell. *Science,* **147:**1294–1295.

Raup, D. M., and Takahashi, T. (1968) Experiments on strength in cephalopod shells (abs.). *Geol. Soc. Amer. Spec. Paper 101,* p. 172–173.

Reid, R. E. H. (1968) Bathymetric distribution of Calcarea and Hexactinellida in the past and present. *Geol. Magazine,* **105:**546–559.

Rensch, B. (1960) *Evolution Above the Species Level.* New York, Columbia University Press, 419 p.

Rhoads, D. C. (1966) Missing fossils and paleoecology: *Discovery* (Peabody Mus. Nat. Hist., Yale Univ), **2:**19–22.

Rhoads, D. C. (1967) Biogenic reworking of intertidal and subtidal sediments in Barnstable Harbor and Buzzards Bay, Massachusetts. *Jour. Geol.,* **75:**461–476.

Richter, R. (1928) Psychische Reaktionen fossiler Tiere. *Palaeobiologica* **1:**226–244.

Romer, A. S. (1948) Relative growth in pelycosaurian reptiles. *Robert Broom Commem. Vol. Special Publ. Roy. Soc. South Africa,* p. 45–55.

Romer, A. S. (1966) *Vertebrate Paleontology.* Chicago, University of Chicago Press, 468 p.

Ross, C. A. (1967) Development of fusulinid (Foraminiferida) faunal realms. *Jour. Paleont.,* **41:**1341–1354.

Ross, C. A., and Sabins, F. F. (1965) Early and Middle Pennsylvanian fusulinids from southeast Arizona. *Jour. Paleont.,* **39:**173–209.

Rowe, A. W. (1899) An analysis of the genus *Micraster,* as determined by rigid zonal collecting from the zone of *Rhynchonella cuvieri* to that of *Micraster coranguinum. Quart. Jour. Geol. Soc. London,* **55:**494–547.

Rowell, A. J. (1967) A numerical taxonomic study of the chonetacean brachiopods. *In* Teichert, C., and Yochelson, E. L., eds. *Essays In Paleontology and Stratigraphy.* Lawrence, University Press of Kansas, p. 113–140.

Rudwick, M. J. S. (1964) The inference of function from structure in fossils. *Brit. Jour. Philos. Sci.,* **15:**27–40.

Rudwick, M. J. S. (1965) Adaptive homeomorphy in the brachiopods *Tetractinella* Bittner and *Cheirothysis* Rollier. *Paläont. Z.,* **39:**134–146.

Salfeld, H. (1913) Uber Artbildung bei Ammoniten. *Zeitschr. Deutsch. Geol. Gesell., Monatsber.,* **65:**437–440.

Sandberg, P. A., and Hay, W. W. (1967) Study of microfossils by means of the scanning electron microscope. *Jour. Paleont.,* **41:**999–1001.

Sanders, H. L. (1956) Oceanography of Long Island Sound, 1952–1954, X, The biology of marine bottom communities. *Yale Bingham Oceanographic Coll.,* **56:**345–414.

Sanders, H. L., Hessler, R. R., and Hampson, G. R. (1965) An introduction to the study of deep-sea benthic faunal assemblages along the Gay Head–Bermuda transect. *Deep-Sea Research,* **12:**845–867.

Savory, T. (1962) *Naming the Living World.* London, English University Press Ltd., 128 p.

Schafer, W. (1965) *Aktuo-Palaontologie, nach Studien in der Nordsee.* Frankfurt, Waldemar Kramer, 666 p.

Scheltema, R. S. (1968) Dispersal of larvae by equatorial ocean currents and its importance to the zoogeography of shoal-water tropical species. *Nature,* **217:**1159–1162.

Schenk, E. T., and McMasters, J. H. (1956) *Procedure in Taxonomy,* 3rd Ed. Revised by A. M. Keen and S. W. Muller. Stanford, Calif., Stanford University Press, 119 p.

Schindewolf, O. H. (1962) Neokatastrophismus? *Deutsche Geol. Gesell. Zeitschr.* **114:**430–445.

Schlanger, S. O. (1963) Subsurface geology of Eniwetok Atoll. *U.S. Geol. Surv. Prof. Paper 260–BB,* p. 991–1066.

Schopf, J. W. (1968) Microflora of the Bitter Springs Formation, Late Precambrian, central Australia. *Jour. Paleont.,* **42:**651–688.

Scrutton, C. T. (1965) Periodicity in Devonian coral growth. *Palaeontology,* **7:**552–558.

Seilacher, A. (1960) Epizoans as a key to ammonoid ecology. *Jour. Paleont.,* **34:**189–193.

Seilacher, A. (1962) Paleontological studies on turbidite sedimentation and erosion. *Jour. Geol.,* **70:**227–234.

Seilacher, A. (1964) Biogenic sedimentary structures. *In* Imbrie, J., and Newell, N. D., eds. *Approaches to Paleoecology.* New York, Wiley, p. 296–316.

Seilacher, A. (1967) Fossil behavior. *Sci. Amer.* **217:**72–80.

Shaw, A. B. (1964) *Time in Stratigraphy.* New York, McGraw-Hill, 365 p.

Shotwell, J. A. (1955) An approach to the paleoecology of mammals. *Ecology,* **36:**327–337.

Shotwell, J. A. (1958) Intercommunity relationships in Hemphillian (mid-Pliocene) mammals. *Ecology,* **39:**271–282.

Simpson, G. G. (1940a) Mammals and land bridges. *Jour. Washington Acad. Sci.,* **30:**137–163.

Simpson, G. G. (1940b) Types in modern taxonomy. *Amer. Jour. Sci.,* **238:**413–431.

Simpson, G. G. (1951) *Horses.* New York, Oxford University Press, 247 p.

Simpson, G. G. (1952) How many species? *Evolution,* **6:**342.

Simpson, G. G. (1953) *The Major Features of Evolution.* New York, Columbia University Press, 434 p.

Simpson, G. G. (1960) Notes on the measurement of faunal resemblance. *Amer. Jour. Sci.,* **258-A:**300–311.

Simpson, G. G. (1961) *Principles of Animal Taxonomy.* New York, Columbia University Press, 247 p.

Simpson, G. G., Rowe, A., and Lewontin, R. C. (1960) *Quantitative Zoology.* New York, Harcourt, Brace, 440 p.

Sohl, N. F. (1960) *Archeogastropoda, Mesogastropoda and Stratigraphy of the Ripley, Owl Creek, and Prairie Bluff Formations.* U.S. Geol. Surv. Prof. Paper No. 331–A, 151 p.

Sokal, R. R., and Sneath, P. H. A. (1963) *Principles of Numerical Taxonomy.* San Francisco, W. H. Freeman and Company, 359 p.

Sorgenfrei, T. (1958) *Molluscan Assemblages from Marine Middle Miocene of South Jutland and their Environments.* Geol. Surv. Denmark, II Ser. No. 79, 503 p.

Spjeldnaes, N. (1951) Ontogeny of *Beyrichia jonesi* Boll. *Jour. Paleont.,* **25:**745–755.

Srb, A. M., Owen, R. D., and Edgar, R. S. (1965) *General Genetics,* 2nd. Ed. San Francisco, W. H. Freeman and Company, 557 p.

Stanley, S. M. (1966) Paleoecology and diagenesis of Key Largo Limestone, Florida. *Bull. Amer. Assoc. Petrol. Geol.,* **50:**1927–1947.

Stanley, S. M. (1968) Post-Paleozoic adaptive radiation of infaunal bivalve molluscs—a consequence of mantle fusion and siphon formation. *Jour. Paleont.,* **42:**214–229.

Stanley, S. M. (1970) *Relation of Shell Form to Life Habits in the Bivalvia (Mollusca).* Geol. Soc. Amer. Mem. 125.

Stehli, F. G. (1964) Permian zoogeography and its bearing on climate. *In* Nairn, A. E. M., ed. *Problems in Palaeoclimatology.* London, Interscience, p. 537–549.

Stoll, N. R., et al., eds. (1961) *International Code of Zoological Nomenclature.* London, Internat. Trust for Zool. Nomen., 176 p.

Sylvester-Bradley, P. C., ed. (1956) *The Species Concept in Paleontology.* System. Assoc. Publ. 2, 145 p.

Tappan, H. (1968) Primary production, isotopes, extinctions and the atmosphere. *Palaeogeogr., Palaeoclimat., Palaeoecol.,* **4:**187–210.

Teichert, C. (1956) How many fossil species? *Jour. Paleont.,* **30:**967–969.

Teichert, C. (1958) Cold- and deep-water coral banks. *Bull. Amer. Assoc. Petrol. Geol.,* **42:**1064–1082.

Thompson, D'A. W. (1942) *On Growth and Form.* New York, Cambridge University Press, 1116 p.

Thorson, G. (1950) Reproductive and larval ecology of marine bottom invertebrates. *Biol. Rev.,* **25:**1–45.

Thorson, G. (1957) Bottom communities (sublittoral or shallow shelf). *Geol. Soc. Amer. Mem. 67,* **1:**461–534.

Trueman, A. E. (1941) The ammonite body-chamber, with special reference to the buoyancy and mode of life of the living ammonite. *Quart. Jour. Geol. Soc. London,* **96:**339–383.

Trueman, A. E. (1942) Supposed commensalism of Carboniferous spirorbids and certain non-marine lamellibranchs. *Geol. Magazine,* **79:**312–320.

Turpaeva, E. P. (1957) Food interrelationships of dominant species in marine benthic biocoenoses. *Trans. Inst. Oceanol. and Oceanogr.,* **11:**198–211.

Valentine, J. W. (1963) Biogeographic units as biostratigraphic units. *Bull. Amer. Assoc. Petrol. Geol.,* **47:**457–466.

Valentine, J. W. (1968) Climatic regulation of species diversification and extinction. *Geol. Soc. Amer. Bull.,* **79:**273–276.

Vaughan, T. W., and Wells, J. W. (1943) *Revision of the Suborders, Families, and Genera of the Scleractinia.* Geol. Soc. Amer. Spec. Paper 44, 363 p.

Waller, T. R. (1969) *The Evolution of the* Argopecten gibbus *Stock (Mollusca: Bivalvia), with Emphasis on the Tertiary and Quaternary Species of Eastern North America.* Paleont. Soc. Mem. 3, 125 p.

Weimer, R. J., and Hoyt, J. H. (1964) Burrows of *Callianassa major* Say, geologic indicators of littoral and shallow neritic environments. *Jour. Paleont.,* **38:**761–767.

Weisbord, N. E. (1962) Late Cenozoic gastropods from northern Venezuela. *Bull. Amer. Paleont.* **42:**1–672.

Weisbord, N. E. (1964) Late Cenozoic pelecypods from northern Venezuela. *Bull. Amer. Paleont.* **45:**1–564.

Weller, J. M. (1960) *Stratigraphic Principles and Practice.* New York, Harper, 725 p.

Wells, J. W. (1956) Scleractinia. *In* Moore, R. C., ed. *Treatise on Invertebrate Paleontology.* Boulder, Colo., Geological Society of America and University Press of Kansas, Part F, Coelenterata, p. 328–444.

Wells, J. W. (1957) Coral reefs. *Geol. Soc. Amer. Mem. 67,* 2:609–631.

Wells, J. W. (1963) Coral growth and geochronometry. *Nature,* **197:**948–950.

Westoll, T. S. (1949) On the evolution of the Dipnoi. *In* Jepsen, G. L., Simpson, G. G., and Mayr, E., eds., *Genetics, Paleontology, and Evolution.* Princeton, Princeton University Press, p. 121–184.

Whittington, H. B. (1957) The ontogeny of trilobites. *Biol. Rev.,* *32:*421–469.

Woodford, A. O. (1965) *Historical Geology.* San Francisco, W. H. Freeman and Company, 512 p.

Woodring, W. P., Stewart, R., and Richards, R. W. (1940) Geology of the Kettleman Hills Oil Field, California. *U.S. Geol. Surv. Prof. Paper 195,* 170 p.

Wright, S. (1932) The roles of mutation, inbreeding, crossbreeding, and selection in evolution. *Proc. Sixth Intern. Cong. Genetics,* 1:356–366.

Zangerl, R. (1965) Radiographic techniques. *In* Kummel, B., and Raup, D. M., eds. *Handbook of Paleontological Techniques.* San Francisco, W. H. Freeman and Company, p. 305–320.

Ziegler, A. M. (1965) Silurian marine communities and their environmental significance. *Nature,* **207:**270–272.

Ziegler, A. M., Boucot, A. J., and Sheldon, R. P. (1966) Silurian pentameroid brachiopods preserved in position of growth. *Jour. Paleont.,* **40:**1032–1036.

Ziegler, A. M., Cocks, L. R. M., and Bambach, R. K. (1968) The composition and structure of Lower Silurian marine communities. *Lethaia,* 1:1–27.

Zohary, D., and Feldman, M. (1962) Hybridization between amphidiploids and the evolution of polyploids in the wheat (*Aegilops-Triticum*) group. *Evolution,* 16:44–61.

Index

Page numbers referring to illustrations are given in **boldface** type.